Statistical Inference and Analysis

Professor Sir Ronald A. Fisher
(Photograph by Antony Barrington-Brown ARPS.)

Statistical Inference and Analysis

Selected Correspondence of R.A. Fisher

EDITED BY

J.H. Bennett

CLARENDON PRESS · OXFORD

This book has been printed digitally and produced in a standard specification in order to ensure its continuing availability

OXFORD
UNIVERSITY PRESS

Great Clarendon Street, Oxford OX2 6DP

Oxford University Press is a department of the University of Oxford.
It furthers the University's objective of excellence in research, scholarship,
and education by publishing worldwide in

Oxford New York

Auckland Cape Town Dar es Salaam Hong Kong Karachi
Kuala Lumpur Madrid Melbourne Mexico City Nairobi
New Delhi Shanghai Taipei Toronto
With offices in
Argentina Austria Brazil Chile Czech Republic France Greece
Guatemala Hungary Italy Japan South Korea Poland Portugal
Singapore Switzerland Thailand Turkey Ukraine Vietnam

Oxford is a registered trade mark of Oxford University Press
in the UK and in certain other countries

Published in the United States
by Oxford University Press Inc., New York

Reprinted 2011

ISBN 978-0-19-855552-0

Printed and bound in Great Britain by CPI Antony Rowe,
Chippenham and Eastbourne

Preface

Perhaps no one has had a greater impact on the methodology of scientific research in the twentieth century than Ronald Aylmer Fisher. From his early work in developing statistical methods needed in the interpretation of experimental data, he went on to recast the entire theoretical basis for mathematical statistics and to initiate the deliberate study and development of experimental design central to the whole process of the Natural Sciences. In clarifying the principles of inductive inference—of reasoning from the particular to the general—which Fisher suggested was the only way known to man whereby new knowledge comes into the world, he greatly enlarged our understanding of the nature of uncertainty and contributed fundamentally to the philosophy of our age. In mid-century, Fisher (*CP* 251) ventured to suggest that 'Statistical Science was the peculiar aspect of human progress which gave to the twentieth century its special character'. Fisher was also a distinguished geneticist. His epoch-making book, *The genetical theory of natural selection* (1930), was the first major work to provide a general synthesis of Darwinism and Mendelism. It marked a turning point in the development of evolutionary thought and led to the emergence of neo-Darwinism.

This volume presents a selection from Fisher's letters on statistical inference and analysis and related topics. It also includes relevant material from the letters (many from distinguished scientists) to which Fisher was replying. It is a companion volume to *Natural selection, heredity and eugenics: selected correspondence of R.A. Fisher with Leonard Darwin and others* (ed. J.H. Bennett, Clarendon Press, Oxford, 1983).

The letters in this volume cover the period 1922–62. There are few letters from the first seven years and those involve just two correspondents, Norman Campbell and Percy MacMahon. In the first five years Fisher wrote virtually all of his letters by hand and only rarely kept copies. (Fisher's earliest letter, dated 22 July 1922, to Campbell (p. 58) is one of the few of which he kept a copy in his own hand (see p. 60).) Occasionally, when a correspondent had complained that he could not decipher Fisher's handwriting (e.g. MacMahon, p. 276), a typewritten letter would be sent. However, from about 1928 onwards, Fisher's letters were generally typed out by a secretary and then carbon copies were filed away. These copies have provided the basis for the reproduction of most of Fisher's correspondence included in this volume.

Fisher's practice was, wherever possible, to reply to letters immediately after reading them, generally by dictation to his secretary. He had an extraordinary capacity to draw quickly upon his vast mental store of knowledge and ideas; without hesitation, he would recall requisite facts and marshal the many elements of a complex argument and then express himself with great care and precision. His letters composed in this way were rarely changed later. Sometimes, as when his reply required an extensive mathematical argument or perhaps a very long development, he would write this out by hand for his secretary later to copy on a typewriter. Occasionally he would make small changes to a typed letter before adding his signature and his secretary had then to insert these on the carbon copy before the original was dispatched. In the last few years of his life, when he was often travelling and without secretarial assistance, Fisher resorted again to sending handwritten letters without keeping any copy for his files. In such cases where he filed away a relevant incoming letter, it has sometimes been possible to trace and then obtain photocopies of Fisher's manuscript letters through the generosity of his correspondents.

The correspondence presented in this volume has been arranged in six chapters according to subject matter. Inference (Chapter 1) makes up two-thirds of the volume and Statistical Method (Chapter 2) about one-sixth. The other chapters are History of Statistics, Teaching of Statistics, History and Philosophy of Science, and Scientists and Scientific Research. Within each chapter the letters are ordered alphabetically by the correspondents' surnames. This seemed preferable to a simple chronological ordering because the exchanges with some individuals extend over many years and are best read and studied in their entirety. (In only one instance did it seem necessary to break up a letter with different parts placed in two separate chapters; cross references have been given where this occurs.) My object when selecting correspondence has been to include material of scientific or historical interest, avoiding unnecessary repetition and personal references of no scientific interest. Where a word or passage has been omitted, this is indicated by the symbol . . . . Editorial insertions in the correspondence are shown within square brackets [ ]. References to books and papers by Fisher have been inserted wherever it was thought this could be helpful. The numbering system used for references to Fisher's papers is that adopted in the *Collected papers of R.A. Fisher.*[*] Readers seeking further understanding of Fisher's work may be referred to the article by F. Yates and K. Mather in *Biographical memoirs of Fellows of the Royal Society of London*, **9**, 91–120 (1963)[†] and Joan Fisher Box's biography of her father: *R.A. Fisher, the life of a scientist* (Wiley 1978). L.J. Savage's article 'On re-reading R.A. Fisher' (*Annals of Statistics*, **4**, 441–500, (1976)) is also recommended.

An important part of the correspondence in Chapter 1 was produced in 1954–55 whilst Fisher was preparing the manuscript of *Statistical methods and scientific inference*. At various stages he invited Frank Yates and George Barnard to Cambridge to discuss what he had written and some of the correspondence in Chapter 1 stemmed from those discussions. There is much of interest in Fisher's correspondence with all of the distinguished statisticians represented in Chapter 1. However, special mention might be made of (i) the

[*] Copies of this publication (in five volumes) are available from the Secretary, Department of Genetics, University of Adelaide, Adelaide, South Australia 5000.

[†] Reprinted in Volume 1 of the *Collected papers*.

long series of exchanges with Jeffreys in Cambridge, covering virtually the whole of Fisher's period at the Galton Laboratory and ceasing when Fisher returned to Cambridge in 1943 and (ii) the remarkable series of exchanges on fiducial inference (a) with Darmois and Fréchet in Paris in 1940, (b) with Savage and Tukey in USA in the period 1952–55, and (c) in the last months of Fisher's life, with Barnard, Finney, Fraser, A.T. James, Kyburg, Rao and Sprott. Chapter 2 contains many delightful examples of Fisher's eager approach to finding innovative and elegant solutions to various problems involved in the handling of experimental data. We may also note in Chapter 2 his interest in and concern with computation and in the following chapters his comments on the origin of some statistical terms and symbolism as well as his interesting views on more general questions such as the organization of scientific research and freedom in science. His writing throughout is sharp-edged and stimulating. Whilst lightened by many good-humoured comments, it needs to be read very carefully.

As indicated in the earlier volume of Fisher's scientific correspondence as well as in *R.A. Fisher, the life of a scientist*, Charles Darwin's son Leonard (1850–1943) had a profound influence on Fisher's life and work. Leonard's advice was often sought by Fisher early in his career and, when given, it was always greatly respected. After Fisher's major paper on the correlation between relatives had been effectively rejected by the Royal Society of London and then the Royal Society of Edinburgh had said it could only publish a short abstract, it was Darwin's intervention that led to publication of the full paper. When some of Fisher's early papers were refused publication in *Biometrika* and the *Journal of the Royal Statistical Society*, Darwin gave Fisher wise counsel and assistance. It was in such circumstances that Darwin wrote to Fisher in 1923 urging him to 'push on quietly avoiding as far as possible all controversy'. Darwin strengthened his appeal by adding that his father had believed controversy with individuals was a great waste of time and should be avoided. However, Fisher found himself unable to follow this advice for very long. As many of his innovative ideas and developments became controversial, he could not resist being drawn into the fray. With his great intellect and ready wit, he could have a devastating effect in scientific controversy, though it must be said, as Fisher himself recognized, that he sometimes became an irascible protagonist. In his correspondence also he was sometimes inclined to be rather testy but it seems that generally, if he believed a correspondent genuinely shared his aim of getting at the truth or of making some progress in understanding, then he would happily make a great effort to help, although sometimes this might be through prodding the correspondent to think for himself, possibly along suggested new lines.

In his obituary notice on 'Student', Fisher remarked on his clear-headedness, his custom of forming independent judgements, his concern with and responsibility for the practical interpretation of experimental data, his constructive imagination and pertinacity, and his ability to perceive and to solve his problems. These characteristics were also typical of Fisher and they can be seen in his correspondence. Fisher was, in addition, a most able mathematician with deep insight into the subject, but he never lost contact with real data. He only published mathematical work that bore on statistical or genetical theory and only when it could be used to analyse data or solve practical problems. His profound mathematical insight sometimes caused difficulties for readers of his work because he could take a short cut using the geometry of generalized Euclidean space which was very elegant but not easy to follow. He considered that the breadth of logical outlook essential in inductive inference was only easily acquired by long contact with experimental research and it was his belief that mathematicians often lacked this that led him to be highly critical of the influence of mathematicians on statistics. This is a sentiment that appears in a number of his letters.

I am greatly indebted to three of Fisher's colleagues and friends for their invaluable assistance in the preparatory work associated with this volume. Frank Yates, David Finney, and George Barnard have made many helpful suggestions and kindly provided answers to various questions that arose in the course of this work. David Finney visited Adelaide, with support from the University of Adelaide Distinguished Visiting Scholar Scheme, to assist with the final selection of correspondence and its general arrangement. I am grateful to Graham Wilkinson for his help especially with the correspondence on inference and to Alan James for his advice on many questions relating to this volume. Special thanks are due to the authors of correspondence with Fisher, or their literary executors, for permission to quote material from letters addressed to Fisher. I am indebted to Donald Fraser, Henry Kyburg, and David Sprott for providing photocopies of manuscript letters they received from Fisher in 1962. I am indebted to George Barnard for drawing the Fisher–Sprott correspondence to my attention; Sprott's letters to Fisher, some of which were posted to India, did not get into Fisher's files in Adelaide and were apparently lost. I thank the University of Adelaide for permission to quote extracts from *R.A. Fisher Correspondence* held at the University and the Royal Society of London for permission to quote part of a report that Fisher wrote for the Society in 1934. I also wish to thank Miss Georgette Psaltis and Mrs. S.E. Suter for the care with which they prepared the typescript.

J.H.B.

Adelaide, South Australia
February 1989

Contents

Editor's Abbreviations

CP	*Collected papers of R.A. Fisher* (ed. J.H. Bennett). Volumes 1–5. University of Adelaide (1971–4). (Numbered references to papers can be identified from the reference list on p. 369.)
CMS	*Contributions to mathematical statistics* by R.A. Fisher. John Wiley, New York (1950).
DOE	*The design of experiments* by R.A. Fisher. Editions 1–8, Oliver and Boyd, Edinburgh (1935–66). Eighth edition reprinted by Hafner Publishing Company (1971).*
GTNS	*The genetical theory of natural selection* by R.A. Fisher. Clarendon Press, Oxford (1930). Revised edition, Dover Publications, New York (1958).
SMRW	*Statistical methods for research workers* by R.A. Fisher. Editions 1–14, Oliver and Boyd, Edinburgh (1925–70). Fourteenth edition reprinted by Hafner Publishing Company (1973).*
SMSI	*Statistical methods and scientific inference* by R.A. Fisher. Editions 1–2, Oliver and Boyd, Edinburgh (1956–9). Third edition, Hafner Publishing Company (1973).*
ST	*Statistical tables for biological, agricultural and medical research* by R.A. Fisher and F. Yates. Editions 1–6, Oliver and Boyd, Edinburgh (1938–63). Sixth edition reprinted by Longman (1974).

Page numbers given for Fisher's books in the editorial notes in the present volume refer to the most recent editions unless otherwise stated.

*Reprinted by Oxford University Press (1990) in one volume, *Statistical methods, experimental design and scientific inference*, with a Foreword by Frank Yates.

Dates and Events in the Life of R.A. Fisher

1890	Born on 17 February in London.
	Educated at Stanmore Park and Harrow Schools.
1909–13	Student at University of Cambridge.
1912	B.A. (Camb.), having passed the Mathematical Tripos Part II as a Wrangler.
1913–14	Statistician in Mercantile and General Investment Company, London.
1914–18	Assistant Master at Rugby and other schools.
1917	Married to Ruth Eileen Guinness.
1919–33	Statistician, Rothamsted Experimental Station, Harpenden.
1920	M.A. (Camb.).
1921–27	Fellow, Gonville and Caius College, Cambridge.
1925	Publication of *Statistical methods for research workers.*
1926	Sc.D. (Camb.).
1929	Fellow of the Royal Society.
1930	Publication of *The genetical theory of natural selection.*
1933–43	Galton Professor of Eugenics, University College, London.
	Editor of *Annals of Eugenics.*
1934	Honorary member, American Academy of Arts and Sciences.
1935	Publication of *The design of experiments.*
1937	Honorary Fellow, Indian Statistical Institute.
1938	Publication of R.A. Fisher and F. Yates's *Statistical tables for biological, agricultural and medical research.*
	Royal medal of the Royal Society.
1943–57	Arthur Balfour Professor of Genetics, University of Cambridge.
1943–62	Fellow, Gonville and Caius College, Cambridge.
1946	Guy medal of the Royal Statistical Society.
1947	With C.D. Darlington, founded the international journal *Heredity*; co-editor, 1947–62.
1948	Darwin medal of the Royal Society.
	Foreign associate, U.S.A. National Academy of Sciences.
	Foreign member, Royal Swedish Academy of Sciences.

1949	Publication of *The theory of inbreeding*.
1950	Foreign member, Royal Danish Academy of Sciences and Letters.
	Publication of *Contributions to mathematical statistics*.
1952	Knight Bachelor.
1955	Copley medal of the Royal Society.
1956	Publication of *Statistical methods and scientific inference*.
1956–59	President, Gonville and Caius College, Cambridge.
1958	Silver medal of the Linnean Society.
1959–62	Research Fellow, CSIRO Division of Mathematical Statistics, University of Adelaide, South Australia.
1961	Member of the Pontifical Academy of Sciences.
1962	Died on 29 July at Adelaide, South Australia.

Ronald Aylmer Fisher†

The frailties of Ronald Fisher's mortal existence have succumbed, but his scientific spirit will live forever in his published researches, indelibly inscribed across the pages of 50 years in the history of science.

To do full justice to the man and his accomplishments is a task for another time and place, and I have thought it fitting for this occasion to emphasize what seem to me to be outstanding points.

During the latter half of 1918, he was offered two posts almost simultaneously: the first, as senior assistant to the Professor in the Galton Laboratory, University College, London, a centre with an established reputation in mathematical statistics and biometry, and the second, at Rothamsted Experimental Station, where he would have to make his own way.

It must be remembered that up to that time he had only a few papers to his credit, including two which were most important, one in mathematical statistics, and the other in genetics—the two fields in which he ultimately made his greatest contributions.

In characteristic fashion, young Fisher decided to stand on his own feet, and accepted the Rothamsted post. He knew exactly what he was doing, because he foresaw the opportunities for independent research. It was a momentous decision. At that time the experimentalist, in particular the biologist, struggled with research under the severe handicap of biological variation, with only small samples and limited facilities for increasing his observations.

Within five years, Fisher had solved all major problems, and placed in the hands of the experimenter both the techniques for conducting the experiments and the mathematical and arithmetical procedures for making sense of the results.

These researches were gathered together in a book first published in 1925. The effect was electrifying. Word rapidly encircled the globe, and a stream of men converged on Rothamsted to learn at first-hand something of these new developments for application in their respective countries. I emphasize how widely these ideas were spread, and to indicate how deeply they penetrated, I may mention that they even provided material for examination questions as presented to students of agriculture in the University of Melbourne. All this in the space of as many years as can be counted on the fingers of one hand! This is no exaggeration; it is fact, because I was one of those students.

The book, *Statistical methods for research workers*, is without doubt a classic, and made a tremendous impact on science. In 37 years it has gone to

† The address given by Professor E.A. Cornish at the funeral of Sir Ronald Fisher in St Peter's Cathedral, Adelaide, 2 August 1962.

the 13th edition, with a complete reprint of one edition, making the equivalent of 14. Moreover, there are translations in French, German, Italian, Spanish, Russian, and Japanese.

In mathematical statistics, success mounted on success, until today every phase of the subject bears his mark, and in most of them he laid the foundations. Two more books followed on the theory of statistics—the first now in its seventh edition, with translations to other languages, and the second (his latest) in its second edition. And all this had its origin in the fact that he was the right man, at the right time and in the right place.

In genetics his total contribution falls not far short of that in mathematical statistics, but it did not hit the scientific world with the same impact, nor did it spread in the same dramatic manner. There was, however, one very great moment. For many years he had been working on a series of researches of which the outside world knew nothing until publication, in 1930, of the book *The genetical theory of natural selection*, which can be most suitably appraised by stating that it provided a quantitive basis for the Darwinian theory, just as Clerk–Maxwell's electromagnetic theory of light provided the quantitative basis for Faraday's work in experimental physics.

Recognition came with numerous honours; honorary degrees from universities in Britain, USA, India, and Australia; election to national academies in Britain, USA, Sweden, Denmark, and the Pontifical Academy, and a Knight Bachelor from Her Majesty the Queen.

As to the man himself, an outstanding characteristic was his immense capacity for work—aside from his books, his personal contribution totalled between 300 and 400 research papers, but to this must be added an incalculable contribution to the research of literally hundreds of individuals, in the ideas, guidance, and assistance he so generously gave, irrespective of nationality, colour, class, or creed.

This was made possible, first, by his extraordinary insight and acumen which immediately stripped each problem down to its essentials, and secondly, by the innate ability—intuition if you wish—to proceed unerringly to the solution. Two characteristics, I submit, which mark the touch of genius.

The breadth of his knowledge beyond his chosen fields seemed boundless—he could speak with authority on literature and ancient history (in particular ancient civilizations), and in a sense at the other extreme, on continental drift and palaeomagnetism.

His English, both written and spoken, was impeccable; when he chose he was an excellent lecturer; he was always brilliant in conversation, a deadly opponent in argument, completely intolerant of humbug, and like all of us, he had his little idiosyncrasies.

But he was a kindly soul, most inspiring and encouraging to those who genuinely sought his aid, and intensely loyal to his staff and students—as witness the tremendous fight he waged for the scientific reputation of a young Indian cotton researcher.

He had an abiding interest in nature in all its forms. He loved children and they were always attracted to him. And on this note I should like to conclude by relating a touching little incident. We are all familiar with the uninhibited vocal efforts of children at play. The pupils of the school in the immediate vicinity of the Queen Elizabeth Hospital will never know how their cries delighted him during last week as he lay fighting to keep alive the flickering candle of life.

He must go down in history as one of the great men of this our twentieth century.

1
Statistical Inference

A.C. Aitken to Fisher: 22 January 1936

Very many thanks for the offprints, which are safely to hand. I shall return the ones you mention. I had of course read your alternation of papers with Dr. H. Jeffreys in the *Proc. R.S.*[1] . . .

The trimming up of the mathematics of probability, to which you allude, is of secondary importance compared with a correct grasp of the basis, and a recognition of what the variables are when encountered in practice. Several interesting books in the last three years have appeared, founding probability very prettily on the theory of measure and integration of sets of points; but refraining from mentioning what these sets are in such cases as the tossing of an inhomogeneous, irregular and biassed die, etc. No subject is so perennially interesting, or uselessly controversial.

[1] See *CP* 109 and Jeffreys, H. (1934). Probability and scientific method. *Proc. R. Soc.* **A 146**, 9–16.

Fisher to A.C. Aitken: 23 January 1936

I am very glad to have your letter, and agree with you entirely as to the position in mathematical probability. In some form or other the subject must sooner or later be introduced, or re-introduced, into the teaching of mathematics in Universities, and, in this respect, the question of the nomenclature and affiliation of the subject really deserves some consideration. In his great book it was clearly Laplace's intention to enlarge the meaning of the term 'theory of probability' so as to include a wide range of mathematical studies, equations of finite differences, for example, which are cognate but do not involve uncertain inference or its mathematical specification in any form. The same tendency extended to the present day would make the subject include all the various topics vaguely associated with statistics. It is now clear, however, and it was, I think, clear to Gauss, that this was bound to happen, that the old theory of probability would come to mean a group of rather academic topics, studied as a preliminary to statistics. If it had not any other association, therefore, the word 'statistics' might be used for the whole subject which is now opening up, and I used to think that this would be the appropriate modern position, but I now rather doubt it, and would be glad to know your opinion on the way in which these studies may best be furthered. To many, 'statistics' means Government publications, and in Universities it is often absurdly confused with economics. I was greatly attracted by Whittaker's term, 'the calculus of observations', which, if it were shorter, would cover the ground admirably. I suppose the term 'theory of errors' could properly be extended to cover the distributions of statistical estimates, and might have been used for the whole theory of estimation, had I not been concerned with the temporary necessity of making my own approach quite explicitly clear, and to avoid the assumption that I was accepting previous formulations of analogous problems. Even here, though, confusion has been introduced through Pearson and Neyman using the term 'theory of estimation' for the views they have developed on tests of significance, without reference to the results of the original theory.

From my point of view the important point is that the original concept of probability is not adequate to specify the nature of the uncertainty inherent in many forms of inference from observations. From this point of view it is almost unfortunate that a group of cases has been found in which inductive inference may properly be expressed in terms of probability, using the fiducial mode of argument; for this has tempted some mathematicians, and will, I fear, tempt more, to imagine that this type of argument is more widely applicable than is really the case, and to avoid enlarging their imaginations sufficiently to grasp the cases where no probability statement is adequate. This is, in my view, a decisive reason against enlarging the meaning of the theory of probability so as to cover all types of inductive inference, since the word 'probability' must be tied closely to one quite defined mathematical concept.

Pray excuse this long dissertation.

G.A. Barnard to Fisher: 14 *October* 1945

I am enclosing a reprint of my original letter to *Nature* on 2 × 2 tables,[1] together with a copy of a reply[2] to your remarks [*CP* 205] which I have sent in.

As one whose training prior to the war was exclusively mathematical, and who is only too conscious of the fact that his practical experience of experimental work is virtually restricted to development work during the war, I should like to say that there is no one with whom I should less like to disagree than yourself. I am therefore hoping that the apparent disagreement that now exists is principally due to the excessive brevity of my original letter to *Nature*.

May I add a few more points of explanation which I have omitted from the letter to *Nature* for the sake of brevity?

First, I think the first two paragraphs of my first letter are somewhat misleading, in so far as I did not draw a clear distinction between the test of association and the test of homogeneity. Your test is primarily, (though not exclusively) to my mind, a test of association—analogous to a test of correlation—while my test is a test of homogeneity, analogous to a test of equality of means. If X and Y are two random variables representing the events in question (and taking only the values 0 or 1), as I see it the test of association is a test of the correlation coefficient

$$\rho_{XY} = \{M(XY) - M(X)M(Y)\}/(\sigma_X\sigma_Y) = 0? \quad (1)$$

while mine is simply a test of equality of means, $M(X) = M(Y)$? (2)

Ordinarily there would be no confusion between these two questions, but when X and Y represent events (and not measured quantities) and so take the values 0 and 1 only, we have $X^2 = X$ and $Y^2 = Y$, and then *provided neither* σ_X *nor* σ_Y *is zero*, we can transform (1) to make it look like (2).

Because, if neither variance vanishes, (1) reduces to the question, does

$$M(XY) = M(X)M(Y)?$$

which may be put

$$M(XY) - M(X)M(XY) = M(X)M(Y) - M(X)M(XY),$$

i.e.

$$M(1 - X)M(XY) = M(X)M(Y - XY),$$

or

$$M(XY)/M(X) = M\{(1 - X)Y\}/M(1 - X).$$

Now XY is a random variable representing the event Y, given that X has happened, and $(1 - X)Y$ is a random variable representing the event Y given that X has not happened. If we represent these two new events by A and B, then our last equation reduces to

$$M(A) = M(B)$$

which looks like (2). The division respectively by $M(X)$ and $M(1 - X)$ is done in order to make the number of experiments involved equal in the two cases.

Now if all the animals die, this is the case where the variances do vanish, and in this case (1) ceases to have a meaning, while (2) still makes sense.

The other point I should like to make is that my use of the adjective 'powerful' was simply for the sake of brevity, and it in no way means that I accept Neyman and Pearson's theory of testing hypotheses. In fact, what I tried to do with my test was to base it on what might be called 'common-sense' notions. Apart from the condition of 'validity', the condition of 'symmetry' arises directly from the fact that the table

$$\begin{Bmatrix} a & b \\ c & d \end{Bmatrix} \text{ is really the same as the table } \begin{Bmatrix} b & a \\ d & c \end{Bmatrix}$$

while the condition of convexity arises from the fact that, if we consider

$$\begin{Bmatrix} a & b \\ c & d \end{Bmatrix}$$

to be 'significant', then we must (on common-sense grounds) consider

$$\begin{Bmatrix} a-1 & b+1 \\ c & d \end{Bmatrix}$$

to be significant (provided $a<b$). The condition of 'maximum number of points' is perhaps the most doubtful one; I base the argument for this (a) on the requirement that we must have a unique test, and then (b) on the idea that we have no means of choosing, on the data given, one point of the lattice diagram rather than another—which amounts to saying we are testing the null hypothesis, but we have not specified any particular alternatives against which we are testing it. This last idea is certainly in flat contradiction with Neyman's ideas. (see P.S. below).

If I may say so, it has for a long time struck me that Neyman and Pearson's ideas have caught on widely, because they are based on an explicit theory of probability (the neo-classical theory), and they are therefore more easily put in 'clear' mathematical language than your own ideas. It is only too true, however, that 'clear' mathematical language often presents such an abstract picture of the true state of affairs, that 'clarity' is gained at the expense of truth. As far as I know (and I should be very grateful for correction), your own ideas on *probability*, as distinct from statistical testing, have not been set out explicitly in a complete form—I have found suggestions in your early papers in *Proc. Camb. Phil. Soc.*, but they do not seem to have been developed in full.

During this summer I have been trying to work out a theory of mathematical probability which goes beyond the usual 'additive set function' notions, and tries to account for the *origin* of the notion of distribution. The theory is algebraic in character, and it appears that the notion of invariance under the symmetric group, or under sub-groups of the symmetric group, plays an essential part. The theory is in a sense purely mathematical, and it seems to have purely mathematical points of interest, but in addition I venture to think that the role of the symmetric group in this theory parallels the role of 'randomisation' in your theory of significance testing; so that I speculate that this theory may make it possible to 'formalise' some parts of your own theory.

I have nearly finished writing out a preliminary account of this theory, and I should be very grateful for the privilege of sending you a copy for criticism.

P.S. If we make another 'common-sense' condition, that the 'significance' of the result does not depend on the unit of measurement chosen, we can derive certain cases of the *t*-test in a similar way.

[1] Barnard, G.A. (1945). A new test for 2 × 2 tables. *Nature* **156**, 177.
[2] See *Nature* **156**, 783.

Fisher to G.A. Barnard: 30 *October* 1945

I am sorry I have not answered before your letter with your entirely courteous reply to my comments in *Nature* [*CP* 205]. I think I now understand your position much better, though I do not altogether agree with it. You say[1] 'In case (1) we do not have in mind any proposal to do away with blue-eyed people in order to reduce colds', but we surely do have in mind proposals of this kind, e.g. with people bearing in one of their chromosomes the gene for Huntington's Chorea, in order to reduce the incidence of this distressing form of madness. In any case, if blue-eyed people did have colds more than brown-eyed people, the scientific fact would be that a generation with fewer blue-eyed would have fewer colds, supposing the conditions to remain otherwise unaltered, and the question always seems to be whether there is or is not such an association. The all blue-eyed population *is* a possible world, as doubtless every good Japanese would agree.

My quite general point is that even when we fix the total volume of data, it is sometimes more informative than at other times, and it is the general function of ancillary information to notify us of how good or informative our data actually are in all cases where they may be variable in this respect.

Expecting a ratio 1:1, I may have 14 plants, four of one kind and ten of another. The obvious convention is to compare this with a frequency distribution of the various ways in which 14 would be subdivided, given by the binomial $(\frac{1}{2} + \frac{1}{2})^{14}$; but repeated experimentation may not always give 14. Even if I fix the number of seedlings planted out, there may be variation due to some other factor, e.g. perhaps about half the plants are white-flowered and so not classifiable as between original classes pink and purple. I put out in all 25 plants, and I may get 14, but I may get only 11 to be classified. I believe the appropriate test of significance disregards the total of 25 plants which is constant from trial to trial of the kind I am making, and regards the actual randomly distributed number 14 of plants which supply information relevant to my problem.

How do you think this goes?

[1] *Nature* **156**, 783.

G.A. Barnard to Fisher: 12 *January* 1949

Two years ago you were kind enough to write to me about a letter I had sent to *Nature*, about 2 × 2 tables. I did not reply, because your letter gave me so much food for thought that I hesitated to commit myself to paper. I am now sending you the fruits of the thoughts suggested by your letter. I am afraid they are rather long; but I should be most grateful if you could find time to read them.

The duplicated paper is to be read before the Research Section of the Royal Statistical Society on March 10th,[1] but it does not have to go to the printer for another month or so. If you have any alterations to suggest—particularly to the passages in which I suggest that the theory put forward is essentially a development of your own—I should be glad to make them, with proper acknowledgement.

[1] See Barnard, G.A. (1949). Statistical inference. *J. R. Statist. Soc.* **B 11**, 115–39.

Fisher to G.A. Barnard: 21 *January* 1949

Thank you for your letter and enclosures. The typescript was of course of particular interest to me, for there you have, so to speak, developed a general calculus of likelihood applicable more widely than to those situations for which I first introduced the idea.

This seems to me particularly worth doing in view of the fact that, whereas exhaustive estimation leading to [a] probability statement may or may not be possible, the likelihood function becomes well defined as soon as distinct theoretical ideas are formed at all.

Consequently I congratulate you on your enterprise and on your interesting paper. . . .

G.A. Barnard to Fisher: 13 *October* 1953

First, may I thank you for the magnificent dinner on Thursday evening. I do hope your train was not so late back in Cambridge as to make you resolve never to take it again, for I hope you will do me the honour of allowing me to be the host to you next time there is a Council[1] meeting at that time of day.

I am afraid I shall always be a bad correspondent with you, because what you have to say sets up in me a train of thought which is liable to go on for months—even years, as with the remark, which you have probably forgotten, about what happens to the 'fixed' sample size when someone treads on one of the plants. . . .

Coming now to another point we discussed, the definition of 'sufficient statistic' which has now gained widest currency among *mathematicians* goes as follows:

(I) A statistic t is sufficient for a parameter θ if the distribution of any other statistic t', given t, does not involve θ.

This is the way, for example, the thing is put in Wilks.[2] I had assumed that it was your definition, until I looked the matter up in your collected papers. There, the nearest to a definition seemed to be given in 'The mathematical foundations of theoretical statistics' [*CP* 18], (p. 316) where we have:

(II) 'That the statistic chosen should summarise the whole of the relevant information supplied by the sample.

This may be called the *Criterion of Sufficiency*'.

You elsewhere show that (I) is a *true statement* (not a definition) when there are no unknown parameters other than θ, but you do not (so far as I can find) give it as a definition.

I would not bother you with my confusions on this point if I did not know that such confusions are quite widespread. For example, I remember Fieller telling me that the sample correlation coefficient r was not sufficient for the population ρ, because the distribution of the sample means $(\bar{x},\bar{y})$ given r, also involved ρ. I replied that r was 'in a way' sufficient, because you could obviously not make use of the information in $(\bar{x},\bar{y})$ without knowing the population means (μ_1,μ_2). But I thought at the time that Fieller was strictly right according to the definition. Now it seems not. If we take definition (II), r is sufficient for ρ, in the absence of knowledge of μ_1,μ_2, and it only fails to be sufficient if we take (I) as a definition, rather than as a special consequence of the definition.

In fact, so far as I know, you have not given a mathematical definition (as 'opposed' to a logical definition) of the phrase 'sufficient statistic for a parameter θ'. May I now suggest the following:

(A) If $x_1,x_2,\ldots x_n$ are independent observations on X, whose probability function is $\phi(x \mid \theta,\theta')$, the statistic $t = t(x_1,x_2,\ldots x_n)$ is a *statistic for the parameter* θ if, for any $Y = f(X)$ which has a probability function of the same form $\phi(y,\xi,\xi')$, the condition $\theta = \xi$ implies that

$$t(f(x_1),f(x_2),\ldots,f(x_n)) = t(x_1,x_2,\ldots,x_n).$$

(B) A statistic t for a parameter θ is sufficient for that parameter if the conditional distribution, given t, of any other statistic t' which is a statistic for θ, is independent of θ.

The definition (A) differs from what I suggested the other night in that one does not need to have a *group* of transformations—one transformation may be enough. . . .

[1] Council of the Royal Statistical Society. Fisher was President and Barnard a Vice-President of the Society.

[2] Wilks, S.S. (1947). *Mathematical statistics*. Princeton University Press.

Fisher to G.A. Barnard: 17 *October* 1953

Thanks for your letter explaining those of your statements the other night that I did not understand. About sufficiency, at various times in the past I have tried out on classes and discussions a system of ideas much like the following:

(a) A set of functions of the observations, the joint distribution of which in whatever is to be called the population of random samples is independent of all parameters, is a set of ancillary statistics.

(b) A set of functions of the observations such that, given all members of this set, the distribution of any functionally independent function of the observations is independent of all parameters, is an exhaustive set of statistics.

Scholium: By throwing out unnecessary members there is, I suppose, a minimal exhaustive set, but the possibility of such a minimal set not being unique probably needs investigating.

(c) A set of functions of the observations and of the parameters, the simultaneous distribution of which is independent of all parameters, is termed a pivotal set of functions.

Scholium: One rather likes problems in which the maximum number in the pivotal set is equal to the number of parameters and in which only the minimal exhaustive set of statistics is involved.

Probably I have mis-stated some of this, as it is a long while since I have tried to expound it to anyone.

Fisher to G.A. Barnard: 21 *November* 1953

I have just received the enclosed from E.T. Williams of the *D.N.B.*[1] Perhaps you might forward it to [R.F.] George as R.S.S. business.

I am inclined to suggest that you might write an appropriate biographical notice calling attention to the omission in the *D.N.B.* and to their rather surprising policy of not making such omissions good. If we published in the Society's journal it would not only come within reach of the very large membership of the Society, but you would be able to secure offprints for your own distribution.

If you think well of this suggestion, you might enclose my letter also to George.

[1] Fisher had sent the following letter to the Editor, *Dictionary of national biography*, on 31 October 1953 and E.T. Williams had replied advising that earlier omissions are not made good.

re The Reverend Thomas Bayes, F.R.S.

I understand that it is your custom in making up supplementary volumes for the *Dictionary of National Biography* to include, not only those who by recent death have become eligible for inclusion, but also some others of earlier date whose distinction can now be more clearly appreciated.

The Council of the Royal Statistical Society, of which I am this year President, wishes to draw your attention to the case of the Reverend Thomas Bayes (1697–

1762)(?) who in the twentieth century has become one of the best known figures in the history of the development of our understanding of inductive reasoning, a central theme in the study of statistical methodology. Biographies already exist in your *Dictionary* of his father and grandfather who were distinguished as dissenting Ministers, but not of their descendant who was Fellow of the Royal Society for twenty years, and has gained perhaps a more lasting celebrity.

Bayes was cited by Laplace in his *Theorie Analytique*, and obviously exercised a profound influence on this great French writer. In 1838 he is referred to by Augustus De Morgan[1] in the terms 'This was first used by the Rev. T. Bayes, in *Phil. Trans. liii.* 370.; and the author, though now almost forgotten, deserves the most honourable remembrance from all who treat the history of this science'. It is, however, only in the present century, in connection with the theory of experimental design, and of inductive inference in general, that widespread attention has been focussed on Bayes' important contribution.

It would not be difficult at the present time, although many particulars have doubtless been lost, to obtain a competent notice of this distinguished mathematician. To my knowledge, for example, Professor G. Barnard of The Imperial College of Science, and a Vice President of this Society, has been gathering particulars of his family and career. My Council would all be gratified if, through their mediation, it were possible to add to the national series a biographical notice worthy of so remarkable a subject.[2]

[1] De Morgan, A. (1838). *An essay on probabilities and on their application to life contingencies and insurance offices*. Longman, London.

[2] An omissions volume of the *Dictionary of national biography* is being prepared in 1989 and this will include Thomas Bayes.

Fisher to G.A. Barnard: 9 February 1954

. . . I find, looking up the old papers, that I can now understand, much better than before, the early work of Neyman, or Neyman and Pearson, in the light of what you said the other afternoon, for it now seems clear to me, as it did not before, that Neyman, thinking all the time of acceptance procedures, was under the misapprehension that my own work on estimation had only the same end in view.

A subsidiary trouble, which has emerged again and again since, is his non-recognition of the fact that rigorous inductive inference must include the totality of the available information. A good example of the latter I can quote from a paper of 1933, [*Proceedings of the*] *Cambridge Philosophical Society*, volume 29, page 492. In the middle of the first paragraph he says, 'In dealing with the problem of statistical estimation, R.A. Fisher has shown how, under certain conditions, what may be described as rules of behaviour can be employed which will lead to results independent of these probabilities; in this connection he has discussed the important conception of what he terms fiducial limits'.

The points to notice here are (i) the change from rules of inference to rules of *behaviour*, and (ii) the failure to distinguish between inferences valid when no Bayesian *a priori* knowledge is available, and inferences valid whatever may be the nature of such information *a priori*. He ignores completely my warning of three or four years earlier that the fiducial distribution would be invalid to any one possessing knowledge *a priori* in addition to the observed sample.

G.A. Barnard to Fisher: 14 *June* 1954

I enclose the notes on the 'Creasy–Fieller paradox'[1] which I promised. Of course, my elaborate argument is not intended to replace the simple one used by Bliss and Fieller, but to show that if we start by the roundabout route taken by Miss Creasy we still arrive at the same result, provided we proceed with proper care.

I would very much appreciate any critical comments you may have time to make.

On the other paradox, 'Mauldon's paradox',[2] while the Jacobian has a constant sign, one of the sub-Jacobians changes sign. I wonder whether one should require not only that $\partial(u,v,w)/\partial(x,y,z)$ should remain of constant sign, but also that the sub-Jacobians $\partial(u,v)/\partial(x,y)$, $\partial(u,v)/\partial(x,z)$, . . ., $\partial(u,w)/\partial(x,y)$, . . etc. should all remain of constant sign?

The paradox arises from the general Wishart distribution, but the simplest case can be put as follows:

Define $Q = Q(x,y,z|\alpha,\beta,\gamma)$ by

$$Q = \{(x/\alpha)^2 + (y/\beta)^2 + 2(x/\alpha)(y/\beta)\sin z \sin \gamma\}/2,$$

and consider the trivariate distribution

$$dF = (\cos^2\gamma/\pi\, \alpha^2\beta^2)xy\, e^{-Q}\, dxdydz,$$

with $0 < x < \infty$, $0 < y < \infty$, and $-\pi/2 < z < +\pi/2$, where the variates are x,y,z and the parameters are α,β,γ, with $\alpha > 0$, $\beta > 0$, and $-\pi/2 < \gamma < +\pi/2$.

If we make the transformation

$$u = (x/\alpha)\cos \gamma,\ v = (y/\beta)\cos z,\ w = (x/\alpha)\sin \gamma + (y/\beta)\sin z,$$

we find that

$$Q = (u^2 + v^2 + w^2)/2.$$

and the Jacobian is

$$\partial(u,v,w)/\partial(x,y,z) = \begin{vmatrix} (1/\alpha)\cos\gamma & 0 & 0 \\ 0 & (1/\beta)\cos z & -(y/\beta)\sin z \\ (1/\alpha)\sin\gamma & (1/\beta)\sin z & (y/\beta)\cos z \end{vmatrix}$$

$$= \{y/(\alpha\beta^2)\}\cos\gamma$$

giving the distribution

$$dF = (1/\pi)u\, e^{-(u^2+v^2+w^2)/2}\, du dv dw,$$

where $0 < u < \infty, 0 < v < \infty, -\infty < w < +\infty$.

Since this is independent of all parameters, the set (u,v,w) are pivotal quantities, and given the observations (x,y,z) they therefore appear to determine the fiducial distribution of (α,β,γ) as

$$dF = ((1/\pi)x^3y \cos z)\, \alpha^{-4}\beta^{-2} \cos\gamma \cdot e^{-Q} d\alpha d\beta d\gamma.$$

But if instead we had taken the transformation

$$u' = (x/\alpha)\cos z,\ v' = (y/\beta)\cos\gamma,\ w' = (x/\alpha)\sin z + (y/\beta)\sin\gamma,$$

we would have arrived at the apparent fiducial distribution

$$dF = ((1/\pi)xy^3\cos z)\, \alpha^{-2}\beta^{-4} \cos\gamma \cdot e^{-Q}\, d\alpha d\beta d\gamma$$

which is different from the other one, in that α and β are interchanged.

There does seem to be an internal 'twist', associated with the angular variable z, in the distribution, and this may be connected in some way with the change of sign of the sub-Jacobians $\partial(u,w)/\partial(x,y)$, and $\partial(u,v)/\partial(x,z)$, as z passes through 0.

I hope when examinations and other distractions are finished to chew this over thoroughly.

[1] See Fieller's letter of 10 January 1954 (p. 86).
[2] See Mauldon, J.G. (1950). Pivotal quantities for Wishart's and related distributions, and a paradox in fiducial theory. *J. R. Statist. Soc.* **B 17**, 79–85.

[Enclosure]

THE CREASY–FIELLER PARADOX

1. For simplicity we consider the case of a pair (x,y) of independent observations from normal populations with unit variances and means (ξ,η). Given x, the fiducial distribution of ξ is normal, centred at x, and that of η is normal, centred at y, and these fiducial distributions are independent and each has variance 1. The joint fiducial distribution of (ξ,η) is therefore circular normal, centred at the point (x,y). We suppose, for definiteness, that both x and y are positive.

2. The Creasy–Fieller paradox is concerned with the fiducial distribution of the ratio $\xi/\eta = \alpha$. Fieller considers the pivotal quantity $(x - \alpha y)/(1 + \alpha^2)^{\frac{1}{2}}$, which has a standard normal distribution, and in effect says that the fiducial probability that

$$\alpha_1 < \alpha < \alpha_2 \text{ is } \phi\{(x - \alpha_2 y)/(1 + \alpha_2^2)^{\frac{1}{2}}\} - \phi\{(x - \alpha_1 y)/(1 + \alpha_1^2)^{\frac{1}{2}}\}, \qquad (1)$$

where $\phi(u) = (1/\sqrt{2\pi}) \int_{-\infty}^{u} e^{-t^2/2} dt$. As α ranges over $(-\infty, +\infty)$,

$|(x - \alpha y)/(1 + \alpha^2)^{\frac{1}{2}}|$ cannot exceed $(x^2 + y^2)^{\frac{1}{2}}$, and this corresponds to the fact that, at the level of significance $2\phi\{-(x^2 + y^2)^{\frac{1}{2}}\}$, we have no evidence any linear combination of ξ and η differs from zero, i.e. we have no evidence at this level of significance that the ratio ξ/η exists at all.

3. Miss Creasy takes the joint, circular normal, distribution of (ξ,η) and to obtain the fiducial probability that $\alpha_1 < \alpha < \alpha_2$ she integrates this distribution over the sectors BOD, AOC, where AOB is the line $\eta = \alpha_1\xi$ and COD is the line $\eta = \alpha_2\xi$. Her result differs from Fieller's.

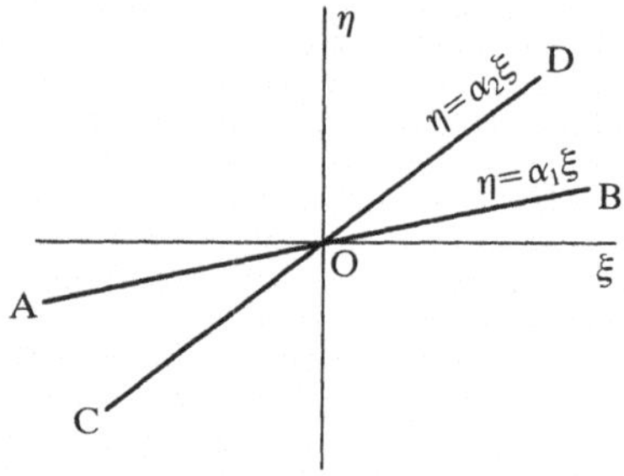

4. The fallacy in Miss Creasy's argument seems to lie in that when we transform from the pair of variables (ξ,η) to a pair (α,β), where $\alpha = \xi/\eta$ and β is any function of ξ and η, the origin is a singular point of the transformation. We may contrast this position with that arising, for example, in the Fisher–Behrens' problem, where we are concerned with the difference $\xi - \eta$ instead of the ratio ξ/η, and where a non-singular transformation is possible.

5. It may be objected that Miss Creasy's method is the same as that ordinarily used to find the distribution of the ratio of two normal deviates. Why should this method not work when we are dealing with fiducial probability?

6. The answer to this objection is that strictly speaking the ordinary method does not give us the distribution of the ratio of two normal deviates unconditionally. It does so only on the assumption that the ratio exists. We must assume that the case (0,0) can be excluded, on some ground or other. Now in most applications of the theory of probability we can interpret events of probability zero as being impossible, though a consistent application of this rule in all cases would lead to contradictions. Because in most cases we can thus ignore possibilities whose probability is zero, we argue here that the case (0,0) can be ignored, as having probability zero. This means that the origin can be removed from the plane we are considering, and then our transformation becomes non-singular.

7. The principle that propositions of probability zero can be ignored, while it is often, but not always, applicable when the propositions are concerned with observations, seems hardly ever to be applicable to propositions concerned with parameters. With observations, we are most often concerned with series of independent trials, and with events which occur in such series with finite relative frequencies. With parameters, the concept of a series of independent trials has not the same interpretation. We may say, roughly, that when we are dealing with a parameter, such as ξ, we have to reckon with the fact that, if by chance ξ were zero, it would forever retain this value, since it is a constant. While if an observation were found to have this value, we could be

sure the next observation would have a different value. The observation with direct probability zero could occur, if at all, only transiently; a parameter value with fiducial probability zero, if it occurred, would remain so for ever.

8. In transforming the joint fiducial distribution of (ξ,η), therefore, special attention must be paid to the possibilities $\xi = 0$, $\eta = 0$. Perhaps the most natural way of doing this is to interpret the fiducial distribution, in relation to the ratio of the means, in the light of the fact that we normally restrict our considerations to the ratios of quantities both of which have the same sign. A very general method of expressing a ratio is as a percentage, but one does not hear of one quantity being −50 per cent of another.

9. If we interpret the fiducial distribution of (ξ,η) in this way, we are led to interpret points lying to the left of the η-axis as corresponding to a non-significant departure of x, and for such cases we may regard the ratio ξ/η as having the 'improper value' 0 — a value improper in the sense that it does not, in conjunction with the value of ξ, determine that of η. Similarly, we may interpret points below the ξ-axis as corresponding to $\eta = 0$, and the improper value ∞ for the ratio ξ/η.But if we thus interpret points to the left of the vertical axis, and points below the horizontal axis, we shall have counted twice the points lying in the third quadrant. To restore the balance, in considering the points lying between the lines AOB, COD, corresponding to values of the ratio between α_1 and α_2, the sector AOB, lying in the third quadrant, must be taken negatively, while that in the first quadrant is taken positively. We are thus led to take, as the fiducial probability that the ratio ξ/η lies between α_1 and α_2 the difference of the probabilities associated with the areas underneath the lines AOB and COD. But the probability associated with the area under AOB is evidently that associated with a normal deviate equal to the distance from the point (x,y) to the line AOB, i.e. to a normal deviate of

$$x - \alpha_1 y/\sqrt{(1 + \alpha_1^2)},$$

and taking the difference of this probability and that corresponding to α_2, we evidently obtain Fieller's result.

Fisher to G.A.Barnard: 16 *June* 1954

Thanks for your letter; I hope I may understand it!

I think there is a distinction of data which may be worth attending to.

If we have a number of observations (x,y) which we interpret as independent shots of some hypothetical pair (ξ,η) with generalised bivariate error of normal specification, I suppose one can calculate $\bar{x}$, $\bar{y}$ and the quadratics

$$A = S(x - \bar{x})^2/\{N(N - 1)\}, \text{etc.},$$

and infer a bivariate fiducial distribution with density

$$\{N/(2\pi)\}\ (1 + r^2)^{-\frac{1}{2}(N+2)}$$

where $r^2(AC - B^2) = C(\xi - \bar{x})^2 - 2B(\xi - \bar{x})(\eta - \bar{y}) + A(\eta - \bar{y})^2$; then the integrated total probability beyond any line will agree (will it not?) with the corresponding 'Student' test for e.g.

$$\lambda(\xi - \bar{x}) + \mu(\eta - \bar{y}) > v,$$
$$v^2 = (A\lambda^2 + 2B\lambda\mu + C\mu^2)\, t_{N-1}^2.$$

On such an interpretation of the data the integral over any chosen region seems to give the appropriate probability statement.

But, in the example in *Statistical Methods* § 26.2, I suppose the two drugs to have effects on any chosen patient in a fixed ratio. I am concerned to set limits to this ratio, not to estimate the properties of any population from which the patients might be supposed to be chosen. This hypothesis gives a series of points, one for each patient, distributed in some unknown manner, but all on one and the same straight line through the origin.

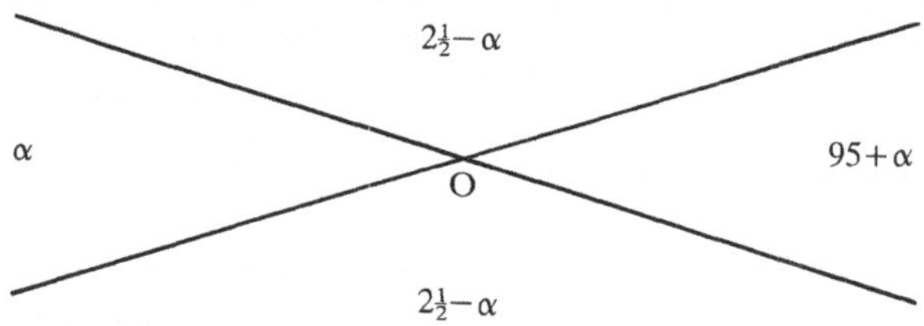

The t tests exclude lines outside the V on the right, and exclude also the V on the left, so that fiducially the probabilities are as above with α undetermined.

If one were to receive an accretion of data to the effect that the patients were normally distributed about some population mean, I suppose one could ease the V in so as to make it include exactly 95%. But this would answer a different question.

Fisher to G.A. Barnard: 15 *July* 1954

I am just writing to see whether, by any chance, a letter of mine of June 16th has not yet reached you.

If it has duly arrived, do not think that I am in a hurry for a reply, though, of course, I should be glad to clear up any point that might have been sufficiently puzzling to delay you in answering.

G.A. Barnard to Fisher: 18 *July* 1954

Thank you for your letter of June 16th and for your further note of July 15th. I must apologise for not having acknowledged earlier your first letter. The fact is, I did not understand it fully at first reading, and so kept it to meditate on. . . . .

I did not write to ask you about my difficulties because my past experience has been that such difficulties have usually been of my own making. It has so often happened that I have found something you have written to be obscure at first reading, only to see later on that this was because I had not understood the problem properly; and I have usually ended by seeing that what you have said was not only correct, but also the simplest and most direct way of stating the truth. And the process of untying the knots in my own mind has been enjoyable as well as salutary.

However, since you encourage me, may I say how far I have progressed up to now?

If I understand you aright, you point out that to assume that

$$z = x - \alpha y$$

is normal, as you do in *Statistical Methods*, Section 26.2, in no way implies that either x or y is normal, nor that they have a joint bivariate normal distribution. For example, if x had a rectangular distribution with range $(-\delta,+\delta)$, and y had the density

$$(1/\sqrt{2\pi}) \int_{-\delta}^{+\delta} \exp\{-(y-u)^2/2\sigma^2\}\, du$$

then z could be normal. The situation is exactly as in regression theory, where the independent variable need not be distributed in any particular way.

If all we know is that z is normal, or can be taken so, for some unknown α, then the method of Section 26.2 gives the only way to proceed. No question of a bivariate fiducial distribution of the means (ξ,η) of x and y can arise in this case.

But if we are given the further information that x and y are each normally distributed, then we have a bivariate normal distribution from which we can infer the joint fiducial distribution of (ξ,η) as on the first page of your letter. The fact that, with this extra information, we can make statements qualitatively different from those we can make without it is not surprising, when we remember the peculiarities of inductive inference, as opposed to deductive inference, which you have so often stressed.

So far so good. But if we consider now the case where we know x and y separately to have independent normal distributions, then we can proceed to discuss the ratio $\alpha = \xi/\eta$ of means *either* by using the single pivotal quantity

$$(x - \alpha y)/(A - 2B\alpha + C\alpha^2)^{\frac{1}{2}}$$

or by using the two pivotal quantities

$$(x - \xi)/\sqrt{A} \text{ and } (y - \eta)/\sqrt{B}$$

and we do not get the same answer. Which method should be used? Here I feel inclined to think that my argument about the singularity at (0,0) shows that the first method should still be used. In other words, although there has been an accretion of data, we are unable to use it.

This last paragraph is where I am now stuck. Of course, in the light of your point about the distinction of the data, the question becomes largely theoretical, since the cases that occur in practice seem all to be covered by Section 26.2.

In all the discussion at the Research Section, no one made your point about the distinction of data, and it would, I think, be a great pity if the discussion went into print without mention of the point. I wonder if you would allow me to mention the matter to Fieller, so that he could refer to it (in terms to be approved by you) in his reply to the discussion? From the practical point of view, what you say certainly disposes of the question.[1]

P.S. I think what I am trying to say in the last paragraph but three above could be put in this way: If, *per impossibile*, we had information that the value (0,0) for (ξ,η) was excluded, but all other positive or negative values were possible, then Miss Creasy's argument would be correct. We could never, of course, have such information; but we might approximate to it in cases where x and y were confined to positive values (cf. the people with negative heights we once discussed)—if, for example, the ratio with which we were concerned were some anthropometric index—and then, to the extent that we were still justified in taking x and y normal, we would have something like this situation. But precisely in this case the means would have to be large compared with the standard deviations, and it is here that Miss Creasy's argument leads to the same answer as that of *Statistical Methods*, Section 26.2.

[1] Shortly after receiving this letter from Barnard, Fisher invited Barnard and Fieller to Cambridge for a discussion of the questions raised.

Fisher to G.A. Barnard: 20 *September* 1954

Thank you for thinking of sending me the little Dutch cigars with which I shall experiment with much pleasure. The container certainly is charming.

I am glad you are thinking of taking up again with me questions involving the use of fiducial probability. I suppose it is agreed that statements of fiducial probability are the only kinds of probability statements that we can make on empirical evidence about the real world.

G.A. Barnard to Fisher: 15 *November* 1954

May I take up yet again the questions on fiducial distributions you were kind enough to discuss with Fieller and me in the summer?

First, you may remember saying that some Russian mathematician had undoubtedly proved a theorem, with many lemmas, to the effect that, if we know, for all α,β, the distribution of $U = \alpha X + \beta Y$, then we know the joint distribution of X and Y. The Russian in question is Romanovsky, who proved in 1928, in the bivariate case, that the characteristic function uniquely determines the distribution function. For the c.f. of U is ξ exp $itU = \xi$ exp $i(\alpha tX + \beta tY) = f(\alpha t,\beta t)$, say. The joint c.f. of X and Y is then ξ exp $(iuX + ivY) = f(u,v)$. Since we know $f(\alpha t,\beta t)$ for all α,β, we know $f(u,v)$ for all u,v.

Now some questions. They are general ones, because I cannot expect you to recall the details of what we discussed so long ago, and also because I think I can see the points better in the more general cases.

Question 1: If we have a sample of n, with mean $\bar{x}$ and s.d. s, and we find the fiducial distribution of μ in the usual way, it is a t-distribution centred at $\bar{x}$. Now suppose that for some reason we can rule out negative values of μ, as being impossible. Can we then take the fiducial distribution to be a 'mixed' one, having the same t-distribution form to the right of the origin, but having the probability which formerly was to the left of the origin all lumped at the origin? This seems to be reasonable. Am I wrong?

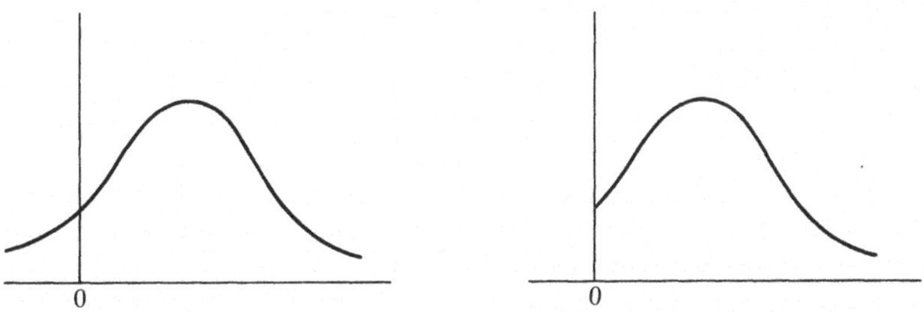

All values of μ possible

Only positive values of μ possible

A mixed type of fiducial distribution like this arises in your paper on 'Dispersion on a sphere' [*CP* 249], where there is a lump of fiducial probability corresponding to the possibility that the vectors are randomly oriented.

The bearing of this on the question of the ratio of two means is, that in the case where (x,y) are known to be jointly normal, and independent, we have, as you say, the option of choosing sectors so that the difference A − C is α, or we may slightly alter them to make the sum A + C = α. Now what should we do? It seems to me that when we can rule out negative values of the means, as impossible, we should lump as above, and then it is appropriate to make A − C = α.

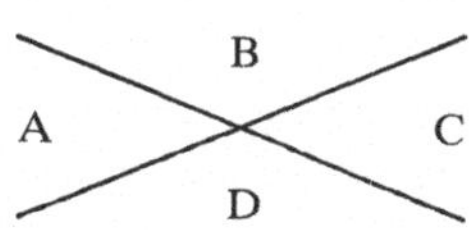

But if we cannot rule out negative values, and still wish to discuss the ratio of the two means, then we should make the *sum* A + C = α. I cannot think of an example for this latter case, and am inclined to think that it could not arise in practice; but the theoretical possibility seems there.

Question 2. In deriving pivotal quantities u,v,w, from the sufficient statistics x,y,z, for the parameters α,β,γ,

$$\begin{aligned} u &= f(x,y,z;\alpha,\beta,\gamma) \\ v &= g(x,y,z;\alpha,\beta,\gamma) \\ w &= h(x,y,z;\alpha,\beta,\gamma) \end{aligned}$$

you have said that the Jacobian must remain of constant sign, and I note George Owen as saying (in his 1948 *Sankhyā* paper) that 'certain monotonicity conditions', e.g. constancy of sign of the Jacobian, are required.

My question is, should we not be cautious whenever f, g, and h are not monotonic in x,y, and z? In other words, should we not require, in general, that not only the Jacobian, but also the sub-Jacobians (e.g. of (u,v) with respect to (x,y)) should also be of constant sign? If this stronger condition is violated, funny things seem able to happen. The possibilities are more easily pictured than written about, I think.

This question relates to a point raised by Mauldon, of Oxford, who obtains two distinct 'pseudo-fiducial' distributions from the same data, by using a transformation in which the sub-Jacobians change sign, although the Jacobian itself stays positive.

Third (and last!) question: Can we not say, whenever we have a given body of data, relating to parameters θ,ϕ, etc., that the likelihood function represents an inference from the data? And can we not deduce, from the likelihood function, that e.g. this pair of values for θ,ϕ is ten times as likely as that pair of values? In this sense, can we not make inferences in terms of likelihood, even when fiducial probability statements are not possible? I would not be happy to say that fiducial probability statements are the only possible ones respecting unknown parameters which it is possible to infer from samples. Fiducial probability statements are the only possible *probability* statements which it is possible to infer; but likelihood statements are always there too. Is this right? . . .

Fisher to G.A. Barnard: 16 *November* 1954

I had been hoping to hear from you, and your letter is even a greater pleasure than I had expected.

i) Your treatment of the uniqueness of the two-way distribution function when the probability dichotomy is given for all lines on a plane is delightfully simple and direct.

ii) I should think, as you do, that there is a condensation of fiducial probabilities in the case you mention, and in a fair number of other cases, of which one of practical importance arises from the fact that we can assume an unknown variance to be non-negative. I have not, however, so much experience of cases of this kind arising in practice that I should be shaken if anyone could show that this type of argument leads to unreasonable conclusions, but then I feel this way about all self-evident axioms. For reflection, however, consider the case where negative values of zero measure, e.g. all numbers of the form $-N/2^n$, were admissible, but not other negative numbers.

iii) In applying this method to the case of (x,y), known to be jointly normal, you say also 'and independent' so are not discussing the case I took in my note [*CP* 257] on the general shemozzle between Fieller and Miss Creasy, but the one I mentioned later in a letter.[1] I agree with your first proposition, but would modify your second from 'still wish to discuss the ratio of the two means', by the addition, 'and to ignore whether for a positive ratio they are both positive or negative', for it seems that the data do also discriminate these cases.

iv) I am very glad that you are thinking of the question of pivotal uniqueness which I feel sure resembles, even if it is not strictly isomorphic with, the question of mapping one set on another, or rather an infinite series of sets. I have never got my head quite clear about this, but it would be exceedingly interesting to demonstrate that the cases are, or are not, strictly isomorphic.

v) I agree with you entirely that there are cases in which we have validly determined likelihood ratios for all theoretical possibilities, and no probability statements, and so far as I can see, this must be the general and prevailing case in respect of scientific data in general. From this point of view it might be thought an unfortunate coincidence of a mathematical nature that so many practically important questions should lead (trifles having been ignored) to statements of fiducial probability, since it has led the world of mathematical statistics to ignore the manifest fact that on some realistic types of data the condition of our rational uncertainty is not expressible in terms of mathematical probability at all. . . .

[1] See Fisher's letter of 16 June 1954 (p. 13).

Fisher to G.A. Barnard: 2 December 1954

It was good of you to turn up to the discussion on Braithwaite's talk,[1] which, indeed, had not much to do with the theory of testing hypotheses. I tried to catch you afterwards without success, so I am writing now to know whether you had ever accomplished the work I rather threw at you last year, of preparing a biographical notice of Thomas Bayes.[2] I believe you had sorted out his family better than had been done in the *D.N.B.*, at least, if I understood you aright, in the definitive edition first published, I believe, in

1900, though I suppose such things can always be corrected in supplementary volumes.

Anyway, I should very much like some time to see Bayes done justice to.

[1] On 22 November 1954, Professor R.B. Braithwaite gave a talk entitled 'Choosing between statistical hypotheses' to the Cambridge University Philosophy of Science Club.
[2] See Fisher's letter of 21 November 1953 (p. 8).

G.A. Barnard to Fisher: 14 *December* 1954

Thank you so much for your letter. I enjoyed the trip to Cambridge very much, particularly because you made quite clear to me the difference of principle between the random allocation of treatments to plots, and the randomisation of an inference. Of course, one feels instinctively that the one is sensible, while the other is, to say the least, suspect. And you had said to me before that the difference lay in the fact that in the second case one *knows* the value of the random component. But until you explained it again, at the meeting, I did not have it quite clear.

The point about randomisation has more importance, I think, than Prof. Braithwaite seemed inclined to admit. I enclose two quotations from Blackwell and Girshick, in the first of which they make such a virtue of randomisation that it becomes a reason for rejecting Bayes principles. (The fact that they virtually contradict this in the second quotation merely shows, I think, that the first passage was written by Girshick and the second by Blackwell). Actually, the kind of randomisation to which Bayes principles fail to lead is the second kind, which, as you have pointed out, is different in principle from randomisation in sampling.

The two quotations are interesting, I think, as indicating a schizophrenic attitude to Bayes principles. Girshick, the more practical and less mathematical, seems to feel instinctively that he must avoid subjective conclusions; while Blackwell, the better pure mathematician, sees quite clearly that the decision function approach leads to subjective conclusions, unless the prior distribution is given and he does not have the practical experience to make him appreciate that this is objectionable.

I shall make a point of coming on January 17th, and look forward to a still more pleasant and profitable evening.

About Thomas Bayes, there was little response to the notice in the Journal.[1] I had a note from the Equitable Life Assurance Company, to say that they believed they had, in their muniment room, a notebook of Bayes, and their man put me on to someone who had made a study of Bayes from a local antiquarian point of view. This man had looked up the copy of Bayes' will in Somerset House, but apart from confirming the impression that Bayes must have been a kindly man, it was of little interest. He had seen the notebook, and I gathered it contained little of importance. However, I must make time to go over to the Equitable offices and look at it myself.

Apart from this, there seems little hope that more information than I have already circulated will come to hand. I hope to be able, after Christmas, to visit the Equitable, after first looking at the Canton MS. in the Royal Society's Library, to check Bayes' handwriting, and then it seems we shall have to close the account. I am afraid the biographical information will hardly do Bayes justice. . . .

[1] See *J. R. Statist. Soc.* **A 117**, 120 (1954).

[Enclosure: from Blackwell, D. and Girshick, M.A. (1954). *Theory of games and statistical decisions.* Wiley, New York]

p. 115: 'An objection often raised to the minimax and minimax loss principle . . . is that they do not take into account any information the decision maker may have about ω (the true state of affairs—G.B.). . . . If he can describe his information about ω by a probability distribution ξ over Ω (the set of all possible ω—G.B.), so that ξ(ω) represents the probability, based on his information, that ω is the true state of the system, then the utility of action (rule — G.B.) *d* is simply

$$U(\xi,d) = \sum_{\omega} \xi(\omega)u(\omega,d)$$

($u(\omega,d)$ is the expected gain if ω is the true state and decision rule *d* is used—G.B.).

Thus he maximises his utility by choosing *d* so as to maximise $U(\xi,d)$. This principle of choice, based on ξ, is called a *Bayes principle*. . . . Bayes principles have the objection that, in most statistical games, ξ is simply an expression of the personal judgement of the decision maker, so that two decision makers, facing the same decision problem and using Bayes principles, might well reach different conclusions from the same data. . . . A further objection to Bayes principles is that they never require randomisation; many statisticians consider that random sampling, which is a form of randomisation, is useful'.

p. 119: 'The above theorem . . . indicates rather clearly that, without assuming an *a priori* distribution on Ω, there is in terms of the present theory no adequate principle of choice for choosing a strategy in a statistical game'.

Fisher to G.A. Barnard: 13 *January* 1955

Thanks for your letter and quotations from Girshick and Blackwell. How remote from the original this talk about Bayes' principle is! . . .

When we meet you must tell me the exact date of Bayes' death, which I have not been able to check, with a view to an opening chapter on history to what may be a book on statistical logic. I suppose there is no longer any hope of your unearthing Bayes' original introduction from which Price quotes.

As news, I believe I now understand what Laplace was saying in his definition of probability. I suspect a definite semantic transference from instances (événemens) to theories, or hypotheses, or states of things, as Boole puts it. Anyway, I shall like to show you what I have written.

Fisher to G.A. Barnard: 19 *January* 1955

I had thought of putting my paper on scientific inference in Series B of the Royal Statistical Society, of which I suppose Irwin is still effectively editor.

I have just now, however, been approached by Linder of Geneva, who was given the tip by Mahalanobis, who would like to secure it for a newish journal called *Dialectica*, now in Volume 8, published jointly by the Neuchatel firm and 'Presses Universitaires de France'. It seems, however, to be mostly in German.

I wonder if you could let me know your own opinion as to whether it ought preferably to be published in England? . . .

G.A. Barnard to Fisher: 23 *January* 1955

. . . About *Dialectica*, I think it rather resembles *Mind*. Few statisticians, or even natural scientists of any sort, read it, I think—at any rate in this country or the U.S.A. So I would urge in favour of Series B of the *Journal*. It would then certainly be easily available to those who would profit most by it.

A thought does occur to me, that the last Ordinary meeting of the R.S.S. for this session is not yet fixed up, and I wonder if you would think fit to give it there. That would have the advantage of allowing a London audience to hear it; but presumably the disadvantage that in the ordinary way the paper would be printed in the *Journal* Series A, which is not quite so widely read as Series B. On the other hand, the written discussion and reply would help a good many, I think.

I think this meeting would be in May. If you like, I would be glad to get in touch with R.F. George about this. But if you prefer Series B, Irwin is still the Editor.

Fisher to G.A. Barnard: 25 *January* 1955

Thanks for your letter. I will see what Irwin thinks.

I do not much want to give it to a meeting of the Society, as it means a long and tiring afternoon with a very wearisome and unprofitable discussion usually to follow. The Society's method of persuading a number of weighty authorities to express themselves at length immediately following a communication is, I think, a bad one in regard to the Society's function of promoting mutual understanding.

However, I suppose there is nothing to prevent Series B printing it[1] without its presentation at a meeting.

[1] See *CP* 261.

Fisher to G.A. Barnard: 18 *February* 1955

As you know I have been putting together ideas on scientific inference, and I believe you have seen most or all of the chapter on the history of the subject in England, which is effectively the history of the subject. I have now completed a first draft of the chapter on the forms of quantitative inference, being particularly concerned to stress the primitive nature of simple tests of significance, and to distinguish them from inferences in terms of mathematical probability in the strict sense, and from inferences in terms of mathematical likelihood.

If you have time I should very much like you to see this chapter, of which I am sending you a copy, together with one of the first chapter, in case you wish to refer to it. The excuse for all this burden on your time is that I very much want personally to have your reaction to what I have said, and to *how* I have said it, with a view later to a little discussion among like minded people who share my belief that we need to develop means of self-expression in the natural sciences more appropriate, and for our purposes more accurate, than have been offered by Neyman, Wald and their followers.

To be successful we shall inevitably have to form firm opinions about a great deal of miscellaneous and inconclusive discussion on probability, logic, and what not. . . .

G.A. Barnard to Fisher: 25 *February* 1955

Thank you so much for your letter and enclosures . . . I shall write again after chewing over, but for the present would commend the enclosed extract, in connection with the historical chapter—you may like to quote it—which seems to me to embody the earliest published significance test, by Daniel Bernoulli, writing in 1734. It seems to help also to stress the primitive nature of tests of significance; though why such a mode of reasoning should remain in virtual eclipse for something approaching 200 years is a bit of a mystery.

One other remark. In connection with the fiducial argument, I have the feeling that many people have found this hard to grasp because they are misled by mathematical symbolism. In real scientific work, an observation is something real, and so is what we represent by a parameter. When we regard an observation as one of a series, or a parameter value as one of a series, we become able to speak of them as 'random variables'; but if we choose to regard them by themselves, alone, they will not be 'random variables', in the mathematical sense. Thus if T denotes a function of the observations, and θ the parameter, T, or θ, or both, may denote random variables, or not, according to the point of view we are, at a given moment, taking. What confuses people, among other things, I think, is that in pure mathematics, where we do not reason on things having objective existence, but on pure symbols, we must *fix* our attitude, so that once T is a random variable, it must always be so; and once θ is regarded as a fixed constant, not a random variable, so it must also remain.

I am afraid I do not make myself clear, and seem to be writing very loosely. I will think some more, to see if I can put things more precisely, in writing if possible, and then perhaps in discussion. I have in mind the distinction, in logic, between what is called the semantic approach and what is called the syntactical approach.

I look forward to seeing you.

[Enclosure]

'—Pièce de M. Daniel Bernoulli.
—Recherches physiques et astronomiques sur le problème proposé pour la seconde fois par l'Académie Royale des Sciences de Paris: Quelle est la cause physique de l'inclination des Plans des Orbites des Planètes par rapport au plan de l'Equateur de la révolution du Soleil autour de son axe; Et d'où vient que les inclinaisons de ces orbites sont différentes entre elles.—Paris, Academie Royale des Sciences—*Recueil des Pièces qui ont remporté les prix de l'Académie.* Tome troisième (1734–37), Paris 1752.

III. Avant que d'entreprendre ces deux points, il ne sera pas hors de propos d'examiner plus particulièrement ce que nous avons posé en fait; savoir, *que les Orbites célestes s'approchent de trop près pour ne point affecter quelque plan commun situé au milieu d'elles, et que ce n'est que par une circonstance particulière, que les mêmes Orbites ne sont pas entièrement unies dans un même plan.* Sans cet examen, on pourroit attribuer à un hasard le Phénomène qui fait le sujet de notre question, et regarder tout notre raisonnement comme superflu, ou peut-être même chimérique.

Voici comme je m'y prendrai: Je chercherai de toutes les Orbites planétaires les deux qui se coupent sous le plus grand angle; après quoi je calculerai quelle probabilité il y a, que toutes les autres Orbites soient renfermées par hasard dans les limites de ces deux Orbites. On verra par-là que cette probabilité est si petite, qu'elle doit passer pour une impossibilité morale.

IV. Après avoir comparé chaque Orbite avec chacune, et calculé les angles sous lesquels elles s'entre-coupent, j'ai trouvé se couper sous le plus grand angle l'Orbite de Mercure, et celle de la Terre ou l'écliptique: car leurs plans font un angle de 6° 45′: pendant que l'Orbite de Saturne ne fait, avec celle de Mercure, qu'un angle de 6° 24′; et l'Orbite de Jupiter, encore avec celle de Mercure, un angle de 6° 8′. Toutes les autres Orbites, de quelque manière qu'on les combine, se coupent sous des angles beaucoup plus petits. Je parle ici des Orbites des Planètes principales.

(Il est facile de voir qu'on peut trouver lesdites intersections par la simple Trigonométrie; . . .).

Je m'imagine donc toute la surface sphérique ceinte d'une zone, ou espèce de Zodiaque, de la largeur de 6° 54′. (Car telle est la plus grande inclinaison de l'Orbite de Mercure avec l'écliptique.) Cette zone contiendra à peu près la dix-septième partie de la surface sphérique. Si l'on considère donc les Orbites planétaires comme placées par un pur hasard, il sera question de déterminer quel degré de probabilité il y a, pour que toutes les Orbites tombent dans une zone donnée de position, faisant la dix-septième partie de toute la surface sphérique. Mais la position elle-même de la zone se détermine par une des Orbites, quelle qu'elle soit, puisqu'elles ne diffèrent guère entre-elles; ce qui fait qu'il n'y a plus que cinq Orbites qui entrent en ligne de compte: cela posé, on trouvera par les règles ordinaires, le nombre des cas, qui fassent tomber les 5 Orbites dans ladite zone, au nombre des cas contraires, comme 1 à $17^5 - 1$; c'est-à-dire, comme 1 à 1419856.

(Je ne donne pas à cette méthode toute la précision géométrique, ce que le Lecteur n'aura pas manqué de remarquer; mais je m'en suis contenté, parce qu'il ne s'agit ici que d'avoir quelque idée générale de la chose. Un nombre considérablement plus grand ou plus petit, ne nous feroit pas envisager autrement le point de la question. On voit pourtant assez que notre proportion ne peut-être fort éloignée de la véritable. Mais, me demandera-t-on, quelle est donc la véritable? Je réponds à cette demande, qu'on ne sauroit le déterminer, à cause du mouvement des noeuds qui changent à tout moment les limites des Orbites: j'ai donc simplement considéré une zone, hors de laquelle aucun point des Orbites, quoique changeantes de position, ne sorte jamais, et j'ai comparé cette zone avec la surface de la sphère dont elle fait à peu près la dix-septième partie, tantôt plus, tantôt moins, à cause de la variabilité des limites. Dans cette zone, il n'y a aucun point qui ne soit sujet à être touché par une des Orbites; et hors de la même zone, il n'y a aucun point qui puisse jamais l'être; d'où l'on voit assez le fondement de ma solution. Si tous les noeuds étoient constamment dans un même point commun, il auroit fallu avoir égard au plus grand angle d'intersection de 2 Orbites que nous avons vu être de 6° 54′: et comme cet angle auroit pu aller jusqu'à 90°, si le hasard l'avoit formé, il faudroit comparer ces deux angles, et dire que le premier fait environ la treizième partie du second; d'où l'on tireroit le degré de probabilité, (pour qu'aucune des Orbites ne fit avec une autre Orbite un angle plus grand que de 6° 54′) égal à 1: $(13^5 - 1)$ qui donne une proportion environ quatre fois plus grand, que dans la première solution; savoir, celle de 1 à 371292. Enfin, la meilleure manière de calculer le degré de probabilité seroit de considérer le plan au milieu des Orbites, (qui, selon toutes les apparences, est le plan même de l'Equateur solaire) avec lequel chaque Orbite, quoique mobile, fait sans doute un angle constant, ou presque constant. Si ce plan étoit donné de position, il faudroit calculer quelle Orbite fait le plus grand angle avec ce plan, et quelle est la grandeur de cet angle; et comme dans l'hypothèse des Orbites fortuitement placées, cet angle auroit pu monter jusqu'à 90 degrés, on auroit encore eu à considérer le rapport dudit angle avec celui de 90°, et, posé ce rapport être de 1 a m, le degré de probabilité cherché, seroit maintenant comme 1 à $m^6 - 1$. Je mets ici l'exposant 6 au lieu de 5, que j'ai mis dans les deux exemples précédents, parce que le terme fixé n'est pas ici une des Orbites, mais l'Equateur solaire. Cette méthode me paroîtroit la plus juste de toutes, si la détermination de l'Equateur solaire étoit un peu plus certaine; suivant ce que M. Cassini rapporte dans les Mémoires de l'Académie Royale des Sciences de Paris, de l'année 1701, c'est l'Orbite de la Terre qui fait le plus grand angle avec l'équateur solaire, et cet angle doit être de 7° 30′, cela donneroit $m = 12$, et $m^6 - 1 = 2985983$. Si donc toutes les Orbites étoient placeés fortuitement par rapport à l'équateur solaire, il y auroit à parier 2985983 contre 1, qu'elles n'en seroient pas toutes si proches. Toutes ces méthodes, quoique fort différentes, ne donnent pas des nombres extrêmement inégaux. Cependant je m'attacherai au nombre donné en premier lieu, et n'ai fait cette addition que dans le dessein de faire voir au Lecteur quel fond on y peut faire.)'

Fisher to G.A. Barnard: 26 February 1955

Thank you for your letter and its interesting enclosure. The latter I will cap with a quotation from de Moivre,[1] which I believe, but have not yet checked, to be from his first edition of 1718, and to have an error of 10000 in the calculation, which has been corrected in the third edition by adding four noughts after the first eight digits. My memory being bad, I might be mistaken on some of this, but it seems likely that had he done the thing right, he would have used six figures only and said 'millions' one time more.

You will notice that de Moivre assumes wrongly that he has arrived at a probability statement.

I found your example exceedingly interesting, especially the fact that the Academy had posed for a prize essay competition the problem of the angles between the planes of revolution of the planets, a bone that was still being worried over 100 years later in Boole's *Laws of Thought*. It is curious that Bernoulli does not simplify his problem by considering the poles of the respective planes of revolution, and the probability of them, or their antipoles, being grouped as closely as they are found to be. It seems to me irrelevant that his zone should contain only a 17th part of the sphere, for that seems not to mean the same as that the chance of a great circle chosen at random, lying entirely within the zone, should be one part in 17. However, I may be wronging the great Daniel. . . .

[1] There is no record of the quotation in Fisher's file, but Professor Barnard has kindly provided a copy of the relevant passage, shown below, from page *v* of the Preface to the third edition of A. de Moivre's *The doctrine of chances* (1756).

'Further, the same Arguments which explode the Notion of Luck, may, on the other side, be useful in some Cases to establish a due comparison between Chance and Design: We may imagine Chance and Design to be, as it were, in Competition with each other, for the production of some sorts of Events, and may calculate what Probability there is, that those Events should be rather owing to one than to the other. To give a familiar Instance of this, Let us suppose that two Packs of Piquet-Cards being sent for, it should be perceived that there is, from Top to Bottom, the same Disposition of the Cards in both Packs; let us likewise suppose that, some doubt arising about this Disposition of the Cards, it should be questioned whether it ought to be attributed to Chance, or to the Maker's Design: In this Case the Doctrine of Combinations decides the Question; since it may be proved by its Rules, that there are the Odds of above 263130830000 Millions of Millions of Millions of Millions to One, that the Cards were designedly set in the Order in which they were found.

From this last Consideration we may learn, in many Cases, how to distinguish the Events which are the effect of Chance, from those which are produced by Design: The very Doctrine that finds Chance where it really is, being able to prove by a gradual Increase of Probability, till it arrive at Demonstration, that where Uniformity, Order and Constancy reside, there also reside Choice and Design.'

G.A. Barnard to Fisher: 16 *April* 1955

I enclose the draft chapters, in reply to your letter[1] which I found on returning from Paris, where I have been giving some lectures about foundations. . . .

There are just a few detailed points on the drafts:

1. p.1: On Bayes' mathematical attainments, I would suggest adding to 'his mathematical contributions to the *Phil. Trans.*' 'and his (anonymous) tract on fluxions'.
Also, I wonder if you think fit to make the point that not only was the scientific atmosphere ripe (after Newton and Boyle), but also the philosophical atmosphere, after Locke and Hume. . . .

Perhaps it might be mentioned that Cournot, in France, expressed views similar to those of Boole in England; though Cournot suffered even more neglect from his countrymen than did Boole from his.

On the general point, I think the only one is that made by Frank Yates, about considering the sensitivity of a test. And on this, perhaps it might be pointed out that a good test of significance needs to be both sensitive and insensitive — sensitive to the kind of departures from the null hypothesis which are of interest, and insensitive to the kind of departures not of interest. Thus *t* is sensitive to departures of the mean, but not to departures from normality. But Bartlett's test for homogeneity of variances is sensitive to departures from normality almost as much as it is to departures from equality of variances.

I am still thinking hard about the question of axiomatising ignorance!

[1] Fisher's secretary had written on 1 April asking about the two draft chapters which Fisher had sent Barnard on 18 February.

Fisher to G.A. Barnard: 18 *April* 1955

Many thanks for your long letter. I am afraid I knew nothing about the 'anonymous tract on fluxions'. I should like to know more, but I do not think I can use this knowledge without making an unnecessary display of learning which I do not possess. This applies also to Locke and Hume.

By the way, have you completed the biographical notice which I think you were invited by the Statistical Society to give about Thomas Bayes? If that is published, I could meet your points by referring to such a notice.

Perhaps you could give a reference to the work of Cournot you refer to, and if you have any such things in your mind, to quotations showing the state of French opinion at the time. On the question of axiomatics, you might be interested in a little section which I have inserted somewhere in the second chapter. I enclose, also, a similar section from chapter four, as a further step in the asymptotic approach to intelligibility!

Fisher to G.A. Barnard: 21 *May* 1955

I was very glad to receive the invitation to your inaugural lecture with a title embodying one of my own special fads, namely that statistics should be taught as a technological subject.

I hope to be able to come up and shall look forward to seeing you, perhaps at tea.

I have written as much of my book now as was burning to find expression, though I suppose I could be flogged into writing some more if any yawning chasms are pointed out. If you want to know how it has developed, let me know and I will send or bring a copy of some part of it.

G.A. Barnard to Fisher: 11 *June* 1955

I am writing to thank you for going to so much trouble to come to my inaugural lecture last week. I only wish I could feel that what I had to say was of some interest. . . .

You wrote some time ago to say that your six chapters were in the hands of Frank Yates, and that I might see them later. I would be very glad to do this, if you would be so kind as to send them.

Fisher to G.A. Barnard: 15 *June* 1955

It is I that should have written to you to thank you for the most enjoyable lecture, and subsequent Dinner, at Imperial College. . . .

I have sent a complete typescript to the printers in Edinburgh, and of the second copy here, chapter five, 'Some simple examples of inferences involving probability and likelihood', is, I think, in Yates's hands. I should guess that it and number six are the only two you have not seen, and number six is chiefly concerned with connecting the ideas of the rest of the book with those of the theory of estimation, and with the difficulties that have been felt about fiducial distributions with many parameters.

Why not send Frank a line and tell him you are ready for chapter five, and I am ready for his comments on it, and let this office send you chapter six when you want it, probably while I am in Brazil?

Fisher to G.A. Barnard: 27 *July* 1955

I think I remember your wanting to see chapter five. I have now got this back from Frank, and could let you have it whenever it may be convenient.

Fisher to G.A. Barnard: 27 *September* 1955

I wonder if the long working hours of a family man on holiday have left you any time to look at my fifth chapter. Anyway, when you are through with it, if possible without hurrying, it would be nice to have it back.

On your instigation I have looked at the Pearson and Hartley Table 11, and for the case of the utmost simplicity ($n_1 = n_2$, and, in my sense, $s_1 = s_2$) I was quite taken aback to find that what Sukhatme and I have called the value of d (here called v) is, as given by Welch, or perhaps Mrs. Aspin, (for Pearson takes a lot of trouble to distribute the responsibility) actually less than t, at the level of significance chosen, for $n_1 + n_2$ degrees of freedom; so you get a sharper test if you can say 'I really am not quite sure that these variances are equal' than if you have reluctantly to admit that they must absolutely be exactly equal, in which case 'Student's' test would have *faute de mieux* to be applied!

Actually, it is not difficult to demonstrate that if d were less than t for the total number of degrees of freedom, then whatever might be the true variance-ratio of the populations sampled, the number of claims to significance, at any level, would be invariably greater than the fraction defining the level chosen; and this to a really rather large extent.

I enclose a short note[1] of a demonstration of this kind, though, of course, I have many years ago given effectively the same analysis; and wonder whether you think that in the atmosphere of concreteness which should be produced by the production of actual tabular values, it would do anything to prevent all but the wildest asses from deceiving themselves when they have data of this sort.

[1] See *CP* 264.

Fisher to G.A. Barnard: 7 October 1955

I have been hoping to hear from you about the miscellaneous material of mine in your hands, for which the paper on Welch's table will not take you long, and I hope you will send it soon to Frank Yates.

The draft of chapter five I should also be glad to have back, so as to revise it adequately before it is set up in print.

G.A. Barnard to Fisher: 12 *October* 1955

I have been a long time answering your letter about Welch's test, because I found myself disagreeing with you. And experience teaches me that when I do this I am wrong.

But it seems to me that what you have proved is that in the conditional set in which s_1^2/s_2^2 is fixed at unity, the relative frequency of judgements of significance, at the 10 per cent level, is greater than 10 per cent, when the null hypothesis is true. This does not exclude the possibility that in other conditional sets, where s_1^2/s_2^2 is fixed at some other value, the relative frequency of such judgements may be less than 10 per cent; so that overall, when s_1^2/s_2^2 is allowed to vary, the relative frequency may have the required value of 10 per cent.

I have looked up the literature and found a paper of Aspin and Welch (*Biometrika 36*, (1949) pp. 290–296) in which Welch gives the following figures for the probabilities of significant (according to his test) positive values of d, depending on $\gamma = \sigma_1^2/(\sigma_1^2 + \sigma_2^2)$:

γ:	0.1	0.2	0.3	0.4	0.5
P:	0.0501	0.0500	0.0500	0.0498	0.0498

These were obtained by coarse-meshed quadrature, so that the last digit is not reliable. (For two-tailed tests, the figures should be doubled.)

However, using Welch's test, we are left with the paradox that, with $s_1 = s_2$, we can judge that a given result significantly contradicts the hypothesis of equal means, while the same result fails to contradict the hypothesis of equal means and equal variances. We therefore seem to be in a situation where we can believe A, but not B, although B is a logical consequence of A.

If we require that a fixed level of significance should correspond to a fixed frequency in repeated samples, of judgements of significance when the null hypothesis is true, then such a paradox as this one inevitably arises whenever we have a test of a wider hypothesis (B), and of a narrower hypothesis (A), unless these two tests happen to be identical.

We might try to get round the difficulty by saying that B, being a wider hypothesis, should be tested at a higher level of significance than should A. But this does not seem to me at all convincing.

I am sending your Chapter V by separate, registered post. I found it *very* meaty. . . .

Fisher to G.A. Barnard: 13 *October* 1955

Thank you for looking at my note on Pearson's table, and for your comments thereon. Unless I misunderstand you, to a user who has observed $s_1 = s_2$, and has calculated his value of d, and later complains that the value tabulated by Mrs. Aspin on Welch's formula would make him claim significance with a frequency of more than 12%, the reply is, if I have you right, 'Don't you worry about that! If s_1/s_2 had been some other value, the table would have led you to reject the hypothesis in less than 10% of trials'.

To the experimenter who actually hopes to learn from his data, this is just as much as to say 'The value tabulated for v, or d, is too low in the case you have encountered, but in other parts of the table this is compensated by tabulating a value too high'.

What Welch and his colleagues have failed to realize is that the table is entered with the ratio s_1/s_2, so that each tabular value must be judged right, or wrong, according to the subset of possible experience specified by that tabular entry. The rider that it is not entered with the ratio σ_1/σ_2, this being unknown, might, of course, suggest that the subset of possible experience represented by a fixed ratio σ_1/σ_2 is irrelevant to a significance test.

I have tried to make simple points simply, because as between people who wish to understand each other it is important, so far as possible, to take such things one at a time.

I hope you will have another shot if you think I am missing anything. Of course, I do not believe there really is compensation in other parts of the table, but it might be nice to know in what parts to look for undue insensitiveness, which is supposed to counteract (though to the user it does not) the undue laxity in those parts of the table to which I have pointed.

[P.S.] I suppose you have now sent my note to Frank.

G.A. Barnard to Fisher: 13 *October* 1955

I enclose Chapter V, and would like to say I feel privileged to have been able to see it at this stage. I also hate to let if go, because I am finding so much food for thought in it. Would you please let me see a copy as soon as you can spare one for good?

One thing I found especially salutary was the interpretation of fiducial statements about parameters as limiting forms of fiducial statements about observations, serving to emphasise the lack of valid distinction between fiducial probability and other 'kinds' of probability.

I do just wonder whether the fiducial distribution of μ, given on p. 10, may mislead. You say (near the bottom) 'the equation

$$\mu = \bar{x} - st \tag{76}$$

thus specifies the fiducial distribution of μ. . . . . .'.
In inserting 'thus', it seems to me you suggest that the deduction is immediate; whereas, if I have rightly grasped p. 15, the rigorous step-by-step derivation here would require, first the distribution of σ, then the joint distribution of μ and σ, and thence the marginal distribution of μ. Am I right?

Other thoughts will take time to find a brief expression.

Fisher to G.A. Barnard: 14 *October* 1955

Thank you for your second letter, about chapter five. After equation (76), $\mu = \bar{x} - st$, I am inclined to say 'supplies a basis for a rigorous fiducial argument giving the distribution of μ'. In a few years time I hope this may sound over-formal, but at present, with people like Tukey ready to get deeply confused, I suppose one cannot be too careful.

I am inclined, also, to follow the three 'unsolicited testimonials' to Welch's solution by the additional comment:

I have no doubt that the error of these calculations consists in ignoring the fact that the Table is entered with s_1/s_2, or some equivalent function, and therefore that each tabular entry can come into use only in that selection of cases in which this ratio is realized. The frequency distributions of v (or d) at other ratios, or mixtures of them, are therefore misleading for each particular entry used. [*CP* 264, p. 59.]

Re-reading your letter I do not think the marginal distribution of μ requires to be based on the joint distribution of μ and σ, for the inequality

$$\mu > \bar{x} - st_p$$

has a definite probability of being realized in the set of all values of μ, $\bar{x}$ and s appropriate to any sample, and if it is admitted that no subset can be recognized having a different probability, and to which the observed sample certainly belongs, (as can scarcely be disputed since $\bar{x}$ and s are jointly sufficient and it is postulated that no information *a priori* is available), the distribution of μ follows from that of t, as indeed 'Student' recognized, though naturally enough he was not very easy about the mode of expressing this inference.

G.A. Barnard to Fisher: 17 *October* 1955

Thank you for you letter of the 14th, and that of the 13th, which make all clear, on Welch's test.

Now that you have brought it out, I am reminded that in my 1947 paper I quoted Yates' 1939 paper[1] on this very topic, indicating why one was concerned with fixed s_1/s_2! But, as you say,—and my mistake, I think, proves it—you can't be too careful.

There is a connection, isn't there?, with the 2 × 2 table, where we fix the marginal totals for much the same reason as we fix s_1/s_2.

But now I think that what you have proved proves more—in fact, we can surely get Wilks' result, that if we require fixed frequency of wrong judgements of significance, in the conditional sets s_1/s_2 fixed, then we cannot have this in repeated samples of the same size. For, from the form of the likelihood function, any admissible test criterion must be based on $\bar{x}_1$, $\bar{x}_2$, s_1, and s_2 only; that is, it must be a function of $d = (\bar{x}_1 - \bar{x}_2)/(s_1^2 + s_2^2)^{1/2}$, of s_1/s_2 and perhaps of $\bar{x}_1$ and s_1. (Since any function of the four quantities $\bar{x}_1$,$\bar{x}_2$,s_1,s_2 is expressible as a function of the four quantities d, s_1/s_2, $\bar{x}_1$, and s_1). But by a shift of common origin and change of scale we can make $\bar{x}_1$ and s_1 take any value we please, while d and s_1/s_2 are invariant. Hence, since our test must be unaffected by shift of origin and change of scale, it must depend on d and s_1/s_2 only. And if we fix s_1/s_2, you have shown that the distribution of d depends on σ_1/σ_2, in repeated samples of fixed size. Thus we must either give up the insistence on fixed sample size, or we must give up fixed s_1/s_2, or some other requirement of equal necessity.

I shall write again about the fiducial distribution of μ, given $\bar{x}$ and s, because I am still not quite happy about that. I am wondering whether we ought not to require, in the first place, that there should be a pairing of the sufficient statistics with the parameters, in such a way that one sufficient statistic only is involved for each parameter—like s with σ; then we may get further fiducial distributions by using those we already have to get rid of nuisance parameters. But I will try to formulate my difficulty more precisely before asking you to help me with it.

[1] Yates, F. (1939). An apparent inconsistency arising from tests of significance based on fiducial distributions of unknown parameters. *Proc. Camb. Phil. Soc.* **35**, 579–91.

Fisher to G.A. Barnard: 21 *October* 1955

I am exceedingly glad that the extra paragraph made the point clear, though, of course, it is only a guess that the various people I quoted have actually overlooked the necessity, for the purpose of a test of significance, of calculating the frequency in *pari materia*.

Logically it seems to me, as to you, to be exactly the same criticism as should be made against the use with the 2 × 2 table of frequencies based on varying marginal totals, for judging the significance of a case having marginal totals known. It is rather like mixing in some data from rats in testing the significance of an infection, or prophylactic, applied to mice. . . .

Fisher to G.A. Barnard: 14 *March* 1956

. . . Have you sufficiently wrestled yet with that sixth chapter? The last example should be critical for the understanding of people like Tukey and Savage who have been writing as though any sort of Jacobian transformation would do. I imagine that reflection on the examples I have given of introducing parameters not only one-by-one, but sometimes two-by-two, may suggest to you conditions of the legitimacy of such an operation.

However, probably you are horribly busy.

G.A. Barnard to Fisher: 15 *March* 1956

. . . I have fixed to come to Cambridge for a day or two about the weekend of April 14th, and as you suggested I will ring your secretary a little before then. . . .

G.A. Barnard to Fisher: 24 *April* 1956

I am still unhappy about linear functions of the cell frequencies. In the case of a continuous variate, the sum of squared deviations from the sample mean is expressible as the difference between the sum of squares, and the squared sum divided by N. Both the sum of squares and the sum are linear functions of the cell frequencies, but squaring the sum introduces non-linearity. However, it does not introduce discontinuity. Of course, you allow non-linear functions of a single linear function of cell frequencies. But this example seems to show that you need functions involving more than one linear function.

If one allowed continuous functions of a finite number of linear functions of the cell frequencies, these difficulties would seem to be overcome. But just why one should make precisely this limitation is still not clear to me, though I can see one can't allow entirely arbitrary functions. . . .

Fisher to G.A. Barnard: 25 April 1956

So far as developing a mathematical theory is concerned, I have no objection to offer for the use of non-linear functions of the frequencies, though I think one does require that the functions of the frequencies used shall be so definable as to have a meaning when non-integral, or irrational, expectations are to be substituted for them. I do not suppose, however, that this offers any difficulty.

Of course you are right in saying one may have functions of many linear functions. The stipulation of linearity in the frequencies refers to that part of the statistical process in which the evidence of a *series of frequencies* is to be combined. Once they have been combined in one or several linear functions, I see no inconvenience in combining these in any chosen way. My point about linearity is merely that I see no prospect of any function that is not built up of such linear functions being of any use or interest. I might easily be wrong without having injured the development of the theory, which of course I did develop at first without considering that functions linear in the frequencies had any particular rationality.

For a student, I should say that the frequencies are peculiarly related to one another in that they are counts of mutually exclusive occurrences, and therefore that the whole sequence of frequencies must be considered together in relation, usually, to a fixed total. Measurements may be regarded as frequencies of unity in rather specially defined classes, and though the number of measurements is relevant, they are, as frequencies, a series of frequencies of one, which may, or may not, occupy the same class. I think you only get discontinuity if you were to apply a non-linear function of the frequencies (such as Pearson used long ago in demonstrating, with much sound and fury, that least squares was not the right way of fitting frequency curves, i.e. least squares of frequencies!) to such frequency classes defining measurements. The point is that a frequency distribution is an organic whole corresponding more or less with one or more measurements, but not particularly with N measurements if N is the size of the sample, but really it scarcely corresponds with exact measurements at all.

Fisher to G.A. Barnard: 20 *August* 1956

It is a pleasure that I have been looking forward to, to send you an early copy of this book, which I believe will be on sale in September. I have not noted, so far, a great many misprints, although there doubtless are many more than I have seen. . . .

By the way, I should rather like to know if any of the people, and they are doubtless many, with whom you have argued about fiducial probability, have been misled by the misapprehension that I tried to remove on page 120,[1] i.e. think that a fiducial probability statement about unknown parameters, because it has the same form as it would have if such parameters had been chosen from a super-population, asserts or implies that such a super-population is a historical reality, which would be nonsense seeing that we know nothing of the origin of the population sampled, but we do know something, and that something expressible in terms of probability, about its character. Casting my mind back to a period more than twenty years ago when, under Neyman's influence, Bartlett was attacking the theory, it seems to me that people like Bill Cochran may have been largely influenced by this misapprehension. It may, however, have been quite without influence in more recent discussions in which I have taken no part, for at a fairly early stage I formed the opinion that the number of tolerably good mathematicians who were then in the subject might safely be left to get their thoughts straight by themselves and without my continually yapping, for it seemed to me unavoidable that they would have much to withdraw, and such withdrawals are made most gracefully when made spontaneously. In consequence of this, I have not heard perhaps so much as I should have of the points which may now be thought logically influential. Arguing during the last year with John Tukey, I encountered a most impenetrable mass incorporating every sort of fallacy imaginable, and I cannot tell whether I have had any success in loosening it up.

Anyway, as these discussions are likely to become active, perhaps you will let me know what you can.

[1] *SMSI*, p. 124.

G.A. Barnard to Fisher: 7 *September* 1956

I am writing to thank you most warmly for the copy of your new book, and for the most kind letter enclosed. I have not had the chance to study it in peace as I would like, so far. . . .

I have had some discussion with John Tukey this summer, from which it seems to me you must have cleared his mind a good deal on the fiducial argument. . . .

G.A. Barnard to Fisher: 17 *January* 1957

May I say how glad I am to hear you are back, and to thank you for the very pleasant evening you gave me just before you went away. I hope you and your daughter enjoyed your stay in Calcutta.

A point has come up while you were away which has led me to feel that the notion of group invariance, or some other way of expressing ignorance about a parameter, is essential to the validity of the fiducial argument. You may remember my putting forward the idea some time ago that we could express our ignorance of μ, in relation to the estimation of σ, by requiring that our estimator should be invariant under translations (i.e. changing x to $x + a$).

To apply this idea in a wider context, we might say that, in order to be able to carry through the fiducial argument, we not only need to be in ignorance about the parameters in question, but also we need to be able to give this ignorance a precise mathematical expression. One such way is to be able to point to a group of transformations operating on the set of possible parameter values, under which our analysis of the situation must be invariant.

The specification of a continuous group of transformations on the set of possible parameter values would imply the unique specification of a measure function defined for sets of parameter values. And it would be natural to express the primary likelihood inference from the data by integrating this likelihood function with respect to the measure. In this way we would obtain a probability, which could be regarded as the fiducial probability distribution of the parameter.

The reason why I think some such requirement is necessary is that Lindley has shown that the following is possible; we have a sample S, made up of two sub-samples S_1,S_2. S gives a fiducial distribution $F(\theta)$ for a parameter θ. S_1 gives the fiducial distribution $F_1(\theta)$, and when we use F_1 as a prior distribution, in relation to S_2, we obtain the distribution *a posteriori* (i.e. given S_2) $G(\theta)$. Now we would expect that $F(\theta)$ should be the same as $G(\theta)$, but Lindley shows that this is not the case in general. I can show that it will be the case if we impose the requirements about group invariants.

I am afraid this is a long letter, which you may not find at all clear. I wonder if I might now come to Cambridge to see you about it? . . .

Fisher to G.A. Barnard: 18 *January* 1957

Thanks for your letter, which indeed deserves more consideration than I shall give it before my next lecture at 12 o'clock. I have seen a rather abusive review by Lindley, which is indeed, I think, the second he has written, in which he speaks with great confidence of detecting a mathematical error which overthrows the notion of fiducial probability. My work, like that of others, is liable to contain mathematical errors, and his criticism has led me to a revision of one example, but he is mistaken in thinking that it affects the logical status of fiducial probability, which is, I believe, the only form of probability about the real world which can be inferred from quantitative observational material.

Of course, the overruling requirement of sufficient or exhaustive estimation imposes analytic restrictions on the problems for which fiducial probability can be calculated, and it may be that Lindley has been rediscovering these restrictions under other forms and expressing them in other language. The fallacy of Lindley's argument lies in the fact that the fiducial probability, derived from sample S_1, is only valid if there is no means of recognising subsets in the reference sets used for probability statements. It is certain that a second sample makes such recognition possible, just as information *a priori* might do, and it is in each particular case a matter for investigation whether the fiducial probability distribution is thereby altered. The example on the question of observations of two kinds (Section 6, page 123)[1] is a mistaken one just because there is no exhaustive method of estimation for the two kinds of observation concurrently. I do not think it will be difficult to think up an example free from this defect to exemplify the position (a) that different types of observation lead to statements at different levels of uncertainty, and (b) that the combination of data of these two kinds, when a rigorous small sample combination is possible, will yield statements of probability (i.e. of the higher quality) with the quantitative force of both sets of observations together.

I hope I have not expressed this too obscurely.

No, your letter is not either long or obscure. I do not understand what you say about 'group invariance', but I presume that it is a reappearance of the analytical requirements of exhaustive combination of data.

[1] *SMSI*, p. 127.

Fisher to G.A. Barnard: 22 *January* 1958

I believe some time ago you expressed the intention of answering Lindley's review in *Heredity*[1] of my new book. Some people seem to be expecting that I should do so myself, but I should prefer not, since the review is so full of mis-statements about it, e.g. on his first page:

> 'The fourth chapter is concerned essentially with the problem of what space we should integrate over in forming our significance levels . . .'

which is merely an evasion of the straight exposure in that chapter of the false criterion advocated by Neyman, Pearson, and Bartlett as a ground for doubting the exactitude of Behrens' test of significance.

And of course at the end of this paragraph also the strong definition of consistency and the proof of efficiency (why asymptotic?) of the maximum likelihood estimate are neither of them new.

Of course the rest of the article is equally chaotic, e.g. on page 282 all chromosomes are apparently known *a priori* to be equivalent. The important point, however, is that the second paragraph on page 281 is not impugned by the bad example I took in Section 6 of Chapter V, of which I believe you have already seen the replacement which proves the same point without the original fault. . . .

[1] *Heredity* **11**, 280–3, (1957).

Fisher to G.A. Barnard: 8 February 1958

I have been reconsidering the example in Chapter V, Section 6, to which Lindley took so much exception, and I see that he has put on it an interpretation which the example as it stands will not at all bear, for it treats specifically of the combination of data of two distinct kinds, of which one *ex hypothesi* is incapable of supplying probability statements.

In fact the more I consider it, the more clearly it would appear that I have been doing almost exactly what Bayes had done in the 18th century. As Lindley purports to be a protagonist of Bayes, it would seem that his misunderstanding and confusion goes deeper than anyone could imagine.

In fact I am not inclined in the second edition to alter a word of the section, but to add to it a short further explanation of which I enclose a copy. Anyway, I should be glad to hear from you.

G.A. Barnard to Fisher: 27 February 1958

. . . About Lindley, and his review, I must say I had hoped that some editor would have given me an opportunity to review your book, so that an indirect reply to Lindley would have been possible. But this has not happened, so that one is left to choose between a letter to *Heredity*, and an article or series of articles, perhaps more suitably published elsewhere. My feeling is that the latter would be preferable, because the bases of Lindley's misunderstandings lie too deep for treatment in a short letter, or reasonably short correspondence. And the important thing is that others . . . have fundamentally similar misunderstandings of principle.

That this is so seems to emerge from a series of discussions we have held in our colloquium, in which a series of people, mainly the 'younger generation' have been invited to give their views on your book. Almost everyone agrees with your criticisms of Neyman and Jeffreys, and with at least a part of your general theory. But nearly all become highly convoluted because (I think) their mathematical training has followed the now fashionable trend towards axiomatics to such an extent as to make them sometimes incapable of following a semantic argument.

This makes me think that the next step, for me, is a series of articles, covering the general field as far as possible, in which I would try to go meet these mathematicians, by using their language and notation as much as possible. I gave a course of lectures last term along these lines which Jenkins (who was trained by Pearson) said cleared up a good many things for him. And I hope things will ease a bit here after Easter, so that I could get some sections drafted. I would hope very much to have the benefit of your criticism of them.

Apart from exposition of your ideas, I have in mind a few new things. For one, I have made a detailed analysis of the concept of a 'long run'. According to Neyman's followers, all statistical procedures are to be considered in relation to 'the long run'; but the idea is not at all as simple as they seem to think. It is worth analysis, I think, because it is relevant to some technological applications. And incidentally it seems to throw some further light on Behrens' problem.

Another idea is that one can give a 'consistency proof' for the fiducial argument in a wide class of cases—rather as one could give a consistency proof for the law of excluded middle in mathematical logic, a la Hilbert.

With regard to your second letter, I agree there seems no need to change what is in the book, though your addendum will be illuminating, I think. I should perhaps suggest that after 'The percentile values' in line 10, 'of the distribution *a posteriori*' should be inserted, since this will serve to avoid confusion with the percentile values referred to a little before.

Fisher to G.A. Barnard: 3 *March* 1958

It is extremely nice to have your letter and to hear what you have been thinking. What might have been a storm in a teacup seems to be becoming a world war. . . .

In January of last year I received from Lindley a proposed review of *S.M.S.I.*, which was, I believe, materially different from that which later appeared in that journal (*Heredity*). If you have a copy of the original I should be immensely glad to have it, for in any future discussion it is important to know how his mind works (one can allow for mere brashness but not for all the convolutions of the cortex), and I think his change of attitude during last year, perhaps influenced by discussions with you, might help me a good deal.

I am discussing these problems, of course briefly, with the Kapitza Club here on Tuesday evening.

When are you coming to dine with me?

G.A. Barnard to Fisher: 6 *August* 1958

Several months ago you asked me for a copy of the first draft of Lindley's review of your book. I spent some time looking for it but could not find it till now. I expect it is much too late for your purpose, but I am enclosing it just in case you may have use for it. I marked in pencil some of the places where I hoped to persuade Lindley he was wrong fairly easily; but I was wrong, and he was more impenetrable that I expected. . . .

Fisher to G.A. Barnard: 16 *August* 1958

Thank you for sending me the script, which has been less altered than I had thought. . . .

You might like to see the address I have undertaken to give in Brussels in connection with, I believe, the Société Adolphe Quetelet, if not the International Statistical Institute, more or less entangled with an International Pharmaceutical Congress. You may be interested to see the new look which can be given to Bayes' celebrated theorem.

[P.S.] The *Centennial Review* of Michigan State College has a rather similar paper [*CP* 272] on The Nature of Probability being also, of course, a lecture. I will send an offprint when I have one.

G.A. Barnard to Fisher: 16 *October* 1958

Stuart Hunter, no doubt on the prompting of George Box, has sent me your paper about mathematical probability in the natural sciences,[1] for me to referee as a joint editor of *Technometrics*. I am glad of this because it gives me the chance to write to you and say—as I have had to say so many times before—that on reflection I think you are right. I would like to withdraw completely what I said in Brussels and to apologise for making myself a pest.

I tried to consult Bayes's paper in Brussels, but found that their files of the Royal Society *Transactions* go back only to 1831, so that it was not until I was able to read the printer's proofs of the forthcoming reprint in *Biometrika*[2] that I could confirm the accuracy of your account of Bayes, and the inaccuracy of my own recollections. I also looked up Gauss and here again found myself in the wrong. The only thing I could plead in connection with Gauss is that even in his early work of 1809 he is clearly not entirely satisfied with the use of prior probabilities, since he says later on that the principle of least squares should itself be taken as an axiom.

I wonder if, as assistant editor, however, I might suggest two small changes in your paper. On page 4 you say 'for it seems to be only in mathematical departments insulated from practical research in the natural sciences that confusion and misapprehensions abound'. I feel that this statement, while undoubtedly true, is open to misinterpretation and I wonder therefore if you feel it could be deleted without serious loss. Second, on page 15, line 9 the phrase 'of two types' might be misinterpreted by someone who felt he was not abreast of the latest news on fundamental particles. Perhaps 'by two methods' would be better than 'of two types'.

Hunter has suggested that we might institute the practice in *Technometrics* of printing some discussion along with the paper, rather as the read papers are printed in the R.S.S *Journal*, and he has asked me if I would care to draft something in connection with your paper. With certain sorts of paper I think this would be an excellent idea, but with yours my first reaction is that I could do nothing but gild the lily. If pressed I could I suppose expand the historical references a bit and I might make a few additional comments concerning the definition of probability. I would be very glad to know how you would feel about such a proposal.

[1] See *CP* 273.

[2] Bayes, T. An essay towards solving a problem in the doctrine of chances. (Reproduced from *Phil. Trans. R. Soc.* **53**, 370–418, (1763)). With a biographical note by G.A. Barnard. *Biometrika* **45**, 293–315, (1958).

Fisher to G.A. Barnard: 17 *October* 1958

Thank you for your letter, which was indeed very welcome to me as I did not really want a longish controversy in print, or even in correspondence, involving long quotations in German, or for my part preferably in Latin.

About Gauss, you probably have the paper of mine from the Cambridge Philosophical Society of 1930 in which I say a fair amount about what I think his argument was, though not of course right on the point now in discussion. I should be interested to know if you find yourself in agreement with these earlier thoughts.

Of course I agree with you that Gauss used the word 'probability' to cover what I later called 'likelihood', being influenced, I have no doubt, by the very wide extension of the meaning of the word current in the early nineteenth century on the strength of Laplace's comprehensive definition. This penumbral meaning of the word, though quite unsuited for exact or mathematical thought, has been accepted a good deal too easily by some mathematicians, as by those who argue that we do not lack knowledge *a priori* seeing that we usually have some vague information before we start arguing seriously. What a mathematician should emphasize is that the probability *a priori* required by Bayes' theorem, and without which Bayes' theorem cannot be applied, is an exact and complete specification of the unknown as a random variable, and this no-one thinks that he possesses unless he has found it experimentally.

About your changes, the second is easy, though hard to find since it is on page 15 and not page 12 as you say! I have altered 'types' to 'sorts'. The first one is considerably more difficult. I do not want to be argumentative, but I do regard the position as serious, and it will certainly remain serious unless the self-satisfaction of some mathematical departments is shaken, for, as nineteenth century experience shows, it is not easy to reform anything so highly traditional as mathematical teaching. In my paper the phrase, like that with which the paper ended, was an attempt to warn pharmacists in particular that they could easily be led up the garden path by confident young men apparently well-qualified as mathematicians. I do not like to think of the standardization of drugs, the issue of prophylactic sera, or the aiming of ballistic missiles, being influenced by the arguments of which *Biometrika Tables* No. 11 is a concrete product. I certainly owe a duty to my scientific friends to inform them of what they probably would find unimaginable, of the levity with which make-believe is accepted in some mathematical departments, even when they do not think they are joking.

I do not of course say, or, I hope, seem to imply, that in all mathematical departments the Neyman fog has settled in, but that it has settled in only in those departments which are insulated from practical research in the Natural Sciences, and that the remedy is that, at least in all departments teaching mathematical statistics (other branches of mathematics must look after themselves), people with *real* experience of the fields of application are needed. Remote control consultant work is only a first step to understanding the function of statistical methods and their applications. Hence I should like to think more and discuss further with you your first suggestion.

Of course I should welcome the kind of commentary you suggest to follow my paper. Reviewers of my book have rather curiously found nothing new in it to comment on. . . .

G.A. Barnard to Fisher: 28 *October* 1959

I have been very glad to hear, at third hand through George Box, that you are enjoying life in Australia.

A point has occurred to me, in connection with the fiducial argument, which so far as I know has not been touched on by you in published work. Suppose we are measuring (observing?) a length x which is restricted to discrete values $\pm\, nh$, $n = 0,1,2,\ldots$ and suppose we take the distribution to be the 'discrete normal' with probability function

$$P_r\{x = nh\} = \{hK(h)/(\sqrt{2\pi})\}\exp\{-(nh - \mu)^2/2\}$$

(with $K(h)$ very nearly 1.) And suppose the parameter μ is also restricted to discrete values $\pm\, hr$, $r = 0,1,2,\ldots$ then it seems to me the quantity $(x - \mu)$ is pivotal, with

$$P_r\{x - \mu = rh\} = \{hK(h)/\sqrt{2\pi}\}\exp(-r^2h^2/2)$$

and, if nothing is otherwise known of μ, we can infer, from an observation x, the (fiducial) probability function

$$\{hK(h)/\sqrt{2\pi}\}\exp\{-(rh - x)^2/2\}$$

for $\mu = rh$.

Am I wrong?

I came upon this point after trying to give a rigorous account, for the *continuous* case, of the idea of sufficiency, likelihood, quantity of information, etc. One gets very much tangled up with the continuum hypothesis, non-measurable sets, and so forth; and it seemed to me one way to cut the Gordian knot would be to argue that all observable distributions are discrete. But then one seems to come against difficulties with the fiducial argument, unless one takes the point of view I have indicated.

G.A. Barnard to Fisher: 12 *January* 1960

Thank you for your letter.[1] If ever there is a chance I shall try to get the misprint with 0 and o corrected.

It is most exciting to hear that you have derived the simultaneous distribution of a set of correlation coefficients. I first made the acquaintance of the variance co-variance distribution from Wilks's book and perhaps because I was wanting it at the time I realised very well that there was a good deal further to go before one had the correlation coefficient distribution. I had one or two goes at it myself, especially when prodded by psychologist friends, but made no headway. Is there any way of seeing how nearly jointly normal are the 3 z-transformed correlations between 3 variates? Presumably the approximate joint normality will cease long before the approximate marginal normality, so that we would here have a case where normality of marginal distributions went along with non-normality of the joint distribution. This would have a certain personal interest for me in that the only inaccurate statement I ever found in the later editions of Yule's book was to the effect that normality of marginal distributions implied normality of joint distribution; a theoretical counter-example is easy to find but it is not so easy to find such a distribution in practice.

[1] Fisher's letter (probably handwritten) from Australia has been lost.

G.A. Barnard to Fisher: 8 *May* 1961

I am enclosing the papers which you left with me and have kept copies for myself, for which many thanks.

One point that arises in connection with the case where the observations O are obtained by counting (case B)[1] is that it would appear that an alternative way of converting the observations O to the form required to give a fiducial distribution would be to include the actual times at which the particles arrived within the period of observation. It seems to me one could then treat the interval from the beginning of counting to the arrival of the first particle as corresponding to the detector W, and then the remainder of the observation interval as corresponding to detector O. If I am right it would seem that although the information about the times of arrival of the particles contributes no information about λ in the likelihood sense, it does enable us to convert the likelihood statement into a probability statement. I hope I make myself clear and that I have not misunderstood the situation. I would like to discuss this and the other question when I come to Cambridge.

[1] See *CP* 289.

G.A. Barnard to Fisher: 13 *July* 1961

David Sprott, whom I mentioned to you as having written two papers concerning the connection between the likelihood function and the necessary form for the fiducial distribution, has come over here and we have been discussing a number of points in connection with the fiducial argument. We have reached a stage where we feel very much in need of your help and I wonder if we could come to see you on Monday of next week (the 17th) to discuss it with you.

If I may briefly indicate the nature of the difficulty it is this. At the end of *Statistical Methods and Scientific Inference* you derive the fiducial distribution for the normal bi-variate [parameters] ρ, σ_1, σ_2. At the same time you point out that one could use another set of pivotal quantities, one of which is the variance about the regression line, which might conceivably be misused to derive another 'fiducial distribution', for the same set of parameters. The difference between David Sprott (and incidentally Quenouille) on the one hand, and myself on the other, is that they seem to think that the second set of pivotal quantities are appropriate for a fiducial distribution of the parameters corresponding to them; but that this fiducial distribution is not transformable to a distribution of ρ, σ_1 and σ_2. For my part it seems to me that this second 'fiducial distribution' is just wrong, since when it is transformed in the normal way to its expression in terms of ρ, σ_1 and σ_2 it contradicts the validity principle.

This is just to indicate the kind of issue involved and I hope we may explain it more fully when we see you. . . .

Fisher to G.A. Barnard: March (?) 1962[1]

Your letter Feb. 15 was forwarded from India, and I have just read it. I think we have arrived at nearly indistinguishable conclusions. Let me now formulate my own position.

A pivotal quantity is a function of parameters and statistics, the distribution of which is independent of all parameters. To be of any use in deducing probability statements about parameters, let me add

(a) it involves only one parameter,
(b) the statistics involved are jointly exhaustive for that parameter,
(c) it varies monotonically with the parameter.

As you have observed, and as the last example in *Statistical Inference* was intended to make clear, parameters and their corresponding exhaustive statistics may arrange themselves in strata.

For the normal sample s alone is exhaustive for σ, consequently the marginal distribution of σ is expressible in terms of s, using the pivotal s/σ. Both s and $\bar{x}$ are needed for exhaustive estimation of μ using $(\mu - \bar{x})/s$. In this case it is noticeable and probably essential that the two pivotals together each involves one parameter. *Jointly* they involve a set of statistics exhaustive for both. Each is monotonic in its parameter *uniformly* for variations of the other. Also, I should have stipulated that their simultaneous distribution is independent of all parameters.

That I think is enough. In the bivariate case s_1/σ_1 and s_2/σ_2 have a joint distribution independent of σ_1, σ_2 but not of ρ. However, for arbitrarily assigned ρ, they suffice to give the simultaneous distribution of σ_1 and σ_2 in an array (with ρ constant) and this suffices for the trivariate distribution.

For sets of pivotals then I add

(d) the joint distribution is independent of parameters (of as high or higher stratum)

(e) all are monotonic, uniformly for variations of parameters of as high or higher stratum.

See if you can formulate a justification for ignoring parameters of lower strata once their simultaneous marginal distribution is determined.

Fraser wrote hopefully about the trivariate case, but he may not see all the difficulties, e.g.

$$\frac{r_{12}}{\sqrt{1 - r_{12}^2}}\chi_{N-2} - \frac{\rho_{12}}{\sqrt{1 - \rho_{12}^2}}\chi_{N-1}$$

is distributed in a standardized normal distribution, and is uniformly monotonic in ρ_{12} but is unlikely to be distributed independently of

$$\frac{r_{13}}{\sqrt{1 - r_{13}^2}}\chi_{N-2} - \frac{\rho_{13}}{\sqrt{1 - \rho_{13}^2}}\chi_{N-1} .$$

In fact the distribution of r_{13},r_{23} conditional on r_{12} is not at all alluring.

I expect I told you I had run out the simultaneous distribution of r_{ij} given the system ρ_{ij} for any number of variables.

[1] This letter was published in Barnard, G.A. (1963). Fisher's contributions to mathematical statistics. *J. R. Statist. Soc.* **A 126**, 165–6.

M.S. Bartlett to Fisher: 25 *September* 1933

Though no doubt you have seen my paper in the last Royal Society *Proceedings*,[1] may I take the liberty of sending you a reprint? Since this paper was accepted, Jeffreys' latest paper in the *Proceedings* was published;[2] and perhaps my intrusion into a 'controversy' which is primarily between Jeffreys and yourself can very easily become somewhat superfluous. I do not, however, remember that Jeffreys' paper had in it anything to alter the point of view I have tried to take up in my own paper, though I feel conscious that this view point I may have rather inadequately expressed.

[1] Bartlett, M.S. (1933). Probability and chance in the theory of statistics. *Proc. R. Soc.* **A 141**, 518–34.
[2] Jeffreys, H. (1933). Probability, statistics, and the theory of errors. *Proc. R. Soc.* **A 140**, 523–35.

Fisher to M.S. Bartlett: 26 September 1933

I had not before seen your paper which I am sure will be interesting. It is as you suggest a pity to have published before Jeffreys wrote, as effectively his reply is an attempt to defend a simple blunder, or howler as it is called with boys, by a confusion of definition. Of course it is easy to find ample precedent for such confusion in the controversial history of the subject.

M.S. Bartlett to Fisher: 30 December 1935

In your recent book *Design of Experiments*, I was interested to see your concluding section on the information supplied when the variance σ^2 is unknown, particularly because I had looked at this problem some time ago and had not been satisfied with the solution I reached. It was for this reason that in a paper[1] I wrote at the time (not yet out) in which I was attempting to look at the general problem of small samples with more than one unknown, I classed the problem you discuss in a different category from other problems in which the reduction in the amount of information available when another parameter is estimated is more straightforward.

Looking at the t-distribution problem again in the light of your solution, I feel, though I would agree that we can, if we want to, regard the intrinsic accuracy of the t-distribution we are using to estimate the mean μ as

$$\frac{n+1}{(n+3)s^2}, \tag{1}$$

that the interpretation of this result rather raises the whole question of the interpretation of information for small samples. Thus it is clearly not permissible to imagine any actual set of samples from which we are going to estimate μ, since s^2 would vary for such a set. We might, however, consider such an actual set, s^2 varying as well as x, and our estimate would then, according to my reckoning, contain the information per sample,

$$\frac{n-1}{(n+1)\sigma^2}. \tag{2}$$

This vanishes for $n = 1$, but this simply means that our method of estimation (which is equivalent to supposing we know nothing about σ^2 *for each sample*) fails when s^2 is based on only 1 d.f.

Without going into the question here of the interpretation of the information in small samples generally, I rather favour the consistency of (2) if we want some measure of the information we are using, though it is certainly not quite what we wanted. (1) has the disadvantage that it cannot correspond to any actual procedure; it obviously cannot, for example, be averaged for s^2.

[1] Bartlett, M.S. (1936). Statistical information and properties of sufficiency. *Proc. R. Soc.* **A 154**, 124–37.

Fisher to M.S. Bartlett: 31 *December* 1935

Thanks for your letter of December 30th.

I am afraid I don't yet see where your formula

$$\frac{n-1}{(n+1)\sigma^2}$$

comes from and therefore what it is put forward as meaning. The meaning of (1) is that different experiments for which the formula has the same value supply estimates of equal intrinsic accuracy, though the error curves are different in form.

Are you not perhaps bringing in the different conception of forming a weighted average from samples which are not known to have the same true variance?

I shall be glad if you are able to give more time to examining the meaning of information for small samples, for it is just in this case that the concept seems to me valuable.

M.S. Bartlett to Fisher: 2 *January* 1936

Many thanks for your reply to my letter. I am sorry that I did not make myself altogether clear. The second formula

$$I = \frac{n-1}{(n+1)\sigma^2}$$

I gave is the information obtained from a sample when we are estimating the mean from several samples for which we do not wish to assume the constancy of σ^2. This is admittedly a different conception from the intrinsic accuracy of an estimate obtained from any experiment as given by your formula. My point was that before this type of problem was considered, there had been no need to make such a distinction, intrinsic accuracy in the case of one parameter being defined in terms of the accuracy obtained from a large set of estimates. If we go on to take the case of a regression coefficient (σ^2 assumed known) the information in any particular sample will depend on our value of $\Sigma(x - \bar{x})^2$ but when we consider the corresponding problem of the information in a large number of samples when the x's vary, the information per sample is simply the average value of the information for any one sample. The same sort of problem can occur in some cases where instead of $\Sigma(x - \bar{x})^2$, we have an estimate of another unknown. It does not apply in the problem we are considering, (the reason being that s^2 is not the theoretically complete statistic for σ^2).

I agree that the second formula involves a different conception from the first, and is hardly what we wanted; I referred to it because, given its proper interpretation, it seemed to be immediately consistent with preceding definitions of information. I was somewhat puzzled in connection with the first formula to know how you were regarding the definition of intrinsic accuracy, in view of the fact that the idea of several values from the same t distribution of $x - \mu$ would not appear to have any direct practical procedure to correspond to it.

Fisher to M.S. Bartlett: 8 January 1936

I am sorry, but I have still not got at the origin of your formula

$$I = \frac{n-1}{(n+1)\sigma^2}\,.$$

I suppose n here stands for the degrees of freedom, as it does in the other case.

M.S. Bartlett to Fisher: 9 January 1936

If we confine ourselves to the problem of estimating m from a large set of samples with the same true variance (possibly out of a larger set with differing variances), instead of the equation of estimation

$$\Sigma\ \frac{x-m}{\sigma^2} = 0,$$

which we should have if σ^2 were known, we might anticipate the equation

$$\Sigma\ \frac{x-m}{ns^2 + (x-m)^2} = 0,$$

since the denominator is a (theoretical) sufficient statistic for σ^2 (n = no. of degrees of freedom). If we maximise the likelihood from the t distribution, or from the equivalent distribution of

$$r = (x - m)/\{ns^2 + (x - m)^2\}^{1/2},$$

then the equation

$$\Sigma \frac{\partial L}{\partial m} = 0 \quad (L = \text{log likelihood})$$

is in fact equivalent to the equation above. The only remaining problem is to evaluate the information in the estimate provided by this equation when x and s are allowed to vary at random. There is a snag here in that $\Sigma(\partial L/\partial m)$ does not go to normality. It does go to normality, however, for fixed $ns^2 + (x - m)^2$, so that using the r distribution, for which

$$\Sigma \frac{\partial L}{\partial m} \equiv \Sigma \frac{(n-1)(x-m)}{ns^2 + (x-m)^2},$$

we find:

$$E_r\left(-\frac{\partial^2 L}{\partial m^2}\right) = E_r\left(\frac{\partial L}{\partial m}\right)^2 = \frac{(n-1)^2}{(n+1)\{ns^2 + (x-m)^2\}}$$

of which the average value when we allow $ns^2 + (x - m)^2$ to vary is

$$I = \frac{n-1}{(n+1)\sigma^2}.$$

Fisher to M.S. Bartlett: 14 *January* 1936

Thanks for your letter. I still do not see how you get your formula, unless there has been some confusion of degrees of freedom with numbers in sample. I have put down on the attached sheet[1] a discussion of your problem of a number of different samples supposed to be drawn from populations having the same mean, but, possibly, different variances. The samples are supposed to be of different numbers. I get an equation of estimation rather like yours, with a factor $n + 1$ in each term, but the amount of information in the estimate derived from it seems to be given by my formula.

[1] Not in Fisher's file.

M.S. Bartlett to Fisher: 15 *January* 1936

The two points where we differ are as follows:—

(1) I had the equation

$$\Sigma \frac{(n-1)(x-m)}{ns^2 + (x-m)^2} = 0$$

whereas you had

$$\Sigma \frac{(n+1)(x-m)}{ns^2 + (x-m)^2} = 0.$$

You will agree, however, that if we confine ourselves to samples with the same number of degrees of freedom, the estimate from either equation should contain the same amount of information.

(2) When considering the average value of $(\partial L/\partial m)^2$, I averaged first for t (or $r \propto t/\sqrt{\{1 + t^2/n\}}$) and then for $ns^2 + (x - m)^2$, *which is independent of t.* Your averaging proceeds on the lines indicated in your book, and it is this that I have been doubtful about,—namely, the averaging for t, ignoring s^2 (which is not independent of t), and leaving it in the formula. I do not follow the interpretation of this. Thus it surely cannot represent the information available for any actual problem, such as the one I proposed, since if we considered samples with one degree of freedom, our average information, if the true variance were in fact constant, would appear infinite.

Fisher to M.S. Bartlett: 17 *January* 1936

I am sorry you do not discuss the origin of your equation of estimation, as this might throw light on what you are aiming at. The whole point of my procedure, as I think must be clear in my book, is to retain s^2 as the sole available fact about the precision of the average. I should say it must certainly represent the information available for the actual problem I have in view, namely, one in which s^2 supplies the only available knowledge about the variance. From this point of view it is not appropriate to average the value of $1/s^2$ for different samples from the same population, but to average the value of $1/\sigma^2$ for the different populations leading to the observed values.

M.S. Bartlett to Fisher: 20 *January* 1936

Many thanks for your letter. I'm afraid, however, that I am still not clear on the interpretation of your procedure. It is true that if the fiducial distribution of σ^2, when s^2 is given, were taken, the average value of $1/\sigma^2$ would be $1/s^2$ but I do not see that the fiducial distribution can be used in this way.

It is also true that I considered for simplicity, in order to get a definite answer, a constant value of σ^2. But before averaging by means of any distribution depending on σ^2, I had the formula

$$I = \frac{(n-1)^2}{(n+1)\{ns^2 + (x-m)^2\}}$$

and if σ^2 were not constant, our average information would still be the average value of this. But I do not see why we should not be able to find out what information we should have used if σ^2 were constant after all.

I hope to be up on Thursday for your lecture. Perhaps if you care to make any further comment, you would let me know then. I must apologise for this correspondence becoming so lengthy.

M.S. Bartlett to Fisher: 16 *April* 1937

Cochran has mentioned to me that you do not agree with the discussion I gave at the top of p. 565 of my last *Proc. Camb. Phil. Soc.* paper (Part 4, 1936)[1] in connection with testing differences in means. I refer there to the special (and practically trivial) case where there is only one degree of freedom each for the estimates of the unknown and unequal variances of the two means.

I have not detected any error in my remarks, but I should be glad to examine them again in the light of any comments you care to make.

[1] Bartlett, M.S. (1936). The information available in small samples. *Proc. Camb. Phil. Soc.* **32**, 560–6.

Fisher to M.S. Bartlett: 19 *April* 1937

I have written a note on the discrepancy between your results and those obtained by the fiducial argument, which will appear in a forthcoming number of *Annals* [*CP* 151]. I think the cause of the discrepancy lies in the introduction of fixed values for σ_1 and σ_2 into an argument in which the distribution of these values has already been determined.

M.S. Bartlett to Fisher: 14 *July* 1937

As I was in town yesterday afternoon, I called in at the Galton Lab. on the chance of being able to see you for a few minutes, but found you away. I thought perhaps we might have been able to resolve our apparent difference of opinion on the Behrens' test business, on which you comment in a note in the last number of *Annals of Eugenics*.

With regard to the particular point at issue there (the one degree of freedom case), your representation of my view did not appear to me to be altogether fair. You refer, for example, to the distribution of 'my ratios', whereas the composite statistic $(T\,T')$ (your notation) essentially is not to be split up into T and T'. Granting this, you appear to recognize in your concluding paragraphs that the test would do what I claimed for it, but thought that no experimenter would be likely to use it. With this last remark I entirely agree, but I had no intention of recommending any test based on one degree of freedom—it would be a rotten test anyway!

In the practical example I considered at the end of my own paper (*Proc. Camb. Phil. Soc.*), I used an inequality statement as being a convenient test in the circumstances. Practically, the curious feature of the Behrens' test for small numbers of degrees of freedom is the tendency to get a less significant result for s/s' finite than where it is 0 or ∞. Here Behrens' original table seems a little misleading, for one would infer from it that this was always the case, whereas on examination the significance level for small numbers greater than one appears to change over in comparison with the ordinary t test, (assuming $\phi = \sigma_1^2/\sigma_2^2 = 0$), as the probability value of this level ranges from 1/2 to 0. Whether or not any exact probability interpretation of the Behrens' test can be made, these results appear rather anomalous. . . .

Fisher to M.S. Bartlett: 15 *July* 1937

I am sorry you thought my reference to your paper unfair, for I doubt if you will ever receive fairer treatment from those who differ from you on mathematical points. You say that I appear to recognise, in my concluding paragraph, that the test has done what you claimed for it. Since for paired data this is 'Student's' test of 1908, I imagine there could be no doubt on the subject. The difficulty is that, for data not paired, different possible samples are placed in the wrong order in respect of significance, and therefore, though the distribution is correct, as a test of sigificance it is not available except in the case to which 'Student' applies it. This is quite a different criticism from saying that it is based on one degree of freedom only.

The point has, I think, only this importance, that you adduce your test as disagreeing with Behrens' and as showing, in consequence, that Behrens' test must be wrong.

I do not follow the statement in your last paragraph, e.g. that Behrens' test has a tendency to get a less significant result for s/s' finite than where it is zero or infinity. There must be some meaning behind this, but I do not know what it is. Like any other test, Behrens' accepts certain bodies of data as significant, and distinguishes them from other bodies which, in this test, are non-significant. Do you mean that for a given difference in the means significant results are got less frequently when the ratio s/s' is near to unity than when it is larger or smaller? If so, I cannot see its bearing as a criticism of the test. . . .

M.S. Bartlett to Fisher: 21 *July* 1937

. . . By the statement I made in my letter (3rd paragraph) about the Behrens' test, I meant the following. The statistic $z = (x_1 - x_2)/(\sqrt{s_1^2 + s_2^2})$ (equal numbers of degrees of freedom assumed for simplicity) is used either in the ordinary t-test or in the Behrens' test. In the latter it is considered in conjunction with the value of s_1^2/s_2^2. If the latter value is ignored, the distribution of the statistic z will depend on the unknown ϕ ($\equiv \sigma_1^2/\sigma_2^2$), but we know that, whatever ϕ is, we shall err *on the side of caution* if we consider it zero and consider z as a t with the same number of degrees of freedom as s_1^2 (or s_2^2). On the other hand, if we use Behrens' test and brought in s_1^2/s_2^2, we might (for small numbers of degrees of freedom) consider z as even *less* significant than we thought it when we knew we were already erring on the cautious side. I think I should prefer the 'inequality test'. For the case of $n_1 = n_2 = 1$, the significance level given by Behrens' test is P_1 where

$$P_1(s_1^2/s_2^2) < P(\phi = 0), \quad \text{whatever } s_1^2/s_2^2,$$

($P(\phi = 0)$ being the ordinary t test level for $n = 1$); and the significance level given by the 'inequality test',

$$P_2(\phi) \geqq P(\phi = 0), \quad \text{whatever } \phi.$$

I don't think I am altogether clear yet about your objections to my 'randomisation test', but the above comments appear to me also to require some consideration.

Fisher to M.S. Bartlett: 22 *July* 1937

If we are to get on any further I am afraid you will have to tell me why we know that 'whatever ϕ is, we shall err on the side of caution if we consider it zero and consider z as a t with the same number of degrees of freedom as s_1^2 or s_2^2'. You must be suggesting some objective procedure, but I do not know what it is. We cannot, without ignoring some of the information supplied by the sample, consider σ_1 to be zero if s_1 is not zero.

In the same paragraph I do not know what *the inequality test* means.

M.S. Bartlett to Fisher: 23 *July* 1937

I apologise for being obscure, but since I was referring to a test whose application I had illustrated in my *Proc. Camb. Phil. Soc.* paper (final paragraph) I may have neglected to make myself clear about it in my letter. Perhaps if I enclose a formal mathematical statement of what I meant by the test (which I called for convenience the 'inequality test'), it will make my arguments clearer. The proof[1] I give in this enclosure of the 'inequality' is of course merely a verification that the test is mathematically correct, and is perhaps hardly necessary.

I agree with you that we are ignoring the information supplied by s_1^2/s_2^2, but that does not seem to me to imply that the possible value or relevance of the test is nil.

[1] This proof, not reproduced here, was subsequently included in §7 of Bartlett's 1939 paper, Complete simultaneous fiducial distributions. *Ann. Math. Stat.* **10**, 129–38.

W.U. Behrens to Fisher: 13 *August* 1929

Ich interessiere mich für die Anwendung der Wahrscheinlichkeitslehre auf landwirtschaftliche Probleme. Auf der Tagung der Internationalen Bodenkundlichen Gesellschaft hatte ich Gelegenheit, mit ausländischen Herren hierüber zu diskutieren und ich wurde immer wieder auf Ihr Werk: *Statistical Methods for Research Workers* aufmerksam gemacht. Ich habe es mir daher sofort gekauft und lese es jetzt mit grossem Interesse. Ich bedaure ausserordentlich, dass ich es nicht früher gekannt habe, ich hätte mir sonst viel Arbeit ersparen können. Aus meiner Arbeit:[1] 'Ein Beitrag zur Fehlerberechnung bei wenigen Beobachtungen', von der ich Ihnen gleichzeitig ein Separatum schicke, werden Sie ersehen, dass ich zum Teil zu Ergebnissen gekommen bin, wie sie schon längst den englischen Statistikern bekannt sind. Ich hatte die Arbeit vor der Drucklegung an einen der bedeutensten deutschen Wahrscheinlichkeitstheoretiker geschickt und angefragt, ob seines Wissens bereits Arbeiten mit ähnlicher Problemstellung vorliegen. Er antwortete mir damals, dass ihm keine derartigen Arbeiten bekannt seien. Ich werde selbstverständlich bei der nächsten sich bietenden Gelegenheit darauf aufmerksam machen, dass ich *nicht* die Priorität habe.

Auf Seite 820–822 meiner Arbeit komme ich zu anderen Ergebnissen als die englischen Statistiker resp. Sie. Ich wäre Ihnen sehr dankbar, wenn Sie hierzu Stellung nehmen und mir Ihre Ansicht mitteilen könnten.

Grüssen Sie bitte Herrn Dr. Crowther von mir.

[1] *Landw. Jbr.* **68**, 807–37 (1929).

Fisher to W. U. Behrens: 5 September 1929

I received your letter with very great interest during my holiday, and since my return have studied the reprint of your paper. It is a great pleasure to see that you have arrived independently at 'Student's' distribution, the importance of which I have now for many years been endeavouring to make clear to the English speaking statisticians. Your attack upon the more complex problem of comparing the means of samples having different variances seems to me wholly original.

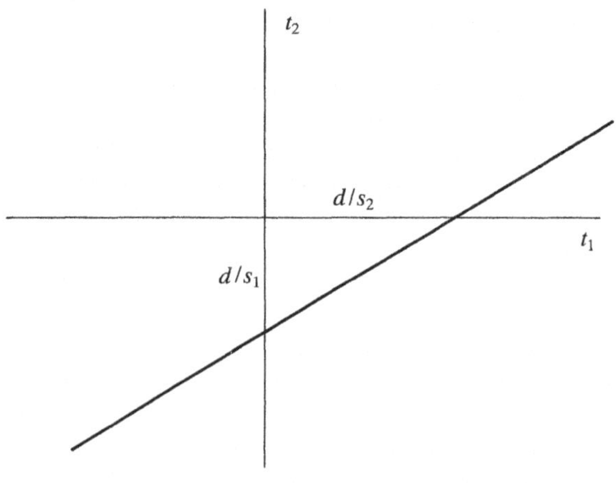

The integration of the simultaneous distributions of two values t_1 and t_2, dependent from [? upon] two different samples of sizes n_1 and n_2, on one side of the line, making intercepts d/s_1 and d/s_2 respectively, would be of very great interest. Unfortunately this probability depends from [? upon] four parameters, n_1 and n_2, in addition to d/s_1 and d/s_2 or as you have it $d/\sqrt{s_1^2 + s_2^2}$ and s_1/s_2. The numerical values you give are confined, I believe, to the case $n_1 = n_2$; the more general case is, however, of considerable interest for experimentalists. The tabulation would be simplified and the results expressed in a form convenient for tests of significance if one could tabulate the 5 per cent and 1 per cent values of $d/\sqrt{s_1^2 + s_2^2}$ for different values of n_1, n_2 and s_1/s_2, replacing the last modulus by its arc tan for purposes of interpolation. Probably 10 values at intervals of 10° would then be sufficient, while for n_1 and n_2 I should utilise the harmonic series 6, 8, 12, 24, ∞, which works excellently for 'Student's' distribution.

I am glad that you are calling attention in Germany to recent advances in the theory of errors, which besides their intrinsic importance are having an increasing influence upon the design of biological and agricultural experi ments.

I enclose two recent papers which I hope may be of interest to you.

W. U. Behrens to Fisher: 10 *September* 1929

Verbindlichsten Dank für Ihren werten Brief vom 5.9. und Ihre interessanten Separata. Aus Ihrem Brief ersehe ich, dass Sie es für nützlich halten, grössere Tabellen für die 'tests of significance' von Beobachtungsdifferenzen aus zurechnen. Ihr Plan findet meinen Beifall, ich möchte vorschlagen, auch die 2 per cent values von $d/\sqrt{s_1^2 + s_2^2}$ zu berechnen. Sollten Sie auf meine Mitarbeit Wert legen, so bin ich gern bereit, mich an einer gemeinsamen Arbeit zu beteiligen, und ich erwarte dann von Ihnen nähere Vorschläge. Allerdings bin ich jetzt durch andere Arbeiten in Anspruch genommen, aber von Dezember dieses Jahres bis Marz 1930 hoffe ich mehr Zeit zur Verfügung zu haben. Die Ergebnisse könnten wir dann gemeinsam publizieren, Sie könnten vielleicht einen Aufsatz für eine englische Zeitschrift schreiben, ich würde den Text für eine deutsche Publikation in den Landwirtschaftlichen Jahrbüchern schreiben.

Fisher to W. U. Behrens: 9 *February* 1957

I am sending you herewith a recent offprint [*CP* 265] on the test of significance which bears your name, at least in English-speaking countries.

I do not know if it has ever come to your knowledge that a number of attacks have been made on the validity of this test by J. Neyman, and others influenced by him, such as Pearson and Bartlett; indeed, tables of a rival test have been published—Table 11 in the *Biometrika Tables* of Pearson and Hartley. I have recently criticized these tables, which are, in my opinion, grossly inaccurate, and make the observed difference significant on much too easy terms.

The fact is that the problem which you first discussed, I think about 1929, is a good test case for the validity of the rather academic exposition of tests of significance put forward by Neyman and Pearson more than twenty years ago, of which the key thought is the frequency with which a statement would be found to be correct in 'repeated samples from the same population', which is in fact very far from being a measure of the strength of the evidence provided by the data against some specific and well defined hypothesis or group of hypotheses.

I hope, at least, you find these two offprints [*CP* 264, 265] interesting.

I have been wondering if we shall see you at the Stockholm meeting of the International Statistical Institute. I feel sure an invitation could be obtained if this would enable you to get any necessary financial assistance.

W. U. Behrens to Fisher: 22 *September* 1957

I beg your pardon that I have not answered you before today. I had planned to come to Stockholm, but I had not the time. The money was not missing, but the time. My occupation is the agricultural chemistry in the fertilizer industry, the biomathematics are my hobby. So I have not the time for study of the international literature as it is necessary, and I thank you very much for your information about the so-called Behrens–Fisher test. . . .

Fisher to C.I. Bliss: 4 *October* 1956

. . . I shall be much interested to learn how you, and others among your countrymen, bear with my efforts [in *SMSI*] to draw the necessary logical distinctions, and to use words accurately, in a subject which has been so deeply entangled and knotted together as the theory of probability had become early in the century, and the theory of testing hypotheses since about 1930.

I believe, now, I should have stressed early and loudly that Keynes was mistaken in defining probability as 'the measure of rational belief', in that whereas the phrase 'measure of rational belief' was a penetrating one and needed in the subject, yet it is not the classical and mathematically defined concept of probability that fills this bill.[1] This does appear, indeed, gradually and by stages in the course of Chapters II and III, but perhaps you could give me an opinion as to whether the reader might like to have fair warning of so large a semantic change. The effort to find probability statements appropriate to every case in which belief, or disbelief, in some degree can be supported on good and communicable reasons, is, of course, the main cause of the numerically erroneous values arrived at on the basis of Pearson and Neyman's theory of testing hypotheses, of which I give an example in the current *J.R.S.S.*, Series B [*CP* 264], which I hope you may see before I have an offprint to send you. . . .

[1] A review by Fisher of J.M. Keynes's book *A treatise on probability* (1921) was published in *Eugen. Rev.* **14**, 46–50, (1923).

O.K. Buros to Fisher: 16 *February* 1959

In your *Statistical Methods for Research Workers*, you describe biased intraclass correlation coefficients, methods for correcting for the bias, and unbiased intraclass coefficients obtained by means of analysis of variance. Have you written more extensively on these topics elsewhere?

An article, 'Unbiased estimation of certain correlation coefficients' in the March 1958 issue of *Annals of Mathematical Statistics* seems (I am unable to follow the proof) to indicate that your unbiased estimates of the intraclass correlation coefficient are biased. Would you be willing to comment on this? . . .

O.K. Buros to Fisher: 17 *February* 1959

Since writing you yesterday, I have received the following statement from a friend:

> 'I raised the question (regarding what has been described as Fisher's unbiased estimate of the population intraclass correlation coefficient) with Professor S.S. Wilks. He has written down a proof that no ratio of linear functions of 'among' and 'between' sums of squares can be an unbiased estimate of the population intraclass correlation. Of course the numerator and denominator taken separately are unbiased estimates of important population parameters, but this does not make the ratio itself an unbiased estimate.'

I have always talked about biased and unbiased estimates of intraclass correlations, thinking that I was interpreting you correctly. Now that Professor Wilks states that such a ratio cannot be an unbiased estimate, I should like to learn what the facts are from you. . . .

Fisher to O.K. Buros: 23 *February* 1959

I have your letters of February 16th and 17th on a point which has needed clarification in statistical teaching for a long while, but as a good many points can be introduced it would be too long to discuss in a single letter. In my recent book *Statistical Methods and Scientific Inference* I say something about the use of 'unbiased' as a criterion in estimation (p. 140)[1] because as used by Neyman, and perhaps other teachers in Berkeley, it seems to have given a great deal of complexity to otherwise simple problems.

Perhaps the simplest interpretation of Professor Wilks's recent demonstration is that he must be using the word 'biased' in a sense which is not usefully applicable by practical statisticians, e.g. the sense in which the estimated standard deviation is biased, although found by taking the square root of an unbiased estimate of the variance.

In Section 2 of the chapter to which I have referred, there is developed with small sample exactitude the notion of Consistent estimation, which fulfils the common sense requirement of unbiasedness by stipulating that the estimate shall be a function of the observed frequencies of such a kind that if expected frequencies are substituted for those observed, the estimate shall be exactly equal to the true parametric value. This criterion, which I put forward I think about as early as 1922 (Mathematical foundations of theoretical statistics: *Phil. Trans.* A, vol. 222) [*CP* 18], is invariant for transformations of the parameter, such as squaring, and so avoids the difficulty which formal bias, as defined by Wilks and others, falls into.

I think the only practical importance of what has become by now a rather intricate subject lies in the avoidance of bias in the methods of sample survey, where it is often desired to total or average numerous small sample estimates. Of course you know Dr. Frank Yates's book *Sampling Methods for Censuses and Surveys* (1949) in which he gives to this point all the care it deserves. If you do refer to my work, however, it might be better, in regard to the current use of language, to speak of the estimates I recommend rather as 'Consistent' than as 'unbiased'. This, of course, is just as strict a criterion but one more suitable for the situation of statistical estimation.

[1] *SMSI*, p. 146.

N. Campbell to Fisher: 18 *July* 1922

On my return from a holiday I have received the reprint of your paper [*CP* 18], for which I am greatly obliged. Just before leaving I had read your paper with the very greatest interest, for it appeared to me a most welcome reversion to realities after the wholly abstract and impracticable dissertations to which we had become accustomed from mathematicians.

As you say, I think our ideas, insofar as they cover the same ground, are essentially similar. But there are differences. The difference in nomenclature is nothing. I prefer to get away from the connotations of the term probability, but even if I used that term *my* probability ('chance') would not be the same as that which you define or describe on p. 312. My chance is something essentially experimental and 'an infinite number of throws' means nothing to an experimenter—you might as well talk of a round square. Moreover randomness is essential to my chance, whereas, so far as I can see, your die would have a 'probability' if it turned up six regularly at the 1st, 7th 13th, . . . throw. But as a matter of fact, the *use* you make of your term seems to me to make it approximate much more nearly to mine than the definition suggests. If a more 'technical' word than chance is thought desirable for my conception, I should suggest 'statistical frequency'—but most European languages have a word exactly corresponding to our 'chance'.

For the rest the problem of statistics seems to me that of identifying a system by means of a measurement of its chances. And that is a problem which (as I believe you recognise) mathematics alone can never solve. Whatever degree of agreement is found between the measured chances and those calculated for some system with which it is proposed to identify that examined experimentally, there is always room and need for the characteristically scientific decision whether, taking everything into account, there is sufficient evidence to permit the identification to be made. That kind of judgement is distinctive of all physical work and by no means confined to statistical study. Still there is room for the mathematician (and it is here that I find your conception of a statistic of maximum efficiency so helpful) in selecting for measurement those chances which provide the best basis for comparison. I make no pretence to have mastered your methods completely and indeed it is probably impossible for one so little practised in statistics to do so, but I am thoroughly convinced that you are on the right lines. The only other method which seems to offer a plausible alternative is some kind of systematic search for regularity in 'residuals'. The less regularity such search discloses, the more weighty the evidence that the system under investigation really has a chance of the kind and of the magnitude which is proposed.

But whether you consider my remarks on such matters sensible or no, pray accept my best thanks and warmest congratulations.

N. Campbell to Fisher: 21 *July* 1922

I hope you will not think that because you were good enough to write to me I am going to drag you into an endless controversy.[1] But I do want to make my position clear; for I am rather alarmed that so sympathetic a critic should seem to miss the essential feature of my contention.

You say that I nominally reduce chance to a synonym for frequency. I protest violently. I maintain on the contrary that the bare knowledge that an event has occurred 15 times out of 100 trials tells me nothing whatsoever about its chance. For that knowledge does not tell me that the event has a chance at all. Even if I know beforehand that the event has a chance, it gives me but the vaguest knowledge what that chance is. It tells me perhaps that the chance is *not* 1 in 1000 and is not 999 in 1000, but it tells me little more. In order to estimate chance I must have a whole series of frequencies, and the chance which I estimate will be determined by the whole of the series and not by a single member of it.

I think it is very useful to have before one's eyes a concrete example. Accordingly on the enclosed diagram[2] I have shown how I should determine the chance that the last figure of a seven-figure logarithm is either 3, 6, or 9. The ordinates give the number of such last figures in the first *n* logarithms, where *n* is the abscissa. After I have plotted 250 points I ask myself whether I am justified in drawing a straight line through the points. If I knew nothing about the events, I should be doubtful in this instance, for the earlier points tend to lie above the mean line, the later below it. I should have to take many more points (or plot the diagram again from a different page of the book) to be sure whether this regularity was real. But if I assume that there is a chance, it is pretty closely defined. I have drawn the 'theoretical' line; and it is clear that any line I could reasonably draw would not differ as much as 10% in slope from it. But that slope, I want you to observe, is determined by all the frequencies taken together; it is not determined by any one of them; in fact at only four points (marked by arrows) do the frequencies coincide with the line that determines the chance.

I suggest that what you mean by the 'true' frequency, or the frequency of a hypothetical infinite population, is simply the frequency corresponding to points on the line. And of course this true frequency can only agree with the actual frequency (in this instance) when the number of trials is a multiple of 10. What you are essentially doing in interpreting statistics is finding a theoretically determined line (such as I have drawn on the diagram) which will give an adequate representation of the scattered actual points. But I won't go further into that. I only want you to be clear what *I* mean, not to discuss what *you* mean.

[1] There is no copy in Fisher's files of his reply to Campbell's letter of 18 July 1922. Fisher's letter of 22 July 1922 to Campbell is the earliest of all his letters included in this volume; the original copy, in Fisher's hand, is reproduced here.
[2] Not shown here.

Fisher to N. Campbell: 22 July 1922

Many thanks for your letter. I do not think there is any danger of an endless controversy, especially as I imagine we are both concerned to make sense of a

Telegrams: LABORATORY, HARPENDEN.
Telephone: 21 HARPENDEN.

Railway Station:
HARPENDEN (MIDLAND RAILWAY).

Lawes Agricultural Trust.

Director:
E. J. RUSSELL, D.Sc., F.R.S.
R. A. FISHER, M.A., *Chief Statistician.*
WINIFRED A. MACKENZIE, B.Sc. (ECON).

Rothamsted Experimental Station
Harpenden.

July 22 1922

Dear Dr Campbell,

many thanks for your letter. I do not think there is any danger of an endless controversy, especially as I imagine we are both concerned to make sense of a rather intricate subject, about which a good many false assumptions have been widely accepted. Your suggestion, in your letter of the 18th, that apart from the difference in nomenclature, your "chance" and my "probability" were intended to designate somewhat different ideas, interested me; the main point, as you say, was for me to get a clear idea of what you meant. I had judged from your remarks in the Phil. Mag., that as the chance of an event occurring would was a physical property of the material system concerned, that we did mean exactly the same thing.

In your letter of the 18th the distinction you drew (apart from the point about randomness in which I fully agree with you) was that "My chance is something essentially experimental and 'an infinite number of throws' means nothing to an experimenter", and again "If a more 'technical' word than chance is thought desirable for my conception, I should suggest 'statistical frequency'". This, of course, would make your chance quite distinct from my notion of probability, but I did not think, after what you had written in the Phil. Mag. that you really wished to identify chance with frequency, and I am glad that you "protest violently" against the suggestion.

A certain number of writers on Probability have tried to define the notion on an objective basis. Thus Leslie Ellis (1843): "If the probability of ~~an event~~ a given event be correctly determined, the event will on a long run of trials tend to recur with frequency proportional to their probability. This is generally proved mathematically. It seems to me to be true *a priori*", and again "I have been unable to sever the judgement that one event is more likely to happen than another, from the belief that in the long run it will occur more frequently."

About the same time Cournot wrote "La theorie des probabilités a pour objet certains rapports numeriques qui prendraient des valeurs fixes et complètement determinées, si l'on pouvait répéter à l'infini

Telegrams: LABORATORY, HARPENDEN.
Telephone: 21 HARPENDEN.

Railway Station: HARPENDEN (MIDLAND RAILWAY).

Laws Agricultural Trust.

Director: E. J. RUSSELL, D.Sc., F.R.S.

R. A. FISHER, M.A., Chief Statistician.
WINIFRED A. MACKENZIE, B.Sc. (Econ).

Rothamsted Experimental Station
Harpenden.

. . . . 192....

les épreuves des mêmes hasards, et qui, pour un nombre fini d'épreuves, oscillent entre des limites d'autant plus resserrées, d'autant plus voisines des valeurs finales, que le nombre des épreuves est plus grand."

These views, which have been called the Frequency Theory of Probability, seem to me to be sound, though perhaps you will jib at Cournot's "répétée à l'infini", and at Ellis' "in the long run". I prefer the former. I do not doubt that the frequency theory would have been universally accepted, if a sufficiently clear distinction had always been maintained between the hypothetical and the experimental value.

Apart from what we actually mean by the probability or chance, when we have decided that there is one, and that we wish to determine it experimentally, your letter of the 21st, deals with the important practical question of testing the homogeneity of data, or from the point of view of a series, of testing if the series is what I should call a changing or an unchanging series. For example it is a sufficiently objective question, to ask if our weather is changing, quite apart from fluctuations. A series of meteorological records may be tested with this in view. Such a question as "is the chance of getting over 40 inches of rain in the year, the same as it was 50 years ago", depends on the same point. Your diagram seems to me a rough but sufficient test of the corresponding question about the log tables. But this is really a separate question. If it be admitted that your material is homogeneous, then to estimate the frequency of 3, 6, 9, I only want to know the end point of your graph, i.e. the fact that ~~these~~ these digits occur 76 times out of 250 trials. If I want to know if your series is consistent with the theoretical probability $\frac{3}{10}$, I merely compare the discrepancy, 1, with its standard error 7.25. Clearly the discrepancy is insignificant; but to test the homogeneity of the table in this respect is quite another matter from defining, or estimating the probability.

Yours sincerely

R. A. Fisher

rather intricate subject, about which a good many false assumptions have been widely accepted. Your suggestion, in your letter of the 18th, that apart from the difference in nomenclature, your 'chance' and my 'probability' were intended to designate somewhat different ideas, interested me; the main point, as you say, was for me to get a clear idea of what you meant. I had judged from your remark in the *Phil. Mag.*,[1] that the chance of an event occurring was a physical property of the material system concerned, that we did mean exactly the same thing.

In your letter of the 18th the distinction you draw (apart from the point about randomness in which I fully agree with you) was that 'My chance is something essentially experimental and "an infinite number of throws" means nothing to an experimenter', and again, 'If a more "technical" word than chance is thought desirable for my conception, I should suggest "statistical frequency".' This, of course, would make your chance quite distinct from my notion of probability, but I did not think, after what you had written in the *Phil. Mag.*, that you really wished to identify chance with frequency, and I am glad that you 'protest violently' against the suggestion.

A certain number of writers on Probability have tried to define the notion on an objective basis. Thus Leslie Ellis (1843);[2] 'If the probability of a given event be correctly determined, the event will on a long run of trials tend to recur with frequency proportional to their probability. This is generally proved mathematically. It seems to me to be true *a priori*', and again, 'I have been unable to sever the judgement that one event is more likely to happen than another from the belief that in the long run it will occur more frequently.'

About the same time Cournot wrote[3] 'La théorie des probabilités a pour objet certains rapports numériques qui prendraient des valeurs fixes et complétement déterminées, si l'on pouvait répéter à l'infini les épreuves des mêmes hasards, et qui, pour un nombre fini d'épreuves, oscillent entre des limites d'autant plus resserrées, d'autant plus voisines des valeurs *finales*, que le nombre des épreuves est plus grand.'

These views, which have been called the Frequency Theory of Probability, seem to me to be sound, though perhaps you will jib at Cournot's 'répéter à l'infini' and at Ellis' 'in the long run'. I prefer the former. I do not doubt that the frequency theory would have been universally accepted if a sufficiently clear distinction had always been maintained between the hypothetical and the experimental value.

Apart from what we actually mean by the probability or chance, when we have decided that there is one, and that we wish to determine it experimentally, your letter of the 21st deals with the important practical question of testing the homogeneity of data, or from the point of view of a series, of testing if the series is what I call a changing or an unchanging series. For example, it is a sufficiently objective question to ask if our weather is changing, quite apart from fluctuations. A series of meteorological records may be tested with this in view. Such a question as 'Is the chance of getting over 40 inches of rain in the year the same as it was 50 years ago?' depends on the same point. Your diagram seems to me a rough but sufficient test of the corresponding question about the log tables. But this is really a separate question. If it be admitted that your material is homogeneous, then to estimate the frequency of 3,6,9, I only want to know the end point of your graph, i.e. the fact that these digits occur 74 times out of 250 trials. If I want to know if your series is consistent with the theoretical probability 3/10, I merely compare the discrepancy, 1, with its standard error 7.25. Clearly the discrepancy is insignificant, but to test the homogeneity of the table in this respect is quite another matter from defining, or estimating, the probability.

[1] Campbell, N. (1922). The measurement of chance. *Phil. Mag.* **44**, 67–79.
[2] Ellis, R.L. (1843). On the foundations of the theory of probability. *Proc. Camb. Phil. Soc.* **8**, 1–6.
[3] Cournot, A. (1843). *Exposition de la théorie des chances et des probabilités*. Paris.

[The following paragraph seems worthy of inclusion here. It is the final part of a 'Note on Dr. Norman Campbell's alternative to the method of least squares', written by Fisher probably about 1925. See Campbell, N. (1924). The adjustment of observations. *Phil. Mag.* **47**, 816–26.]

Finally perhaps a word may be said with an eye to the underlying motive of Dr. Campbell's innovation. He is evidently oppressed by the arithmetical labour involved in the recognised processes of reduction. These processes, like commercial book-keeping, do require some arithmetical labour; and would constitute an intolerable burden on a man whose mind was full of other and more important business. The commercial man gets over this difficulty by hiring such computational assistance as he needs; the alternative of inventing for himself new book-keeping methods, so quick and simple that he could do it all himself during train journeys, would not, I suggest, make a strong appeal. Need the organisation of scientific research be so much behind that of an ordinary small business establishment? In my experience, much of the troublesome arithmetic with which research workers, physicists and biologists, harass themselves, could be done *better* by a boy of 16 of average intelligence, working in a properly organised statistical laboratory. The constant crop of inefficient methods, of which Dr. Campbell's is by no means the worst, which are continually being produced by laboratory workers, who have some but too little experience of statistical problems, is an index of the amount of highly skilled labour which is being wasted in these ways, through lack, it would seem, of organised computational assistance.

H.E. Daniels to Fisher: 29 *January* 1938

Some time ago I wrote to you in regard to the question of intercorrelated observations and the z test, and you were kind enough to send me a most illuminating reply.[1] The following point is now troubling me, and I should be grateful for your advice.

Consider e.g. two samples of size n_1, n_2 respectively and we wish to test whether the means differ significantly. Suppose that the population variances are not identical, but are σ_1^2, σ_2^2. The expected value of

$$\{\Sigma_r(x_{1r} - \bar{x}_1)^2 + \Sigma_r(x_{2r} - \bar{x}_2)^2\}/(n_1 + n_2 - 2)$$

is

$$\{(n_1 - 1)\sigma_1^2 + (n_2 - 1)\sigma_2^2\}/(n_1 + n_2 - 2) = v_1$$

and that of

$$n_1 n_2(\bar{x}_1 - \bar{x}_2)^2/(n_1 + n_2) \text{ is } (n_2\sigma_1^2 + n_1\sigma_2^2)/(n_1 + n_2) = v_2$$

so that

$$v_2 - v_1 = (\sigma_1^2 - \sigma_2^2)\,(n_1 + n_2 - 1)\,(n_2 - n_1)/\{(n_1 + n_2 - 2)\,(n_1 + n_2)\}.$$

When

$$\sigma_2 > \sigma_1 \text{ [and] } n_2 > n_1, \text{ [then] } v_2 \text{ is less than } v_1.$$

In case[2]

$$\begin{aligned} \sigma_1^2 &= 4, \sigma_2^2 = 24, n_1 = 5, n_2 = 25, \\ v_1 &= 21, v_2 \cong 7 \end{aligned}$$

and while the expected value of t^2 will not quite equal v_2/v_1 it seems very unlikely that it should exceed unity. If correct, this would appear to contradict your remark[3] (*Statistical Methods* p. 121) that a real difference in variance would always enhance t. One might argue that only those cases which do enhance t are important since when significance is not obtained, no positive conclusion is drawn. But the test would seem to suffer as a test of the null hypothesis that the samples arose from the same normal population.

[1] See p. 253.

[2] Daniels had written $n_1 = 25$, $n_2 = 5$, $v_1 \simeq 82$ and $v_2 = 21$, but the values shown here are evidently those intended.

[3] '. . . a difference in variance between the populations from which the samples are drawn will tend somewhat to enhance the value of t obtained.' (*SMRW*, 5th edition, 1934, Section 24.1, p. 121). However, with the 6th edition in 1936, Fisher replaced 'somewhat' in the above sentence by 'sometimes'. See *SMRW*, p. 124.

Fisher to H.E. Daniels: 18 *February* 1938

Your letter of January 29th was given me on my return, and I agree with it, except that I do not think I ever said that a real difference in variances would always enhance t. Perhaps you will let me know whether it was I or some one else who gave you this impression. I think Stevens mentioned that the possibility that the populations sampled differed in their variances does not affect the exactitude of the test of the null hypothesis that they are drawn from the same population. Such a case as you mention, supposing t to be non-significant on 'Student's' test, would seem to be nicely treated by the supplementary z which I suggest for the differences between the variances, for this would show that the variances were very significantly different, although on the data it could not be said that the means were different.

Now in practice this would advance the investigation as much as in the standard case; for, supposing the test were made between two varieties of the plant, the fact of a real difference in the variances shows that in some circumstances one variety is the better and in other circumstances that it is the worse. The situation is thus proved to be more complicated than perhaps the experimenter originally thought. He will see now that it is useless to compare the means unless he has some specification of the circumstances, or range of circumstances, in which the test is to be made. His preliminary enquiry 'Are the samples from the same population?' is answered definitely in the negative and, this being so, it will depend entirely on the circumstances of the case whether any comparison of the means is desired at all.

The point has, I think, received the rather large amount of theoretical attention that it has chiefly through lack of contact with the practical experimental situation. Some years ago Behrens published a test of significance appropriate to the rather academic question 'Might these samples have been drawn from different normal populations having the same mean?', and more recently Sukhatme has been preparing tables needed for Behrens' test. I have, however, always doubted whether the test has any real importance.

Fisher to G. Darmois: 5 June 1936

I write to thank you most heartily for your excellent little book on *Methods of Estimation*, which you were kind enough to send to me. I do not know any other publication of the same length which gives so good an introduction to the subject.

There is one point to which perhaps I may draw your attention, in respect of the propositions you ascribe on p. 19 to J.L. Doob. I have little doubt that the estimate obtained by the method of maximum likelihood always converges, and, if so, it certainly is consistent, or correct. It is not, however, always true that its limiting sampling distribution, when the sample is increased, is Gaussian. This is only true in general when the amount of information, g_{ii} in your notation, is finite, i.e. neither zero nor infinity. For particular values of the parameter it may always be made zero or infinity by an appropriate transformation of the parameter, but such failures of the measure of information are only trivial. When, however, the likelihood is a discontinuous function, as it is for the probability distribution

$$e^{-(x-\mu)}dx$$

for which the least observation in the sample supplies a sufficient estimate of μ, we have, nevertheless, no measurement of the amount of information, and the limiting distribution is not Gaussian. These cases, therefore, where the measurement of information is not a convergent integral for all values of the parameter, deserve special study to ascertain whether any measure of the value of the estimate can be found by which such cases might be compared.

On another point, on which I think you do not touch, I believe it must be true that the maximum likelihood estimate has always a higher intrinsic precision than any other estimate which can be made, but I have only proved this as true in the limit for large samples, where the loss of information is certainly minimised by using the method of maximum likelihood.

I notice you use the term 'exhaustive estimation' for what I speak of as 'sufficient statistics'. I should like, if you approve, to use the term 'exhaustive estimation' in a somewhat wider sense, including not only the cases in which sufficient statistics exist, but also those cases described in my paper 'Two new properties of mathematical likelihood' [*CP* 108], in which the whole of the ancillary information may be utilised by taking account of characteristics of the sample, which, by themselves, yield no information respecting the parameter sought. Estimation in these cases also may be said to be exhaustive, and it seems to me likely that further research may enlarge the field of application of this method. Indeed exhaustive estimation in this sense would be possible if it were always true:- that whenever $x_1, \ldots, x_n$ have a simultaneous distribution depending on a parameter θ it is possible to find $n-1$ functionally independent functions of the variates, the distribution of each of which is independent of θ. It would be interesting to know if you could form an intuitive judgement of the probable truth of this proposition.

Fisher to G. Darmois: 2 *April* 1940

I am very glad to have your letter. . . .

I should be very much surprised in a practical case to get a divergent series[1] by the process which I think is easiest in principle, namely:

If $m_r(\theta_1, \ldots \theta_p)$, when $r = 1, \ldots s$, is the expected frequency in class r for given values of the parameters θ_1 to θ_p, let

$$I_{jk} = \sum_{r=1}^{s} \frac{1}{m_r} \frac{dm_r}{d\theta_j} \frac{dm_r}{d\theta_k}$$

and let V_{jk} be the corresponding element in the reciprocal matrix, then if a_r is the observed frequency, let

$$\sum_{r=1}^{s} \frac{a_r}{m_r} \frac{dm_r}{d\theta_j} = A_j$$

then

$$\theta_j' = \theta_j + \sum_{k=1}^{p} V_{jk} A_k$$

gives a second approximation. The matrices are determined at the trial values, and for more precise values of the variances and co-variances of the values fitted may usefully be redetermined from the improved values. From my own experience I should expect in practice that this would fail to give a convergent series of approximations only if the trial values were exceptionally unfortunate. I think, however, I should have time in the near future to try my hand at any batch of data which has been giving trouble. An approximation based on percentiles is often quicker to obtain than one based on moments, and is likely to be more accurate if the data are heavily grouped. If I ran into a divergent series I should more readily suspect that the theoretical form chosen was unsuitable to the data, i.e. that the goodness of fit was very unsatisfactory, than that my starting point, supposing this to be ordinarily plausible, was the cause of the trouble. . . .

[1] Darmois had asked if the method of successive approximations could lead to a divergent series when used with maximum likelihood estimation.

Fisher to G. Darmois: 27 July 1940

I am sending herewith some of the material you want . . .

We are much looking forward to seeing you on Tuesday and certainly hope that you will meet with no official obstructions, though anything is possible in these days. I am enclosing also what I have just written out, in the hope that you may be interested, an example of ancillary information which I think you have seen before, though perhaps not so fully worked out. I have not, however, delayed to enlarge upon the question of its relation to the problem of the Nile. I believe, however, that this problem is equivalent to the finding of appropriate ancillary statistics in general.

[Enclosure]

Example of ancillary information supplied by characteristics of the data other than the total number of observations

Consider a distribution of two observable values x, y such that the element of frequency is

$$df = e^{-(\theta x + y/\theta)} dx\, dy$$

where x and y each may take positive values from 0 to ∞. The parameter θ is unknown but may be estimated from the observations.

Estimation

If we have a sample of n observations, and apply the Method of Maximal Likelihood, we find

$$\log L = -\theta S(x) - S(y)/\theta,$$

$$\frac{\partial}{\partial\theta} \;'' \quad = - \; S(x) + S(y)/\theta^2.$$

Equating this to zero, we have the Equation of Estimation

$$\theta^2 = S(y)/S(x), \tag{1}$$

and may define our estimate T by the equation

$$T^2 = S(y)/S(x); \tag{2}$$

we shall then be interested in the distribution of T for given θ.

Distribution

If ancillary information is ignored, we consider the distribution of T for different samples of n observations each. If for brevity we write X for $S(x)$ and Y for $S(y)$, it is easy to see that the simultaneous distribution of X and Y is

$$\mathrm{d}f = \frac{1}{(n-1)!^2} X^{n-1} Y^{n-1} \mathrm{e}^{-\theta X - Y/\theta} \mathrm{d}X\, \mathrm{d}Y. \tag{3}$$

Substituting $Y = T^2 X,$
$\mathrm{d}Y = X\mathrm{d}(T^2),$

we have

$$\mathrm{d}f = \frac{1}{(n-1)!^2} \quad T^{2n-2}\mathrm{d}(T^2).\; X^{2n-1}\mathrm{e}^{-X(\theta + T^2/\theta)}\mathrm{d}X \tag{4}$$

and integrating from 0 to ∞ with respect to X

$$\mathrm{d}f = \frac{(2n-1)!}{(n-1)!^2} \; \frac{T^{2n-2}\,\mathrm{d}(T^2)}{(\theta + T^2/\theta)^{2n}} \tag{5}$$

which is the distribution of T given θ and n.

Amount of information available, and amount utilised.

From the original distribution

$$\mathrm{d}f = \mathrm{e}^{-\theta x - y/\theta}\mathrm{d}x\mathrm{d}y$$

we ascertain the mean value or expectation of

$$\{\mathrm{d}(\log f)/\mathrm{d}\theta\}^2.$$

This is the mean value of

$$(x - y/\theta^2)^2$$

which by simple integration is found to be $2/\theta^2$.

The amount of information about θ contained in n observations is therefore

$$I = 2n/\theta^2. \tag{6}$$

For the amount available in the distribution of T for given θ and n, we require the mean value of

$$4n^2\left(\frac{1 - T^2/\theta^2}{\theta + T^2/\theta}\right)^2;$$

using the Eulerian integral of the first kind it appears that

$$E\left\{\frac{1}{(\theta + T^2/\theta)^2}\right\} = \frac{n(n+1)}{2n(2n+1)} \cdot \frac{1}{\theta^2},$$

$$E\left\{\frac{T^2/\theta^2}{(\theta + T^2/\theta)^2}\right\} = \frac{n^2}{2n(2n+1)} \cdot \frac{1}{\theta^2},$$

$$E\left\{\frac{T^4/\theta^4}{(\theta + T^2/\theta)^2}\right\} = \frac{n(n+1)}{2n(2n+1)} \cdot \frac{1}{\theta^2},$$

giving in all

$$4n^2/(2n+1)\theta^2. \tag{7}$$

This is less than the total information available by one part in $2n + 1$; or, when $n = 1$, only 2/3 of the available information is utilised.

Ancillary statistic

Suppose now we propose to ignore n, and consider in its place the ancillary statistic N defined by

$$N^2 = XY.$$

The simultaneous distribution of T and N for given n is found by substituting

$$N/T \text{ for } X \text{ and } NT \text{ for } Y;$$

then

$$\mathrm{d}X\mathrm{d}Y = (2N/T)\mathrm{d}N\mathrm{d}T$$

and

$$\mathrm{d}f = \frac{1}{(n-1)!^2} N^{2n-1}\mathrm{d}N\, \mathrm{e}^{-N(\theta/T + T/\theta)} \frac{2\mathrm{d}T}{T}. \tag{8}$$

Now

$$\int_0^\infty \mathrm{e}^{-N(\theta/T + T/\theta)} \frac{\mathrm{d}T}{T} = \int_0^\infty \mathrm{e}^{-N(z + 1/z)} \frac{\mathrm{d}z}{z}$$

is independent of θ, but is some function $K(N)$ of N. Hence the distribution of N is

$$df = \frac{2}{(n-1)!^2} N^{2n-1} K dN \tag{9}$$

and the distribution of T for given N is

$$df = \frac{1}{K} e^{-N(\theta/T + T/\theta)} \frac{dT}{T} \tag{10}$$

which is independent of n. Now when a sample is observed we know n and equally we know N. If we prefer it we have equally good grounds for regarding (10) as the sampling distribution of T as for so regarding equation (5).

It may now be shown that the information supplied by (10) is the whole of that available.

Since by definition

$$K = \int_0^\infty e^{-N(z+1/z)} \frac{dz}{z},$$

$$\frac{d^2K}{dN^2} = \int_0^\infty e^{-N(z+1/z)} (z + 1/z)^2 \frac{dz}{z}. \tag{11}$$

But $I_{T,N}$ is the expectation of $\{N(1/T - T/\theta^2)\}^2$ which is

$$\frac{N^2}{\theta^2 K} \int_0^\infty e^{-N(z+1/z)} (z - 1/z)^2 \frac{dz}{z}$$

$$= \frac{N^2}{\theta^2 K} \left(\frac{d^2K}{dN^2} - 4K \right). \tag{12}$$

We now evaluate the mean value of this expression for samples having fixed n,

$$df = \frac{2}{(n-1)!^2} N^{2n-1} K\, dN.$$

Now $N^{2n}K = 0$ when $N = 0$ and when N is ∞;

hence $$\int_0^\infty N^{2n} \frac{dK}{dN} dN = -2n \int_0^\infty N^{2n-1} K\, dN.$$

Similarly

$$\int_0^\infty 2n(2n+1) N^{2n-1} K dN + 2 \int_0^\infty (2n+1) N^{2n} \frac{dK}{dN} dN + \int_0^\infty N^{2n+1} \frac{d^2K}{dN^2} dN = 0.$$

But the mean value of $I_{T,N}$

$$\frac{1}{\theta^2}\frac{2}{(n-1)!^2}\int_0^\infty N^{2n+1}\frac{d^2K}{dN^2}dN-\frac{1}{\theta^2}\frac{8}{(n-1)!^2}\int_0^\infty N^{2n+1}KdN$$

$$=\frac{1}{\theta^2}\frac{2}{(n-1)!^2}\left\{-2(2n+1)\int_0^\infty N^{2n}\frac{dK}{dN}dN-2n(2n+1)\int_0^\infty N^{2n-1}KdN\right\}-\ldots$$

$$=\frac{1}{\theta^2}\left\{2n(2n+1)-4n^2\right\}$$

$$=\frac{2n}{\theta^2} \tag{13}$$

which is equal to the whole of the information available.

Whereas the estimate T is not sufficient, on the *convention* that n only is used in its distribution, the pair of values N and T supplies exhaustive information, N being an ancillary statistic capable in this case of replacing the sample number n.

For the same number n the value of N will be sometimes greater sometimes less; the change of convention which makes the estimation exhaustive is that we judge of the precision of the estimate from the value of N observed, and can then totally ignore the size of the sample on which the estimate is based.

In other cases both the size of the sample n and some other characteristics of it will be required as ancillary information. The phrase 'conditionally sufficient' has been used by Bartlett and others for such a statistic as T in this case. I do not think this helpful, since it gives the impression that the sufficiency referred to is an intrinsic property of the estimate *per se*, whereas I do not know that almost any estimate has not the same property when interpreted in relation to some chosen set of ancillary statistics.

The choice of N in this case is suggested by the Likelihood function. It is justifiable from the fact that the distribution of N for given (n,θ) is independent of θ. That is, the series of rectangular hyperbolas specified by constant values of N does in fact divide up the field of variation of x and y, or of X and Y, in such a way that the total frequency expected between any two contours is independent of θ. This is the connection with the Problem of the Nile.

G. Darmois to Fisher: 20 *August* 1940

Je vous envoie quelques remarques sur le problème du Nil et les résumés exhaustifs. Ce que je dis aux pages 1 et 2 est fait depuis assez longtemps, j'y avais réfléchi après votre passage à Paris et votre Conférence à la Société de Biotypologie [*CP* 156]. Le reste provient de la lecture des pages que vous m'avez envoyées. . . .

[Enclosure]

Problème du Nil — Résumés exhaustifs

Le problème du Nil se pose pour une loi de probabilité à deux (ou plusieurs) variables, dépendant d'un (ou plusieurs) paramètres. Soit, pour fixer les idées $f(xy,\theta)\, d(xy)$. Il faut voir si une fonction convenablement choisie $X(xy)$ a une loi de probabilité indépendante de θ. Bien entendu, on peut construire tout de suite toutes les lois ayant cette propriété. Il suffit de prendre

$$A(X)\mathrm{d}X\, B_X(Y,\theta)\mathrm{d}Y$$

$A(X)$ étant la diversité de probabilité marginale de X, indépendante de θ, $B_X(Y,\theta)$ est une fonction de X, Y, θ, qui est la densité de probabilité liée de Y quand X est fixé. Ces deux fonctions peuvent être les plus générales de leur définition. Il suffit ensuite de remplacer X par une fonction arbitraire de xy, et de remplacer Y par y. Dans ces conditions, les courbes $X(xy) = \mathrm{C}^{\mathrm{te}}$ limitent des aires où la probabilité totale est indépendante de θ.

Si l'on envisage l'information qu'une telle loi de probabilité peut apporter, relativement au paramètre θ, on sait qu'on peut le représenter par:

$$i = E\left\{\frac{\partial}{\partial\theta}\log f\right\}^2 = -E\left\{\frac{\partial^2}{\partial\theta^2}\log f\right\}.$$

D'autre part un théorème général relatif à une décomposition de la loi de probabilité sous la forme

$$A(X,\theta)\mathrm{d}X\, B_X(Y,\theta)\mathrm{d}Y$$

où X et Y sont deux fonctions distinctes de xy, indique que l'information i est la somme de l'information qu'on peut déduire de la loi marginale de X, et de l'information qu'on peut déduire de la loi de probabilité liée de Y.

On voit que, dans le cas du problème du Nil, pour toute solution de ce problème, l'information totale est fournie par la seule loi de probabilité de Y quand X est connu (espérance mathématique dans la loi totale). La forme générale de solution que nous avons donnée plus haut permet de former aisément une infinité de solutions particulières. En voici une très simple:

$$\frac{1}{\sqrt{2\pi}}\exp\left(-X^2/2\right)\mathrm{d}X\,\frac{1}{\sqrt{2\pi}\sigma(X,\theta)}\exp\left(-\{Y-m(X,\theta)\}^2/2\sigma^2(X,\theta)\right)\mathrm{d}Y.$$

La loi liée est gaussienne si, pour simplifier encore, on suppose σ indépendant de θ ou aux

$$i = \frac{1}{\sqrt{2\pi}}\int_{-\infty}^{\infty}\frac{1}{\sigma^2(X)}\left\{\frac{\partial m(X,\theta)}{\partial\theta}\right\}^2\exp\left(-X^2/2\right)\mathrm{d}X.$$

On peut trouver des formes explicites de solutions par un procédé moins

général, mais qui donne un mécanisme pour l'étalement de la probabilité dans la bande X, $X + \mathrm{d}X$.
Un moyen très simple d'obtenir le résultat est en effet de soumettre les points de cette bande à une transformation ponctuelle, dépendant de θ, mais laissant invariants, à la fois la fonction X et l'élément de probabilité. On voit qu'on est amené à considérer un groupe de transformations qui échange les éléments de probabilité en conservant leur valeur.
Si l'on prend le groupe de transformations:

$$\begin{aligned} X \quad &= x, \\ Y - t &= y, \end{aligned}$$

on voit que l'élément de probabilité a la forme

$$F(X, Y - t)\, \mathrm{d}(XY).$$

Le problème du Nil est résolu par les bandes $X = \mathrm{C^{te}}$, et les éléments de probabilité se correspondant dans une translation t.

L'exemple de R.A. Fisher. L'élément a pour expressions

$$\mathrm{e}^{-x\theta - y/\theta} \mathrm{d}x\, \mathrm{d}y \qquad \begin{cases} 0 \leqslant x < \infty, \\ 0 \leqslant y < \infty, \\ 0 \leqslant \theta < \infty. \end{cases}$$

Il est clair que si, au point x_1, y_1, on fait correspondre x_2, y_2 tels que:

$$\begin{aligned} x_1\theta_1 &= x_2\theta_2, \\ y_1/\theta_1 &= y_2/\theta_2. \end{aligned}$$

On laisse invariants à la fois la grandeur xy et l'élément de probabilité. C'est la transformation ponctuelle indiquée, laissant invariantes les hyperboles $xy = \mathrm{C^{te}}$.
On peut mettre l'exemple de R.A. Fisher sous la *forme canonique.*
Posons:

$$xy = \mathrm{e}^X, \;\; y/x = \mathrm{e}^Y, \;\; \theta = \mathrm{e}^{t/2}.$$

La probabilité élémentaire devient

$$\tfrac{1}{2} \exp - \{\mathrm{e}^{(X-Y+t)/2} + \mathrm{e}^{(X+Y-t)/2}\}.\, \mathrm{e}^X\, \mathrm{d}X\, \mathrm{d}Y.$$

C'est bien la forme $F(X, Y - t)\, \mathrm{d}X\, \mathrm{d}Y$.

On y trouve en évidence la loi marginale de X

$$\mathrm{d}X \int F(X, u)\, \mathrm{d}u$$

où, évidemment, t ne figure plus. L'exemple de R.A. Fisher possède, du point de vue de l'estimation, d'autres propriétés sur lesquelles nous reviendrons.

Quelques exemples. Une loi de Gauss à deux variables, où la forme de l'ellipse de probabilité est connue, mais où le centre est inconnu sur une droite, peut évidemment se mettre sous la forme

$$K \exp - \tfrac{1}{2} \left\{ \frac{x^2}{a^2} + \frac{(y-t)^2}{b^2} \right\} \mathrm{d}(xy).$$

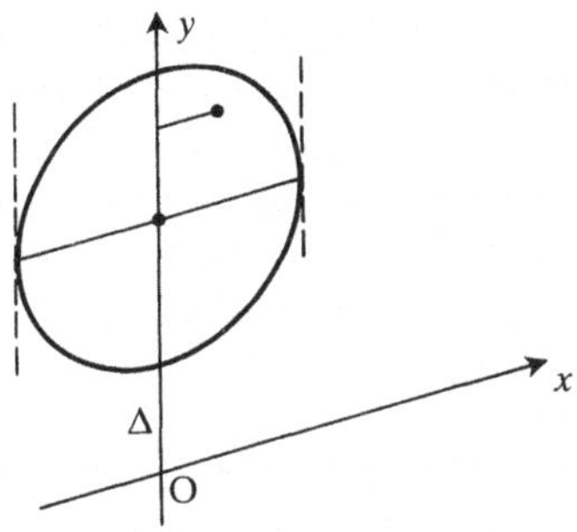

La direction de Ox est conjuguée de la direction de la droite Δ, lieu du centre. Il est clair qu'ici l'information est tout entière dans y. x est inutile.

Exemple de rotation. Une loi de Gauss circulaire, dont le centre seul est inconnu sur un cercle de centre O:

$$K \exp - \tfrac{1}{2}\{(x - r\cos\theta)^2 + (y - r\sin\theta)^2\}\, \mathrm{d}(xy),$$

$$x = \rho\cos\omega,\; y = \rho\sin\omega,$$

$$K \exp - \tfrac{1}{2}\{\rho^2 + r^2 - 2\rho r\cos(\omega - \theta)\}\rho \mathrm{d}\rho \mathrm{d}\omega.$$

On voit bien la forme caractéristique

$$F(\rho, \omega - \theta)\, \mathrm{d}\rho \mathrm{d}\omega.$$

Ici, les bandes qui donnent la solution du problème du Nil sont des anneaux circulaires de centre O.

Information, résumés exhaustifs, estimation. Supposons maintenant que nous considérons n couples x_1y_1, x_2y_2, . . . suivant une même loi de probabilité. Nous voulons en extraire tout ou partie de l'information relative à un paramètre θ. Il y a lieu d'introduire la notion générale de résumé d'une série d'observations, et particulièrement de *résumé exhaustif.* Un résumé sera constitué par une suite d'éléments aléatoires fonctions des observations. Par exemple, centre de gravité des points observés, moments . . ., point de plus grande abscisse . . ., etc.
Par exemple, pour la loi à une variable

$$A\, \mathrm{e}^{-(x-\theta)^4} \mathrm{d}x$$

aux observations $x_1, x_2, \ldots x_n$, nous ferons correspondre le résumé:

$$\Sigma x_i, \Sigma x_i^2, \Sigma x_i^3,$$

formé par les trois premiers moments.
Dans le cas le plus général, le résumé possède une certaine loi de probabilité, et le reste des observations possède une loi de probabilité liée (où le résumé est supposé fixé).
L'information totale est toujours égale à la somme de l'information fournie par le résumé et de l'information contenue dans la loi de probabilité liée.
Si cette dernière loi ne contient plus le paramètre θ, nous dirons que le résumé est exhaustif. Il contient alors toute l'information. Il est clair que ces notions sont absolument générales et que, soit dans l'espace des observations, soit dans l'espace des paramètres, le nombre des coordonnées peut être absolument quelconque.

Exemple. La loi considérée plus haut donne, pour n observations:

$$A^n \exp - (n\theta^4 - 4\theta^3 S_1 + 6\theta^2 S_2 - 4\theta S_3 + S_4)\, dx_1\, dx_2 \ldots dx_n.$$

Il est clair que si l'on prend comme nouvelles coordonnées:

$$S_1, S_2, S_3, \xi_4, \ldots \xi_n, \text{ on aura}$$

$$A^n \{\exp - (n\theta^4 - 4\theta^3 S_1 + 6\theta^2 S_2 - 4\theta S_3)\}\, e^{-S_4}\, \Delta dS_1\, dS_2\, dS_3\, d\xi_4 \ldots d\xi_n.$$

L'intégrale

$$\underbrace{\int \ldots \int}_{n-3} \Delta\, e^{-S_4}\, d\xi_4 \ldots d\xi_n$$

étendu à la multiplicité où S_1, S_2, S_3 sont fixés, est une fonction de S_1, S_2, S_3, $\phi(S_1, S_2, S_3)$, et la loi de probabilité liée est donc

$$\frac{e^{-S_4}\, \Delta(S_1, S_2, S_3, \xi_4, \ldots \xi_n)\, d\xi_4 \ldots d\xi_n,}{\phi(S_1, S_2, S_3)}$$

ou θ a disparu.

Ainsi, l'ensemble S_1, S_2, S_3 constitue un résumé exhaustif, pour tout ce qui regarde la connaissance du paramètre θ.

Estimation. On a évidemment $E(x_i) = \theta$,

$$E(x_i - \theta)^2 = a_2, \quad E(x_i - \theta)^3 = 0.$$

On pourrait construire diverses estimations de θ. La plus simple est S_1/n, dont la précision est donnée par a_2/n. $a_2 \sim 1/3$. L'efficacité $1/12a_2^2$ vaut en gros 70/100.
Ainsi, pour n très grand, l'estimation de l'optimum donne, à une quantité finie près, toute l'information. Le premier moment perd environ 30/100 des observations.
Pour n quelconque, l'information tout entière est comprise dans le résumé

S_1, S_2, S_3. L'estimation de l'optimum serait fournie par:

$$\theta^3 - 3\theta^2 \frac{S_1}{n} + 3\theta \frac{S_2}{n} - \frac{S_3}{n} = 0.$$

Il est facile de voir que cette équation n'a qu'une racine réelle, $\hat{\theta}(S_1,S_2,S_3)$. Vraisemblablement, elle fournit une très grande partie de l'information. Peut être les dérivées suivantes, prise pour la valeur $\hat{\theta}$

$$\left(\hat{\theta} - \frac{S_1}{n}\right)^2 + \frac{S_2}{n} - \left(\frac{S_1}{n}\right)^2,$$

$$\hat{\theta} - \frac{S_1}{n},$$

qui certainement fournissent avec $\hat{\theta}$ toute l'information, en fournissent celles des parties de plus en plus petites.

Forme des lois à résumés exhaustifs. On peut trouver, sous certaines hypothèses, la forme générale de ces lois, qui comprennent évidemment les lois à estimations exhaustives. J'ai donné cette forme dans une note aux *Comptes Rendus T.200* – 1er semestre p. 1265 et dans une communication à la session d'Athènes de l'Institut International de Statistique, en Septembre 1936.

Cas de la loi de R.A. Fisher, solution du problème du Nil
L'élément de probabilité a la forme

$$\mathrm{d}f = \mathrm{e}^{\alpha(xy)a(\theta)+\beta(xy)b(\theta)+\gamma(xy)+c(\theta)}\mathrm{d}(xy)$$

qui convient aux résumés exhaustifs. Le résumé est, dans ce cas

$$\rho_1 = \Sigma\, \alpha(x_i y_i),$$
$$\rho_2 = \Sigma\, \beta(x_i y_i).$$

Dans le cas spécial $\rho_1 = \Sigma x_i$, $\rho_2 = \Sigma y_i$,
ces variables ρ_1, ρ_2 fournissent donc un résumé exhaustif. Si on les appelle ξ,η, leur loi de probabilité est de la forme:

$$\mathrm{d}\phi = \frac{1}{\{(n-1)!\}^2}\ \xi^{n-1}\eta^{n-1}\mathrm{e}^{-\theta\xi-\eta/\theta}\mathrm{d}\xi\mathrm{d}\eta.$$

Toute l'information est donc contenue dans cette loi de probabilité.
Or, il se trouve que la loi en ξ,η est aussi une solution du problème du Nil, puisque la transformàtion

$$\theta_2\xi_2 = \theta_1\xi_1, \quad \eta_2/\theta_2 = \eta_1/\theta_1$$

laisse invariante la fonction $\xi\eta$ et l'élément d'intégrale. Par conséquent, le résumé exhaustif, mis sous la forme

$$X = \xi\eta, \qquad Y = \eta/\xi$$

est tel que toute l'information est fournie par la loi de probabilité de Y liée. On voit que cette loi très simple possède *à la fois* les propriétés de résumés exhaustifs et de solution du problème du Nil.

Recherche d'autres solutions. Il suffit de trouver une solution de la forme

$$e^{\phi(x,y-t)}dxdy$$

où la fonction ϕ permette un résumé exhaustif. On peut en trouver très aisément de deux types:

1) $\phi(x,y-t)$ est un polynome en $y-t$, dont les coefficients sont des fonctions de x. Par exemple

$$A \exp -(x^4+x^2y^2+y^4)\, d(xy)$$

fournit une solution en remplacant y par $y-t$.

2) $\phi(x,y-t) \equiv \Sigma A_i(x) \dfrac{e^{\alpha_i y}}{e^{\alpha_i t}}$.

On voit que la solution de R.A. Fisher appartient à ce deuxième type.
La solution générale du problème qui consiste à mettre $\phi(x,y-t)$ sous une forme capable de résumé exhaustif ne présente aucune difficulté.
Il est sans doute plus difficile de s'arranger pour que le résumé lui-même soit solution du problème du Nil. Toutefois, l'exemple de la page [73] tiré de la rotation, est une solution. En effet, sous la forme

$$\frac{1}{2\pi} \exp -\tfrac{1}{2}\{(x-r\cos\theta)^2+(y-r\sin\theta)^2\}\, dxdy$$

on aperçoit que cette loi comporte un résumé exhaustif Σx, Σy. En posant $\Sigma x/n = \xi$, $\Sigma y/n = \eta$, la loi de ξ,η est évidemment

$$\frac{1}{2\pi}\exp\left(-\frac{n}{2}\{(\xi-r\cos\theta)^2+(\eta-r\sin\theta)^2\}\right) n\, d\xi\, d\eta.$$

Par conséquent, elle a la même propriété que la loi primitive

$$\xi = R\cos\Omega,$$
$$\eta = R\sin\Omega,$$
$$\frac{n}{2\pi}\exp\left(-\frac{n}{2}\{R^2+r^2-2Rr\cos(\Omega-\theta)\}\right) R\, dR\, d\Omega.$$

La loi de R est donc

$$\frac{n}{2\pi}\exp\left\{-\frac{n}{2}(R^2+r^2)\right\} RdR \int_0^{2\pi} \exp(nRr\cos u)\, du.$$

Elle ne dépend pas de θ; par conséquent, la loi de probabilité liée de Ω fournit toute l'information.

On pourrait peut être, sans trop de peine, trouver d'autres exemples.
Il est à remarquer que l'exemple de R.A. Fisher a une autre propriété, c'est que la loi de probabilité liée qui fournit toute l'information est indépendante de n, nombre des observations.

Fisher to G. Darmois: 26 August 1940

I have been reading your letter of August 20th and its extremely interesting enclosure. I am extremely sorry to hear of your accident with the automobile, which must have been most troublesome, though, from your letter, I am glad to see that you can now write perfectly.

With respect to your enclosure, of which I hope you have a copy, I believe it would be helpful as a matter of exposition to bring the size of the sample, let us say n, into the foreground at an early stage. Thus for the amount of information available on the average from a sample of n we have

$$i = E_n\{\partial(\log f)/\partial\theta\}^2$$

but after observing the ancillary statistic X the actual amount of information which our sample contains may be more or less than this expectation, namely

$$E_{n,X}\{\partial(\log f)/\partial\theta\}^2.$$

At this stage I think it is recognisable that the more general expectation is less appropriate to our particular experience than the more limited expectation, and that it was only by a thoughtless convention, or perhaps *faute de mieux*, through our not having discovered the ancillary function X, that the general expectation ever came to be adopted.

If the distribution of our estimate for given X still depends also on n, then n has been replaced by the pair of values n and X. Sometimes, as in the example I gave, and indeed not infrequently, X can be chosen so as to absorb n altogether, so that the knowledge of precision for which we formerly relied on n is now supplied by X only.

In elementary work this occurs for example when a regression coefficient is calculated by the familiar formula

$$b = S\{y(x - \bar{x})\}/S(x - \bar{x})^2.$$

For any set of samples having given values $x_1, \ldots x_n$ it is easily shown that if y is normally distributed with variance v for each value of x, the sampling variance of b is

$$v/S(x - \bar{x})^2;$$

b is normally distributed about the true regression β as mean, with this variance.

From this it follows that if we extend our aggregate of samples to include all having the same value of $S(x - \bar{x})^2$ as we have observed, the same distribution is true of the estimates b we should obtain from them. Thus the number of pairs of values (x,y) observed in the sample is irrelevant, so soon as we have taken note of the value of $S(x - \bar{x})^2$. This is true whatever may be the sampling distribution of the variate x. Even if this were normal, however, the question—'What is the distribution of b for samples of a given size n?' would be not only more complex, but less useful, to answer than the question of the distribution of b for a given value of $S(x - \bar{x})^2$. This quantity completely replaces or supplants n in determining the precision of our estimate. The pair of values (b,n) is not an exhaustive résumé (i.e. b is not a Sufficient estimate), but the pair $\{b, S(x - \bar{x})^2\}$ constitutes an exhaustive résumé. I suggest that whenever n is required it should appear explicitly in the specification of a *résumé*.

In your pretty example of rotation I imagine that $\sqrt{S^2(x) + S^2(y)}$ replaces n in the same way.

The example you take with the quartic exponent

$$df = (1/A) \exp - (x - \theta)^4 \, dx$$

where $$A \text{ is } \tfrac{1}{4}(-\tfrac{3}{4})!.$$

This is of the general form $\phi(x - \theta)\,dx$ for which complete ancillary information is always supplied by the set of differences between successive observations when these are arranged in order of magnitude, these differences having a distribution jointly or severally independent of θ.

In the case you take, the variance of the mean $\bar{x}$ is

$$(-\tfrac{1}{4})!/n(-\tfrac{3}{4})!$$

whereas the average information from a sample of n is

$$12n(-\tfrac{1}{4})!/(-\tfrac{3}{4})! = 4.1138n$$

giving an efficiency 0.70753. This is the fraction utilized by the mean for large samples. If we take the maximal likelihood estimate $T = \bar{x} + u$ where u is the real root of the cubic equation

$$u^3 + 3m_2u - m_3 = 0$$

in which

$$m_2 = S(x - \bar{x})^2/n, \quad m_3 = S(x - \bar{x})^3/n,$$

then T will be efficient and will utilize a fraction of the information tending to unity for large samples. T is not sufficient, the résumé (n,T) supplying on the average less than the total amount of information available, the limiting amount of the deficiency for large samples being given by the formula

$$\frac{S\left\{\frac{1}{m}\left(m'' - \frac{m'^2}{m}\right)^2\right\}}{S\left(\frac{m'^2}{m}\right)} - \frac{1}{n}S\left(\frac{m'^2}{m}\right) - \frac{S^2\left\{\frac{m'}{m}\left(m'' - \frac{m'^2}{m}\right)\right\}}{S^2\left(\frac{m'^2}{m}\right)}$$

where m is the expectation in any class, $m' = \mathrm{d}m/\mathrm{d}\theta$, $m'' = \mathrm{d}^2m/\mathrm{d}\theta^2$, which comes to 4.6231, nearly 1/8th more than the amount of information in one observation. The loss of information in such a résumé is not, therefore, very formidable in practice; but it is, I think, of great theoretical interest to consider the procedure by which it may be recovered.

For samples having a given configuration u is constant, and the amount of information supplied by T is the same as that supplied by $\bar{x}$. For such samples the distribution of $\bar{x}$ is

$$\mathrm{d}f = (1/\phi)\exp\left(-n\{(\bar{x}-\theta)^4 + 6m_2(\bar{x}-\theta)^2 + 4m_3(\bar{x}-\theta)\}\right)\mathrm{d}\bar{x}$$

where
$$\phi(n,m_2,m_3) = \int_{-\infty}^{\infty} \exp\{-n(t^4 + 6m_2t^2 + 4m_3t)\}\,\mathrm{d}t;$$

the amount of information supplied by $\bar{x}$ (together with the configuration) is therefore

$$i = 12n\,E\{(\bar{x}-\theta)^2 + m_2\}$$

of which the average value is seen without difficulty to be

$$12n\,(-\tfrac{1}{4})!/(-\tfrac{3}{4})!.$$

In this case, although you might properly have considered the distribution of T in samples having absolutely the same configuration, it is clear that the sampling distribution depends only on the two elements m_2 and m_3, which together with n and T supply an exhaustive résumé. The particular amount of information supplied by such a résumé may be written explicitly as

$$12n\left\{m_2 + \frac{\int_{-\infty}^{\infty} t^2\exp\{-n(t^4+6m_2t^2+4m_3t)\}\,\mathrm{d}t}{\int_{-\infty}^{\infty} \exp\{-n(t^4+6m_2t^2+4m_3t)\}\,\mathrm{d}t}\right\}$$

If we know anything about the function ϕ, it might be interesting to write this as

$$12nm_2 - \frac{2}{\phi}\frac{\partial\phi}{\partial m_2}, \text{ or } 12nm_2 + \frac{3}{4n\phi}\frac{\partial^2\phi}{\partial m_3^2}\,.$$

Indeed this function is full of interesting points.

Fisher to G. Darmois: 22 March 1955

. . . I am writing to you now primarily to seek your assistance in a literary reference, for I recall about 1946 when I was on a visit to Paris, at your invitation, that you were good enough to show me a paper of Kolmogoroff's in French translation; in particular you wished to show me the curious axiom which Kolmogoroff had excogitated in wrestling with the problem of fiducial inference.

As I am now engaged in setting out more fully, from the point of view of mathematical logic, the bases, as I understand them, of scientific inference, I am wanting to recover the reference of this little essay in axiomatics.

For a time, as perhaps you know, I was doubtful whether Sir Harold Jeffreys and others were not perhaps right in thinking that the form of reasoning, to which I gave the name 'fiducial', required some special axiom, but I am now fully satisfied that this is not so, but that the matter turns on the fact that the word 'probability' was framed by our predecessors in the 17th and 18th centuries not only with abstract and deductive inferences in view, but with the intention of actually applying the idea to the real world, for example, in the advice given to gamblers, and that in consequence the true meaning of the word includes both the specification of what is known, which enters readily into deductive processes, but also a specific requirement as to what is unknown, which is a type of datum we constantly have to use in inductive reasoning, but which is not easily accommodated to the canons of deduction.

All this confirms me in my belief that what is inferred, using the method of fiducial probability, is a classical probability just as conceived by de Moivre or Montmort, and not in any sense a special kind, or species, of probability, as has been diligently insinuated. In fact there are a great many cases in which fiducial inferences could be experimentally verified to any degree of accuracy.

It is in trying to make it clear that I am myself introducing no new axiom that I want to refer to the attempts of Jeffreys and Kolmogoroff to cope with the problem in this way. I hope you will recall the reference without trouble.

Fisher to W.E. Deming: 25 September 1934

Many thanks for sending me your paper with R.T. Birge from *Reviews of Modern Physics.*[1] I think the paper will be found most valuable. It is, I believe, the first attempt to give to physicists, or even to astronomers, a comprehensive account of the ways in which quite modern work has extended and revolutionised the classical theory of errors. You ask me for criticisms, but really I have found very little in substance to criticise. I think the discussion on page 135 is somewhat hard on 'Student's' z test. (By the way, since 1925 'Student' has adopted the transformation I suggested, $t = z\sqrt{n}$, so that he uses the t test as much as I do.) I would not myself admit that 'Student's' test is ever misleading, and it can only be called hazardous in the strict and nonpolemical sense that it lays down and accepts a certain definite hazard. It is the u test which requires guesswork and is, therefore, exposed to objection by those who want their inferences to flow from the data only.

As I expect you know, up to well within the last 15 years writers on statistics were accustomed to be extremely careless in confusing that which is estimated with our estimate of it. The same terms and the same symbols were used for both without distinction. In 1921, in a paper of the *Phil. Trans.* [*CP* 18], aimed at clarifying some of the contradictions and paradoxes of the subject, I introduced two new terms, intended to be antithetical, namely, 'parameter', used to specify the parent population, and 'statistic', calculated from the observed sample. I was quite deliberate in choosing unlike words for these ideas which it was important to distinguish as clearly as possible. That work has now been largely done, so far as concerns the better writers on the subject, and certainly there is no confusion in your paper, where I think you systematically use Greek and Latin letters to distinguish these two classes of quantity; but, perhaps by a slip, you do (Section 3e, line 7) use the expression 'corresponding parameter of a sample', which on consideration you may agree is rather a dangerous one for some classes of student. A population is completely specified by its one or two or more parameters. A sample of n would need n different statistics if these were to be used to specify it. They are, in fact, not used for this purpose at all, but essentially for estimation. To each statistic there corresponds a particular parameter or parametric function to which the value of the statistic tends, as the sample is increased indefinitely, but to each parameter there 'corresponds' in this sense as many different statistics as a cat can have kittens. In fact there is no 1:1 correspondence as suggested by your clause and I am sure it is better not to use the word parameter for one of the fluctuating quantities obtained from samples, which one may call statistical estimates, or something of the kind if that is preferred to the word statistics.

I may say in this connection that I think your exposition in Section 3e of the fiducially related values of σ and s is altogether excellent; the only thing I should add on the logical side is that the statements of fiducial probability obtained should only be taken from distributions, such as s for given σ, where the problem of estimation of the parameters has been completely and therefore uniquely solved, i.e. where s is known to contain the whole of the information contained by the sample. One can see the necessity for this stipulation by considering what would happen if, like the astronomers, we used an estimate of σ based on the mean error, rather than on the mean square error. If s_1 is the estimate, then the distribution clearly will be a function of s_1/σ only and there is nothing but hard work to prevent a misguided astronomer from tabulating the percentile points of the distribution for different sizes of sample. Then, given s_1, it would be possible, apparently, to state the fiducial 5 per cent and 95 per cent points for σ and these would not, of course, agree exactly with the values derived by the mean square method from the same sample. The use of fiducial probability in this precipitate way would, in fact, have led to a definite numerical contradiction, of a kind not unlike those which brought discredit on the use of inverse probability, based on some form of doctrine of insufficient reason.

In the light of the theory of estimation the logical contradiction is easily resolved. An inductive statement (unlike a deductive one) is only true if it is the whole truth; suppression of part of the data and the treatment of the remainder as though it were the whole, although these data are really true and the method of treatment unexceptional if applied to the whole, will, as all statisticians know, lead to very false results. What the theory of estimation is capable of showing is that a definite portion of the information supplied by the sample is omitted or thrown away in using an estimate based on the mean error, but that the whole is retained or conserved in any estimate based on the mean square error. Consequently when faced with such contradictory statements, apparently equally well founded on the fiducial argument, we can with the theory of estimation behind us say that one statement is true and the other is false and why. It is for this reason that I think it worth while to emphasise that the theory of fiducial probability is only an outgrowth or branch of the theory of estimation and that the attempt which Neyman and Pearson have made to make it stand alone without regard to the quantity of information utilised is bound to lead to contradictions and confusion.

Again let me congratulate you most heartily on the completion of a very fine enterprise.

[1] Deming, W.E. and Birge, R.T. (1934). On the statistical theory of errors. *Rev. Mod. Phys.* **6**, 119–61.

Fisher to W.E. Deming: 19 *September* 1935

I am very glad to see from your letter[1] that you only wish to suggest that when σ is known the traditional theory of errors procedure (the u contours of your paper) is appropriate. I got the impression from what you had written that you considered that there was ground for choosing u rather than t, other than the possession of definite knowledge of the value of σ.

If I remember right, 'Student' in putting forward his new test was perfectly clear that he regarded it as a correction of the test traditional up to his time, needed especially for small samples owing to our uncertainty of the true value to be ascribed to the variance of the population.

There is a good deal in the approach chosen by Neyman and Pearson that I disagree with, but so far it seems to have led to nothing more than the conclusion that the tests of significance which I and those who agree with me had previously put forward were the best possible for their purpose; in fact, to use their terminology, the u regions are uniformly the best possible in relation to one class of alternative hypotheses, the population variance being given, while the t contours are uniformly the best possible for another class of alternative hypotheses, the variance being unknown. It is, however, in my opinion, a pity that these writers have introduced the concept of 'errors of the second kind', i.e. of accepting an hypothesis when it is false, seeing that until the true hypothesis is specified, such errors are undefined both in magnitude and in frequency. Their phraseology also encourages the very troublesome fallacy that when a deviation is not significant the hypothesis tested should be accepted as true.

[1] Deming's letter was in response to Fisher's comments on his manuscript that E.B. Wilson had sent to Fisher for an opinion. (See Fisher's letter of 20 May 1935 to Wilson (p. 237).) In his letter, Deming suggested that Neyman and Pearson's papers were the source of his belief that it is *not* a matter of complete indifference as to which samples are rejected — a view that Fisher had criticized.

W.E. Deming to Fisher: 27 *July* 1937

Recently a certain question has come up involving maximum likelihood. It may seem trivial, but the fact is that there is disagreement among people who ought to be able to agree. I appeal to you as the only person who can set us all at ease, hoping that this encroachment on your time will not be irksome.

For the likelihood L of n normal variates having values $x_1, x_2, \ldots, x_n$ we have

$$L = (\sigma\sqrt{2\pi})^{-n} \exp\{-\Sigma(x_i - \mu)^2/2\sigma^2\}. \tag{1}$$

Now $dL/d\mu = 0$ gives $\mu = \bar{x}$ (2)

$dL/d\sigma = 0$ gives $\sigma^2 = s^2 + (\bar{x} - \mu)^2$ (3)

(This s is the S.D. of the sample, as you perceive; not your s.)

Now the question is, what is the maximum likelihood estimate of σ made from the sample of the n observations? Professor Birge and I have taken the estimate of σ^2 to be $s^2 + (\bar{x} - \mu)^2$ as given on the right-hand side of Eq. (3). We say that if you don't know μ then you must go to the distribution of s and apply maximum likelihood to it, the result being of course $s^2n/(n-1)$ for the estimate of σ^2, n being the number of observations. The distinction between the two estimates

$$s^2 + (\bar{x} - \mu)^2 \text{ and } s^2n/(n-1)$$

is that the former contains more information than the latter; knowledge of μ gives us a trifle better estimate than s alone can give.

The point of disagreement comes when some people insist that Eqs. (2) and (3) are to be taken as *simultaneous* in μ and σ. If one does that, he replaces μ by $\bar{x}$ in Eq. (3) and gets simply s^2 for the estimate of σ^2. To Professor Birge and me this seems a highly arbitrary and destructive procedure.

Have you ever published anything on the simultaneous estimation of several parameters, where the estimates by maximum likelihood involve other parameters, as the estimate of σ involves μ? Has anything ever come out in print on this subject? . . .

Fisher to W.E. Deming: 7 August 1937

I have your letter of July 27. It is curious that the point you raise is one of the first that attracted my attention to Statistics. Indeed, I wrote a juvenile paper upon it in 1912 or 1913. It has no particular merit for your purpose.

I notice that with the definition

$$s'^2 = S(x - \bar{x})^2/n$$

you call s' *the* S.D. of the sample. I think this is a very arbitrary nomenclature, since, whereas the standard deviation of the population sampled is unequivocally defined, a sample can provide innumerable different estimates of varying merit, and, without discussing the relative advantages of these, any one of these might equally be called the S.D. of the sample. When the theory of estimation was developed one of the points which I found most surprising is that bias in an estimate, at least bias of the order of $1/n$, where n is the size of the sample, is of no practical importance, whereas its variance and, in the theory of small samples, the form of its distribution curve, is all important. The reason is that, in estimating the value of an unknown parameter θ, you are equally estimating the value of any given function of θ, and if one such estimate is chosen to be unbiassed, the others will generally be found to have positive or negative biases of order $1/n$.

Now there is no denying the fact that, when there is more than one unknown parameter, maximising the likelihood provides simultaneous equations of estimation and these are solved by taking as the estimated variance the statistic s'^2, as defined above. This is not the estimate I use, but it might be used, at some later algebraic inconvenience, to lead to identically the same tests of significance. In fact, in these tests of significance we deliberately make exact allowance for the sampling distribution of s' or s, and avoid the older practices of assuming the true standard deviation to be equal to s' or s; and since s' is a known function of s, to make exact allowance for the sampling distribution of one is to do so equally for the other.

My reason for using an unbiassed estimate, s^2, of the variance, apart from the fact that it simplifies the algebra, of testing the significance of the differences of two variances drawn from samples of different sizes, is that variances are things which one often wants to sum or to average. If this were equally true of standard deviations, or of invariances, it would be equally desirable to use unbiassed estimates of these; and for these, of course, s and $1/s^2$ are not unbiassed estimates. The indifference of bias is brought out most clearly by the property of sufficiency. The fact that the likelihood function involves, apart from a constant factor peculiar to the sample, only the statistics s^2 and $\bar{x}$, shows that these are jointly sufficient for the estimation of the mean and variance of the population. The relation is essentially a joint one. We cannot infer from it in general that $\bar{x}$ is a sufficient estimate for μ and s^2 for the variance. Indeed, as you note, if μ is known, the variance is properly estimated by $s'^2 + (\bar{x} - \mu)^2$. This happens also to be a sufficient estimate. But if, instead of μ being known, there was known only some more general functional relationship between the mean and the variance, the maximum likelihood estimates of these will generally involve both statistics, and not generally be sufficient. The joint sufficiency, however, of $\bar{x}$ and s^2 implies equally the joint sufficiency of any two independent functions of n, $\bar{x}$, and s^2, if such functions can be regarded as estimates at all.

In fact, I think the distinction you are drawing is one without an essential difference, one's choice of an unbiassed estimate being arbitrary in the sense that it is only justifiable by the use to which the estimate is intended to be put.

E.C. Fieller to Fisher: 21 *November* 1946

A copy of the tenth edition of *Statistical Methods*, which we have had on order for some months, arrived here last week, and one of my colleagues promptly pointed out the identity of the argument of the new section 26.2 with the remarks that I published in 1940 and 1944 on the fiducial limits of a ratio. I am enclosing a copy of the latter paper;[1] the former appeared in Vol. VII of the Statistical Society Supplement.[2]

The fact that you do not refer to either of these papers makes me fear that I have inadvertently overlooked an earlier publication of your own; could you let me know, so that I may make due acknowledgment as soon as possible? I could most rapidly do so by adding a footnote to a paper that I have just sent off for publication.

[1] Fieller, E.C. (1944). A fundamental formula in the statistics of biological assay, and some applications. *Q. J. Pharm. Pharmacol.* **17**, 117–23.

[2] Fieller, E.C. (1940). The biological standardization of insulin. *J. R. Statist. Soc., Suppl.* **7**, 1–54.

Fisher to E.C. Fieller: 23 *November* 1946

Thanks for your letter. If you like I could refer to your note in subsequent editions. The general method of substituting values of pivotal quantities such as t and χ^2 at chosen levels of significance in order to make equivalent tests of parameters contained implicitly in these quantities has, I suppose, been creeping into use for a long while. Certainly in the case which you treat I had discussed the quadratic equations to which your method leads with C.I. Bliss, when he was working on toxicological problems at the Galton Laboratory. If anyone has anticipated your note in this application it would be he, but I do not know that he has done so. I seem to remember him using the intercepts of a given ordinate with the Working and Hotelling hyperbola.

E.C. Fieller to Fisher: 29 *January* 1947

I am sorry to have taken so long to answer your letter of November 23rd — as opportunity offered I have been looking up Bliss's publications, and some of them have proved rather inaccessible.

Bliss certainly inverted the Working–Hotelling argument in his 1935 paper, but I doubt whether he then realised the general argument, since he went on using Gaddum's Pearsonian formula in comparative assays until 1944. Paulson gave the fiducial distribution of the ratio of two correlated normal estimates in 1942, but so did I in 1940; what led me to it then was needing to find fiducial limits in an insulin assay after doing a covariance correction; but I tried to write the result in a general form. I would of course take it as a compliment if in subsequent editions of *Statistical Methods* you referred to my note — I agree that the basic idea of equating the significance level of t to the expression (once one has found it) involving the unknown α is a standard one; I think I first learnt it from §62 of *Design of Experiments*. . . .

E.C. Fieller to Fisher: 10 *January* 1954

I believe that Irwin sent you a few weeks ago a draft of a paper by Monica Creasy, in which she seemed to have arrived at an anomalous result concerning the fiducial distribution of a ratio, and mentioned that we were contemplating devoting the March meeting of the R.S.S. Research Section to a discussion of that and related topics.

Of course, we would not want to have such a meeting if you felt that we were likely to confuse rather than clarify the issue, so I am enclosing a revised version of Miss Creasy's paper,[1] and about the first third of the draft that I am preparing to precede her effort. . . . Miss Creasy and I would both very much appreciate your views on our drafts. She is clearly answering a different question from the usual one, but I am not sure that she has yet succeeded in expressing clearly what the difference is; and I would like to be told, if you think that I am committing any howlers.

Miss Creasy is genuinely worried, in particular, to find an appropriate name for her limits; she feels that the title in her draft, 'fiducial distribution limits', is not the right one, but does not like the only alternative, 'apparent limits', that I have been able to suggest.

If you are able to spare the time, what we would both like to do would be to come up to Cambridge one day within the next week or two, and ask you to sort out our minds for us! May we do that?

P.S. I am checking with Chester Bliss that I am not taking his name in vain in my introduction.

[1] Eventually published as Creasy, M. (1954). Limits for the ratio of means. *J. R. Statist. Soc.* **B** **16**, 186–94.

Fisher to E.C. Fieller: 11 *January* 1954

Thanks for sending me the two long screeds by Miss Creasy and yourself which I am afraid I have not time to examine with all the care they doubtless deserve.

I hope Miss Creasy will be persuaded no longer to associate my name with the method which in fact I have never used, for I was always satisfied with Bliss's approach, and was heartily glad to see the last of the awkward forms discussed by Geary, with their infinities appearing within the range of integration.

I wonder who put Miss Creasy on to such an unrewarding job.

Of course I should be very happy if you both came up to have lunch with me but I scarcely think I shall do anything to clarify your minds.

E.C. Fieller to Fisher: 13 *January* 1954

Thank you for yours of the 11th, but I am left wondering what to suggest to the Research Section Committee next week. Perhaps we can have a word about that before or after to-morrow's Council meeting; I am inclined to complete my screed anyway, and propose to the Committee that they should consider it and something of Miss Creasy's for our May meeting, possibly cancelling the one planned for March. I shall still hope to persuade you to read my draft, when I have finished it; for one thing, I would not want to use the word 'fiducial' in any new context without first referring to you.

I understand that Miss Creasy embarked on her investigation of her own free will, regarding it as an exercise in the manipulation of fiducial probabilities that would lead her to the usual result. When she arrived at a different one, she should of course have been told to find the flaw in her argument. Instead, she was unfortunately encouraged by her then colleagues at Oxford to believe that her new limits were somehow more fiducial than the real ones. I think that she is now disabused of that idea, but on the face of it her mathematics leaves the reader wondering why an approach that is valid when we are considering the difference of two means should not also be applicable to a discussion of their quotient. The point seems to me to be an interesting and not unimportant one, and George Barnard told me last night that he thinks that he has an explanation.[1] . . .

[1] See Barnard's letter of 14 June 1954 (p. 10).

Fisher to E.C. Fieller: 14 *January* 1954

Thanks for your letter. Of course I did not wish to interfere with your arrangements. I was only giving my personal reaction.

I imagine that the discrepancy may well be due to the omission, in one treatment or the other, of some latent assumption; e.g. it would seem to me that the T_1, T_2 treatment would be appropriate when x and y are known to be independently distributed, but that your treatment is more general and makes no such assumption. In inductive reasoning generality has not the same meaning as when reasoning is deductive for in the latter we are under no obligation to take into account data which we do not use. Consequently, the more general treatment in a sense includes and supersedes the less general.

This, however, is not at all true in inductive reasoning where we ought to expect a different result whenever we ignore any datum of the problem.

D.J. Finney to Fisher: 28 *May* 1946

I have recently been reading and discussing your reply to E.B. Wilson on the four-fold table (*Science*, *94*, pp. 210–11 [*CP* 183]). While I agree entirely with your solution of the problem which you state, I feel that there is still a difficulty unexplained; I should be very interested to hear your views on this, as the point is one likely to occur quite frequently.

You state the problem: 'Of six treated mice five have died and one lived, while of six controlled mice one has died and five lived.' Clearly you are then only interested in the possibility that treatment improves the survival rate and therefore base your test on one tail of the distribution. The probability is determined as in your letter [to *Science*]. But Wilson's original statement of the problem referred not to treated and control mice but to mice receiving virus A or virus B; he is therefore presumably interested in the possibility either that B is more harmful than A or that A is more harmful than B—i.e. in both tails of distribution. Since he has six mice in each group, the two tails are symmetrical and he can obtain his probability by doubling the result for one tail. What happens if he has eight mice in one group, six in the other? Say:

	Died	Lived	Total
A	5	3	8
B	1	5	6
Total	6	8	14

How is he to test the null hypothesis that A and B are equally harmful, while considering deviations from equality in either direction? Simply to double the total probability for

$$\begin{Bmatrix} 5 & 3 \\ 1 & 5 \end{Bmatrix} \text{and} \begin{Bmatrix} 6 & 2 \\ 0 & 6 \end{Bmatrix}$$

scarcely seems appropriate, as it does not correspond to any discrete subdivision of cases at the other tail such as

$$\begin{Bmatrix} 1 & 7 \\ 5 & 1 \end{Bmatrix} \text{and} \begin{Bmatrix} 0 & 8 \\ 6 & 0 \end{Bmatrix}.$$

Nor does there appear to me any obvious reason for calculating the probabilities for the two most extreme configurations at the other tail (keeping marginal totals unaltered) and adding their total to the appropriate probability for the tail at which the observations occur.

Am I missing something very simple here? I cannot remember having seen this problem discussed, and should be grateful for your views. I hope that I am in time to catch you before you go to the States.

Fisher to D.J. Finney: 31 *May* 1946

Thanks for your letter. It is a good problem, but I believe I can defend the simple solution of doubling the total probability, not because it corresponds to any discrete subdivision of cases of the other tail, but because it corresponds with halving the probability, supposedly chosen in advance, with which the one observed is to be compared. That is to say, one may decide in advance that if the probability is less than one in forty in either direction then we shall consider if, pending further investigation, the viruses are not pathologically equivalent.

How does this strike you?[1]

[1] This letter and the last part of Finney's letter of 28 May 1946 were published in Yates, F. (1984). Tests of significance for 2 × 2 contingency tables. *J. R. Statist. Soc.* **A 147**, 426–63.

D.J. Finney to Fisher: 30 *November* 1949

Suppose that samples from two normal distributions of equal mean give sample means $\bar{x}_1, \bar{x}_2$ and variances of those means ${s_1}^2, {s_2}^2$. Then

$$\bar{x}_1 - \mu = s_1 t_1,$$
$$\bar{x}_2 - \mu = s_2 t_2.$$

If we wish to estimate μ from the combined evidence of the two samples, and to assign fiducial limits to our estimate, I think we can legitimately take

$$\bar{x} = \tfrac{1}{2}(\bar{x}_1 + \bar{x}_2)$$

and use the Fisher–Behrens distribution to give the limits.

We might hope, however, that a weighted mean, $\bar{x}_0$, defined by

$$\bar{x}_0\{(1/s_1^2) + (1/s_2^2)\} = (\bar{x}_1/s_1^2) + (\bar{x}_2/s_2^2)$$

would be more precise. Now

$$\bar{x}_0 - \mu = \{(t_1/s_1) + (t_2/s_2)\}\{(1/s_1^2) + (1/s_2^2)\}^{-1}.$$

Do you see any objection to the use of the Fisher–Behrens argument on this equation? We can integrate

$$(t_1/s_1) + (t_2/s_2)$$

over the region for which it exceeds $d\{(1/s_1^2) + (1/s_2^2)\}^{\frac{1}{2}}$ and it seems to me that the logic should be just as good as in the case you have discussed. I am primarily anxious to be reassured that I am not falling into heresy, before I proceed further. However, I have looked at the next step, in the light of your 1941 paper (*Ann. Eugen. 11*, p. 141 [*CP* 181]). If we modify your transformation at the top of p. 152, writing

$$t_1 = x \cos \theta + y \sin \theta,$$
$$t_2 = -x \sin \theta + y \cos \theta,$$

and $$s_1/s_2 = \cot \theta,$$

I think that the same integral is obtained. If this is so, the Fisher–Sukhatme tables complete the solution to the problem, but with a modified definition of θ. If I have slipped on this last step, presumably the method you used could be adapted to the formation of new tables.

This idea seems very elementary. Is it new? Its usefulness is clear. For example, it gives the complete solution to the estimation problem in an incomplete block experiment, when a final set of means is formed by combination of inter- and intra-block estimates. A weighted mean of k constituents could be tackled in the same way, but $(k - 1)$ ancillary θ-statistics would be needed, and the preparation of tables would be an unpleasant task.

Fisher to D.J. Finney: 2 December 1949

Thanks for your letter. I have not checked the algebra, as I have full confidence that you have done it right. But as the problem is one I have run into before, perhaps I might make a few comments.

I think you assume in your first sentences that the samples are of equal size, but will not wish to limit the discussion to this special case.

In general the simultaneous likelihood of the two samples for varying values of μ when maximised for variations of σ_1 and σ_2 does, with confounded perversity, necessarily involve the difference between the means as a minute source of information for both of these unknowns. So far as I can judge, any competently practical solution will ignore this source of information and use s_1 and s_2 only, just as you have done. In fact I have used, and I believe Frank Yates concurs in using, the Behrens–Sukhatme tables for just this purpose. It might be worth your while to see, however, whether Frank has formed any firm opinion on the matter. . . .

D.J. Finney to Fisher: 13 September 1954

A few days ago we were discussing Fisher–Behrens, and a remark of yours has sent me to re-read your presentation of fiducial inference in *Annals of Eugenics*, *6*, (1935) pp. 391–398 [*CP* 125]. May I trouble you with one or two questions that worry me, not in what you say but in what is left unsaid?

Of course I appreciate that the fiducial argument is strictly based on sufficient statistics, and realize that ambiguities are introduced by trying to base the same type of probability statements on statistics that do not utilize all the information contained in the data. Nevertheless, are there not situations in which a sufficient statistic cannot be obtained from the whole data, and only by discarding some information is it possible or practicable to make any probability statement? Must we then deny the validity of any probability statement, or can we use some kind of approximate fiducial distribution?

Three simple examples occur to me:—

(1) If we have a random sample of n from a normal distribution of unknown mean and variance, we can in the usual way make fiducial statements about these two parameters. Suppose now we have one extra random individual for whom all we know is that his value of x exceeds a particular numerical value (e.g. x is the total leaf area of a plant, and some leaves of one plant were accidentally lost). We can apply a maximum likelihood process to estimate the parameters from the $(n + 1)$ observations, but I scarcely think that we shall have sufficient statistics. If we know the awkward observation to be truly a random one, are we really being illogical as well as slightly wasteful (only very slightly if n is large) by choosing to neglect it and base our fiducial statements upon the n observations as though the $(n + 1)^{\text{th}}$ never existed?

(2) In one of the more complex lattice designs, various types of inter-block mean squares have expectations linearly compounded of inter- and intra-block variances. In addition, there will be an intra-block mean square estimating the intra-block variance alone. Yates and others have shown how to combine the mean squares so as to estimate the two variances. However, if we merely wish to make statements about the intra-block variance and there are rather few degrees of freedom for other mean squares, we might well be content to use the intra-block mean square alone and base fiducial statements on it with the aid of the χ^2 distribution. Are these invalidated by the existence of additional information in other mean squares?

(3) On pp. 296–7 of the *Journal of the Royal Statistical Society*, *B 12* (1950) I suggested the use of the Behrens–Fisher distribution for obtaining a fiducial distribution of a weighted mean of two items, where the items came from distributions with different standard deviations. You had pointed out that the difference between my two items in fact contained a small amount of extra information on the two variances, though I think you will agree that it is not very easily utilized. If x_1, x_2 are means of n_1, n_2 observations, both from distributions with mean μ, and we define

$$\bar{x} = wx_1 + (1 - w)x_2,$$

then

$$(x_1 - \mu)/(s_1/\sqrt{n_1}) \text{ and } (x_2 - \mu)/(s_2/\sqrt{n_2})$$

are both t variates, and my proposal (probably not the first time this has been suggested) was to use Behrens–Fisher on $(\bar{x} - \mu)$ with variance $(w^2s_1^2/n_1) + \{(1 - w)^2s_2^2/n_2\}$. Since $(x_1 - x_2)$ contains information, must we consider this as an entirely heretical use of fiducial argument? I have for a long time been wanting to ask your further advice on this.

Please don't imagine that I am trying to catch you out on this! — except in so far as I am caught myself. For if we refuse to allow any validity to probability arguments along these lines we seem to be denying ourselves access to conclusions that must in some sense be 'very nearly correct', since the information neglected is small. In my example (1), it seems hard that the $(n + 1)$th observation should make us less able to draw a useful conclusion from the data than we were before it was obtained (or than we should be if the experimenter quite legitimately chose not to report it). On the other hand, if we allow an approximate fiducial argument in circumstances where we cannot find a sufficient statistic, or do not want to be bothered with a little extra information, can we deny all validity to the argument based on s' on pp. 392–393 of your paper? Or, perhaps, of more practical importance, if R is the range in a sample of n, the distribution of a 'pseudo-t'

$$(\bar{x} - \mu)/R$$

is known. If computing time is much more precious than time spent in making observations, why should not inferences on the probable position of μ be based on this distribution?

I suspect that I am being very stupid at some point, and I hope that you will put me right. Nothing is further from my mind than to quarrel with the fiducial argument, but I do find a real difficulty here. There's no hurry, as this has been on my mind for a long time.

Fisher to D.J. Finney: 14 *September* 1954

I believe you started writing to me about my old paper on the fiducial argument and then went off to write the rest of the letter about something else. I do not suppose you expect me to deny what I have so often asserted, namely that there are situations in which a sufficient statistic cannot be obtained from the whole data, or the more general proposition that I must have reiterated repeatedly, that the concept of mathematical probability is inadequate to express the nature and extent of our uncertainty in the face of certain types of observational material, while in all cases the concept of mathematical likelihood will supply very helpful guidance, if we are prepared to give up our irrational urge to express ourselves only in terms of mathematical probability.

However, most of your letter is not really on this point but on another. Is there any logical difference relevant to the interpretation of a body of data between such statements as:

(a) I did not get any measurements of that spider because it escaped.

(b) I do not know where I wrote those measurements down, so they are not included in my tabulation.

(c) I could not clearly read what the figures were.

(d) I discarded the observations as untrustworthy.

(e) I could not be bothered to do the calculations all over again.

(f) I had tables based on the range so could ignore intermediate values.

(g) I have found a very good way of calculating my regression line which is sixty times faster than Berkson's,[1] not always quite so accurate (though quite adequately so in most cases!), namely by guessing the number which seems plausible.

I suppose it will be agreed that in all cases a more accurate result could be obtained by giving more care and attention to the problem, and that no line can be drawn between different levels of precision save for specific technological purposes where the amount of tolerance allowable is understood. The point of principle which influences me is not one of greater or less precision, but of misrepresentation to the consumer of the alleged statistical result. He is, as it seems to me, deceived if he is given a value supposedly based on the body of material available but really utilising only a fraction of the information in that body. In fact, the important fact is concealed from him that a more scrupulous examination of the data might have given materially different results.

I conceive that it is possible only by giving students the opportunity of making rather fine distinctions in the logic of the subject, that they can learn to recognize the difference between honest and dishonest work in statistical practice.

In section (3) you seem to be asking my advice but, of course, I was very glad when you showed me what you were doing with the weighted mean from samples of populations with different variances. I pointed out that some information had been neglected, but I do not think I ever encouraged anyone to try to recover what was lost.

Though it may seem hard, it is undoubtedly true that additional information may not only require an adjustment of our conclusions, but also prevent us from making new conclusions in the same form. Of course, in my opinion, we can always ignore any information we like, provided we are scrupulous enough to say what we have ignored. I think a good statistician, in such cases, will wish to present the data in more than one way, these different ways having, if possible, different imperfections.

Some day I hope you will write to me about fiducial inference!

[1] cf. *CP* 256, p. 138.

D.J. Finney to Fisher: 22 September 1954

I am sorry if I wrote about the 'infiducial' argument instead of the fiducial. But am I really as far from the topics of your 1935 paper as you suggest? Of course, the first paragraph of your reply is familiar stuff, but I appreciate this very concise statement. My difficulty is that I cannot see the logical difference between the seven statements you tabulate, at least in respect of any clear dichotomy, yet apparently some lead to fiducial arguments and others do not.

May I vary the statements slightly?

(a) I measure a sample of 20 spiders. From the measurements, I can assign a fiducial distribution to the population mean, by use of the standard fiducial argument in its simplest form.

(b) I intend to measure 20 spiders, but 3 escape. Apart from the risk that escapers are non-random, I can proceed to form a fiducial distribution for the mean exactly as in (a), but with 16 d.f. instead of 19 d.f.

(c) I measure 20 spiders, but accidentally lose a random 3 of my records. I can proceed as in (b), as I am using all the information available to me in the form of sufficient statistics, so that the fiducial argument still applies.

(d) I measure 20 spiders, but wilfully and frivolously discard a random set of 3 measurements. I can make exactly the same calculations as in (c), but is my inference about the population mean still to be classed as fiducial? I have been foolish, but scarcely criminal in thus discarding information.[1] I can state quite honestly what was done, and, after my initial folly, I have used sufficient statistics for all the information that remains.

(e) I measure 20 spiders and use the known distribution of the ratio of '(sample mean – population mean)' to 'sample range' in order to make statements about the population mean. The ratio can be manipulated algebraically as is the *t* value in the fiducial argument, but the argument is no longer fiducial – at least so I have always understood and so I read your 1935 paper. Yet, if (d) is a fiducial argument, why do we make so severe a logical distinction between discarding 3/20 of the observations and discarding a fraction of the information that cannot be associated with a particular subset of observations?

(f) I read a paper in which X reports that he measured 20 spiders, but he records only the mean and the range of the sample. What type of inference can I make about the population mean?

Now it seems to me that (a), (b), (c) correctly employ the fiducial argument, whereas (e) and (f) do not, the confusion I have built up being a consequence of identifying the technical and colloquial usages of the word *information*, with (d) as a borderline quibble that could be classified only after a pedantic insistence on exact definition. This appears to be in line with your 1935 paper, and in particular with your comments on the use of the 'mean error', but surely there *is* then some logical difference between the statements tabulated in your recent letter: (a) and (g) there are at least as distinct as (a) and (f) above. Moreover, the use of the likelihood concept seems to support this.

I want to avoid what you describe as an irrational urge to express myself only in terms of mathematical probability, but I think it is not as irrational to wish to be able to recognize the situations in which such expression is possible and proper.

I am most grateful to you for your last helpful letter, and especially for the valuable statements in the first and last (main) paragraphs. I hope that you may be moved to comment further on this. I know your own reluctance, but I still maintain that you would do a great service to statisticians by producing a new book on inductive inference.

[1] Fisher has written in the margin, 'If stated, no one is misled.'

Fisher to D.J. Finney: 27 *September* 1954

I judge from the letter that you agreed with me that though in all cases information has been lost, yet that the statistician as a professional with his own ethics has to distinguish between cases in which it is lost by his own professional fault, and cases in which he or others have lost it, but not by using inferior statistical methods. The distinction is largely one aimed at discharging the professional duty to giving such services as the client has a right to expect, and, therefore, the fault is largely obviated in case (d) by saying what one has done, or indeed in case (e) if the report makes clear that more accurate and probably stronger statistical statements would have been made by a more sedulous analysis.

In my opinion it is quite arbitrary but very convenient to confine such phrases as 'the fiducial probability' and 'the fiducial distribution', in which the definite article is used, to cases in which these probability statements are unique and as good as could possibly be made from the same data. I would not think it unreasonable to point out that in view of certain difficulties of the data rendering a unique fiducial inference impracticable, yet a fiducial distribution could be inferred from certain portions of the data, or ignoring certain aspects of it, which might be better than nothing. I recur to the point, however, that the mere wasting of information has indeed an economic importance like the wasting of soap, but that what we need to make clear to our colleagues and students is the difference between honest and conscientious work on the one hand, and dishonest and semi-fraudulent work on the other.[1] . . .

[1] For the remainder of this letter, see p. 255.

D.J. Finney to Fisher: 12 *March* 1955

. . . I have a little paradox for you, arising from Section 74 of *The Design of*

Experiments. There you show that if x has a sampling variance s^2 based on n degrees of freedom the precision of x is

$$(n + 1)/\{(n + 3)s^2\},$$

very properly less than it would have been if the variance of x were *known* to be s^2. Suppose now different investigators each obtain an x and an s^2 from the same population. What is the average of their assessments of precision? Since

$$\mathrm{E}(1/s^2) = n/\{(n - 2)\sigma^2\},$$

where σ^2 is the true variance, this seems to be

$$n(n + 1)/\{(n + 3)(n - 2)\sigma^2\} > 1/\sigma^2.$$

Thus on an average the precision is greater than if σ^2 were known!

I suppose the catch is the usual kind of thing in a fiducial argument, namely that the fiducial distribution of μ given s does not relate to sampling from a distribution with fixed σ^2. But what is the right answer to the problem for sampling with fixed σ^2? For example, x is to be the main effect of factor A in a factorial experiment on 4 blocks of 8 plots. If I include only 3 factors, I have s^2 based on 21 d.f. By confounding in a fourth factor, I reduce my d.f. for error to 14, while presumably keeping the same true variance σ^2. In deciding which design to adopt, I want to balance the advantages of studying more factors against the loss of precision to be expected on each of the original main effects and interactions. Ought I to take the ratio of the values of $(n + 1)/(n + 3)$ to represent the relative precision in respect of A? This gives a reasonable figure, but seems to ignore the fact that the two estimates of s^2 have different distributions.

Fisher to D.J. Finney: 15 *March* 1955

. . . I may well have to think further about your paradox, but you might like to know my first reaction.

I have lately been thinking that although I have, for my own part, always made clear that I introduced the word 'fiducial' to qualify a form of argument by which a true or classical probability statement can be arrived at, yet that its use as an adjective to qualify the word 'probability' has been used, not without malice, to imply that there were two or more kinds of probability used in mathematics, and that the fiducial argument led to a special or peculiar kind called fiducial probability.

I have recently been thinking a little about the semantics of this word. What seems to be implied whenever it is used is a distinction of three levels of knowledge, (a) in which nothing whatever is known about some supposed value, (c) in which the exact value is known, and (b) in which a statement can be made in terms of the concept of mathematical probability, in which the case of a stochastic, or random variable, about which a complete set of probability statements can be made, is typical.

Any definition of what is the logical content of a probability statement, therefore, must make plain two things, (1) wherein our factual knowledge differs from complete ignorance, and (2) wherein it differs from perfect knowledge. In respect of the first, it is usual to say that the probability of a member of aggregate A being also a member of B refers to a situation in which the aggregates are measurable, and that fraction of set A which belongs also to B occupies a fraction p of the whole set A, this fraction being termed the mathematical probability. Of course the 'measure' in such a definition is only a mathematical abstraction, or generalized term intended to cover the relationship made familiar by such concepts as frequency in statistics, area in plain geometry, or mass in a mechanical situation.

In mathematical books, however, I do not find any complementary statement which clearly distinguishes the concept of probability from that of complete exact knowledge, and I suggest that the players of the eighteenth century, and the manufacturers of appliances for gambling at the present time, are, or were, perfectly familiar with the second requirement which has been omitted from the mathematical statement; that the gambler throwing a die needs to know not only that there is a set of possible throws, and that one sixth of this set are aces, the aggregate of possible throws being in this respect heterogeneous and with a well defined measure ratio, but that, *and this I take to be the necessary and sufficient condition*, none of the sub-sets of the entire set in which the proportion of aces is different from one sixth can be recognized before the die is cast. Of course there always will be such sub-sets, and a claimant to the faculty of pre-cognition claims to be able to recognize a sub-set, namely that in which he foresees that an ace will be thrown, and that the sub-set so recognized does have a proportion of aces differing from one sixth. To such a pre-cognizer, therefore, the probability is not one sixth, but is greater or less than one sixth according to which set he recognizes the throw to belong to. The same would be true of defects in the gambler's apparatus, or corrupt practices in their use, so that the basis on which the mathematician can interest himself in the theory of probability is that of perfect apparatus fairly used, in which, in fact, the second aspect of the definition of probability, namely that which defines the subjective ignorance needed for a probability statement, is justifiable.

This second side of the definition of mathematical probability is not necessary in the formal pure mathematics in which the concept may be used, in which knowledge and ignorance are not at all distinguished, but is likely to be necessary whenever the concept is applied to the real world in which the distinction between what we know and what we do not know is relevant.

It makes plain, I think, rather simply what has puzzled people about the fiducial argument in which the mathematical steps are as follows:

(i) It is proved that the probability of the inequality $T < \theta$ is equal to some known quantity P for all values of the statistic T obtained by random sampling from the population characterized by the parameter θ. The inequality has then a mathematical probability of being realized in the reference set of all possible samples from a single population.

(ii) It is proved that this relationship holds whatever may be the value of θ and, therefore, that the reference set can be enlarged to include all samples from all populations of this type.

This probability statement will subsume the whole of the information available from the observations on two conditions, namely that there is no given *a priori* distribution of θ, and secondly, that the statistic T is exhaustive or sufficient. Without these two conditions, we cannot be in a position to make a probability statement about a parameter, for we should have not used all that is known about it. If a distribution *a priori* had been in the data, we could have applied Bayes' argument, and utilized the whole of the information, to obtain the frequency with which the probability statement is realized in the sub-set having T constant. In the absence of such information *a priori*, and of more informative statistics, there is no means of recognizing any sub-set, whether characterized by T, or by any other means, having a probability different from that of the entire set specified. The frequency ratio in the entire set, therefore, is the probability of the inequality being realized in any particular case, in exactly the same sense as the frequency in the entire set of future throws with a die gives the probability applicable to any particular throw in view; i.e. it is a statement of classical probability arrived at by an argument with which the eighteenth century writers were unfamiliar, and in which the probability statement would have to be modified if we were to change our pre-suppositions and add new knowledge, such as that provided by a Bayesian distribution *a priori*.

I do not believe that this need to specify specific ignorance as well as specific knowledge affects your problem with s^2. My formula involving the ratio $(n + 1)/(n + 3)$ was, of course, only calculated as that appropriate to the amount of information about the parameter supplied by a single value called x with variance estimated from n d.f. If your different investigators arc certain that their x and s^2 come from the same population, they would, I suppose, pool their estimates of s. But it might be that they thought their true variances might be conceivably in any ratio however high, as in the case of Behrens' samples, and the discussion would then come down to the precision of the estimate based on the combined evidence of the two samples, in which certainly the difference between the two means must be involved as it throws light on the precision of both sets of measurements.

Probably, however, you are not discussing the combination of observations, though you seem to be doing so in considering $1/s^2$ as though it were going to be used as a weight in some such combination.

I hope I have not been too long-winded about probability. I am concerned at the amount of confused rubbish, which seems to have got itself written, about what I should only regard as a rigorous and useful line of argument.

One trouble was that in 1930 [*CP* 84] I was not at all clear that information *a priori* must be positively excluded, although I was primarily writing about such cases.

D.J. Finney to Fisher: 22 *March* 1955

. . . Your discussion of 'fiducial' is most interesting. Let me first admit that, entirely through ignorance and without malice, I for a long time thought that you distinguished fiducial probability from classical probability, and have no doubt myself propagated this heresy. How I first gained the idea, I do not know, whether from false instruction or misreading, but evidently I failed to find explicit statements to the contrary to correct my folly. I mention this in no spirit of controversy, for I think I now see this matter correctly, but only as evidence that malice is not a pre-requisite of misunderstanding! Call it ignorance, folly, or stupidity if you will, but the effect is the same. You say yourself that your own ideas have become clearer since 1930, and I very much wish that you would now produce in book form an integrated account of the theory of estimation. This would be of inestimable value to the plodders like myself, and ought to stimulate much work in the right direction. You may reply that if anyone wants this, he can do it for himself, but there is a great difference between what you as originator of so much of the theory could do by reinterpreting 30 years of work, and what someone else could do by merely re-writing your papers into one long book. I know I have tried you on this before, but I still hope to persuade you.

I like your (a), (b), (c) distinction on pp. 1–2. Should not the 'logical content of a probability statement' also be qualified by specification of other information known or assumed in the course of the analysis? For example, in estimating the median height of Abominable Snow Men from the small sample of 10 collected by the 1959 Himalayan Expedition we might assume (and for a sample of less abominable Englishmen might think we knew) that

(1) The distribution of individuals is normal with S.D. 0.3 feet;
or (2) The distribution of individuals is normal, parameters unknown;
or (3) The distribution looks normal but has a finite range whose limits are at any rate not wider than 0.5 feet, 20 feet;
or (4) The distribution is unimodal, continuous, and always positive;
or (5) The distribution is continuous;
or (6) Nothing.

However, I am not brought much nearer to explaining away my paradox. The idea of combining information from two sources was never in my mind, and my two investigators represented two mutually exclusive possibilities. Perhaps it may help if I describe the precise context in which the trouble arose.

I wanted to make some quantitative assessment of what was lost by introducing a new factor into an experiment, with a view to setting this against the gains. Suppose I have four blocks of 8 plots and I plan a 2^3 experiment on factors A, B, C. Then each of 7 main effects and interactions will be estimated, each with variance $\sigma^2/8$, and σ^2 itself will be estimated by s^2 with 21 d.f. I now want to judge the pros and cons of confounding in a fourth factor, D, on the same plots and blocks. If I do so, I shall presumably still have the same true σ^2, and shall estimate 14 (15 less one confounded) effects with variance $\sigma^2/8$, but s^2, the estimate of σ^2, will now have only 14 d.f. How do I compare the quality of the estimation of, say, the main effect of A by the alternative designs?

It seems to me that I can ask four questions about the information that the experiment provides on the main effect of A:

(1) How much information will the experiment give, if I know the value of σ^2 exactly?

Answer: $8/\sigma^2$.

(2) Having performed the experiment and obtained an estimate, s^2, of σ^2, with n d.f. (or indeed, having such an s^2 from any other source), what is the estimate of the total information that would become available if σ^2 were known?

Answer: $8(n-2)/(ns^2)$, an unbiased estimator of $8/\sigma^2$.

(3) Having performed the experiment as in (2), what is the amount of information that it has given directly, without reference to any other knowledge on σ^2?

Answer: $8(n+1)/\{(n+3)s^2\}$.

(4) What is my expectation under (3) at the start of the experiment? I do not mind this involving an unknown σ^2, as I want it only for comparison of alternative designs *with the same unknown* σ^2.

Answer: Apparently the expectation of (3), evaluated over the distribution of s^2, and therefore $8n(n+1)/\{(n-2)(n+3)\sigma^2\}$.

But (4) is greater than (1), which seems to conflict with common sense!

Moreover, if two designs allow n_1, n_2 d.f. for estimates of variance, the efficiency of the second relative to the first in respect of estimating the main effect of one factor would be, from (3)

$$\frac{(n_1+3)(n_2+1)}{(n_1+1)(n_2+3)}\,\frac{s_1^2}{s_2^2}.$$

If the second design in fact involves taking $(n_1 - n_2)$ contrasts from the first error and assigning them to effects and interaction of new factors, so that

$$n_1 s_1^2 = n_2 s_2^2 + (n_1 - n_2) s_3^2,$$

where s_2^2, s_3^2 are independent mean squares estimating σ^2, the expectation of the efficiency can be integrated out to

$$\frac{n_2(n_2+1)(n_1-2)(n_1+3)}{n_1(n_1+1)(n_2-2)(n_2+3)}$$

which agrees with (4) above. My numerical example had $n_1 = 21$, $n_2 = 14$, whence the efficiency works out at 190/187: thus the 2^4 confounded design appears to be better for estimating the main effect of one of the original factors, quite apart from its merits in respect of extra factors. Of course this can also be seen because my result in (4) *decreases* monotonically to its limit $8/\sigma^2$ as n increases. This again seems to be at variance with common sense.

Now, where do I run off the rails?

[P.S.] I see no reason why you should spend time in putting me right, but I thought that the paradox might amuse you.

Fisher to D.J. Finney: 24 *March* 1955

Thank you for your letter. I do not think anyone would impute malice to you, but, of course, you are not the only one who has referred to these terms. In the process of claiming for Neyman that which he borrowed from me, Pearson, page 75 of *Biometrika Tables*, says:

'The fiducial theory of R.A. Fisher and the confidence interval theory of J. Neyman were developed to meet this situation. Both aim at providing means of calculating from the data intervals within which the unknown parameter θ may be expected to lie within a given measure of probability. While from the practical standpoint there is in most (though not in all) problems no difference between the result of applying the two methods, Fisher's approach introduces the concept of fiducial probability, while Neyman's employs only the classical concept of direct probability. The application of theory described here is in terms of Neyman's approach.'

The description in Pearson's words is as false as it can be, since in both the argument starts with the same probability statement, and in the end Neyman denies the possibility of making any probability statement about the natural facts behind the observations, while the argument I introduced allows for probability statements being made *a posteriori* in the absence of knowledge *a priori*, such as Bayes' argument requires. Where I was mistaken in 1930 was in not making it clear that the absence of knowledge *a priori* is a pre-requisite for the fiducial argument, so that it is not possible, as I wrongly thought at that time, to have both a Bayesian and a fiducial probability statement in apparent conflict.

In the same preface, page 18:

'It will be seen that the two limits are random variables whose values depend only on the observations; their position and distance apart will vary from one sampling to another. The probability statement must be interpreted in the sense that if we determine the limits for σ from the observations by the formulae given above, we are following a procedure which will give in the long run an interval estimate including the true σ in a proportion of about $1 - 2\alpha$ cases. Further, this statement is true even if the successive samples are drawn from populations having different standard deviations. The limits are termed *confidence limits* and $1 - 2\alpha$, often expressed as a percentage, is termed the *confidence coefficient*.'

Pearson takes here a typical example of a fiducial probability statement and takes the trouble to say that its limits, and the probability itself, have the names 'confidence limits' and 'confidence coefficient', so ascribing this novelty of statistical reasoning to himself and Neyman. In this case, however, he clearly differs from Neyman, who would deny that any such probability statement can be made about features of the real world.

I am sorry you did not read my letter further than to give, as you do on page 2, other features which, in your opinion, are inherent in a statement in terms of probability. Here you evidently read a great deal more into the word than I do, for though I said, and though you repeat, the phrase 'the logical content of a probability statement', you seem to want to put into it all that belongs to the framing of a scientific theory. I do not believe that the gambler's belief that the probability of a card chosen at random shall be a king is 1/13 involves more than the two elements, which I was trying to make plain to you, when all these further ideas came into your head, involving what hypotheses should be framed about the distribution of an observed measurement.

As regards your paradox, you ought to consider whether your 'unbiased estimate' gives you satisfactory guidance in the cases of *one* or *two* degrees of freedom.

D.J. Finney to Fisher: 1 *April* 1955

Your two quotations are certainly horrid misrepresentations.

I'm sorry you should think that I did not read your long letter fully, but it is the kind of letter that can be re-read profitably. I think I now see your point more clearly. The situation is more confusing when a continuous variate is involved. For a discrete variate, your two elements suffice to lead to the occurrence of events depending upon a multinomial distribution, or something of that kind.

For a continuous variate (e.g. a random sample of 10 Englishmen measured for height), the nature of the inference that can be expressed probabilistically depends upon the knowledge that we have about the set of possibilities and their relative frequencies. If we know nothing about these except that definite frequencies exist, we are presumably thrown back on rather restricted non-parametric or distribution-free inferences about the population, but this complies with your requirements. If we know that individual heights are normally distributed, that constitutes different information on the possible values and their frequencies. I see now that what I was doing in my earlier letter was to try to list different levels of knowledge about these possibilities, without realizing that all were covered by the first of your two 'elements'. I was not concerned with framing a theory, but with trying to particularize the knowledge with which we start: I now see this to be unnecessary from the point of view of general discussion of necessary and sufficient conditions for making probability statements, though it may become more important in particular instances when we want to know what statements of probability can justifiably be made from certain data.

Meanwhile, my paradox remains with me. I should of course have qualified my previous discussion with '$n>2$'. I think I am right in saying that, for n degrees of freedom,

$$E(1/s^2) = n/\{(n-2)\sigma^2\} \text{ for } n > 2,$$

the integral failing to converge for $n = 1$ or $n = 2$. I don't see why one should not admit the impossibility of estimating $1/\sigma^2$ for an experiment that allowed only 2 d.f. for s^2, but the apparent inconsistencies for larger n remain.

Fisher to D.J. Finney: 18 *April* 1955

I only mentioned the case of $n = 1$ or 2 in order to raise the question in your mind as to the relevance of the expectation of $1/s^2$. I won't admit at all that one knows nothing about an unknown variance when one has one or two degrees of freedom from which to estimate it; in fact I should assert that one can derive in either of these cases an explicit fiducial distribution of the unknown variance.

With Barnard's help I have been wrestling with the relationship between axiomatic assertion of ignorance, inductive reasoning, uncertain inference, etc., and in case you are interested I enclose two little sections from two different chapters of the book I am putting together.

D.J. Finney to Fisher: 6 *May* 1955

I have not replied earlier to your letter of 18 April because I wanted a chance to read your enclosures two or three times. Although I find my original problem something of a mystery still, your own ideas are most stimulating. The preparation of your new book is excellent news. I have been hoping for something like this for years, and these four pages excite my interest greatly. Very many thanks for letting me see them.

Fisher to D.J. Finney: 17 October 1960

. . . I have been working on the distribution of the weighted mean of small normal samples. Like Behrens with an extra parameter. If the normal deviate x defines the level of significance as usual then the correction will be

a cubic in x and $d \div n_1$
5th power in x and $d \div n_1^2$ and $n_1 n_2$
7th power terms $\div\ n_1^3, n_1^2 n_2$.

These last are very troublesome to clean up.

The thing is almost too complicated to be useful, but I should like to see how it tabulates. My impression is that d (Sukhatme's Test) is more important than one would guess, so that approximations taking $d = 0$, or even $d = 1$, would be pretty bad. . . .

D.J. Finney to Fisher: 25 October 1960

. . . Your work on the weighted mean sounds interesting. I have in the past used the standard Fisher–Behrens' tables to give something approximating to fiducial statements about the weighted mean of two quantities where each weight is based upon an ordinary variance estimate. I have described this on page 31 of my book on biological assay, though I do not think that the idea had much originality. I think that you once pointed out to me that this was not a true fiducial argument because, on the hypothesis that the two statistics estimate the same parameter, some information on variance is contained in the difference between them.[1] Am I right in thinking that it is this extra piece of information that you are taking into account in your present study? My guess would be that, unless the degrees of freedom for the two samples are exceedingly small, the information contained in the differences of the two means would be negligible. It will be interesting to compare results based on your new distribution and table with those from the approximation that neglects this information. . . .

[1] See Fisher's letter of 2 December 1949 (p. 90).

Fisher to D.J. Finney: 1 November 1960

A few years ago I came to the conclusion that no one would be able to judge of the value of some approximate approach to the small sample weighted mean problem until he could compare with exact values. Clearly it involves n_1, n_2, θ, d and P, and the practical question is: How important is d?

Using the set-up,

$$\begin{aligned}\bar{x} &= (s_2^2\bar{x}_1 + s_1^2\bar{x}_2)/\{s_1^2 + s_2^2\},\\ 1/S^2 &= (1/s_1^2) + (1/s_2^2),\\ \mu &= \bar{x} + Su,\end{aligned}$$

then if
$$P = \int_x^\infty e^{-\frac{1}{2}v^2}\,dv,$$

$$u = x + \Sigma\{u_{pq}/n_1^p n_2^q\}$$

and the first approximation is given by

$$u_{10} = \tfrac{1}{4}\{(x^3 + x)c^4 + 4(x^2 + 1)dc^3 s + 2x(3d^2 - 1)c^2 s^2 + 4d(d^2 - 1)cs^3\}$$

and u_{qp} may be obtained from u_{pq} by interchanging c and s and reversing d.

I do not believe that we need more than the first term to judge that d is rather important.

Payne gave the adjustments for n_1^{-2} and $n_1^{-1}n_2^{-1}$ though I do not know how far he completed this job. I am trying to perfect the terms for (3) and (21)[1] . . .

[1] cf. *CP* 285.

Fisher to D.J. Finney: 3 April 1962

. . . Have you followed at all the activities of David Sprott of Waterloo, and D.A.S. Fraser of Toronto? They have recently sent me a very ingenious form for the correlation coefficient. Namely,

$$\frac{r}{\sqrt{1-r^2}}\,\chi_{N-2} - \frac{\rho}{\sqrt{1-\rho^2}}\,\chi_{N-1},$$

a pivotal function with a standardized normal distribution, involving the two random variables χ. Choosing any value of ρ, it yields the sampling distribution of r, as I gave it in 1915 [*CP* 4], while if you choose any value of r, you can obtain the distribution of ρ as I gave it in 1930 [*CP* 84], and in some new form such as,

$$\frac{1}{\pi(N-3)!}\quad (1-r^2)^{\frac{1}{2}(N-2)}\,(1-\rho^2)^{\frac{1}{2}(N-3)}\,d\rho\quad \left\{\frac{\partial}{\sin\theta\,\partial\theta}\right\}^{N-3}\frac{\theta - \frac{1}{2}\sin 2\theta}{\sin^3\theta}.$$

They have both been inclined to throw Fiducial methods in the teeth of the Americans; who indeed have been pretty dumb about it. There will probably be an ISI session about Inference at Ottawa next year . . .

D.J. Finney to Fisher: mid April 1962

. . . I have seen a little of the work that Sprott and Fraser have been doing, but I had not seen this extraordinarily interesting proposal for the correlation coefficient. It looks as though they are in the process of doing some very valuable work by putting the formal mathematics of the fiducial argument in a way that will compel the attention of some of those who have previously disregarded it. Am I right, though, in thinking that their treatment of fiducial inference is at present limited to certain classes of distributions and parameters and does not embrace all the uses that you have made? I got the impression from their first paper that they had not subsequently read more, and I am not sure how generally they expect to be able to produce suitable pivotal functions. . . .

Fisher to D.J. Finney: 3 May 1962

. . . I do not suppose that Rao, Fraser and Sprott have cracked the inversion problem, but I suppose the direct simultaneous distribution will appear in *Sankhyā* shortly. I hope also the RSS may soon get around, after more than a year, to publishing my paper on Bayes' experimental procedure [*CP* 289].

I think Fraser must have started with some rather over-simple notions in which the phrase 'continuous group of transformations' was prominent, and exhaustiveness and monotonicity not very visible. Probably he had given little attention to Estimation. Still he extricated himself from the Tukey–Savage sort of rubbish. I do not know whether he has made any *amenda* about the correlation coefficient, in print at least.

My chief intention in the last example in *Statistical Inference* was to illustrate the existence of 3 strata of parameters and their corresponding statistical estimates. This rather complicates the exact statement of the conditions for simultaneous fiducial distributions. I believe I said something about it in my last letter, but I forget how much.

Fisher to D.A.S Fraser: mid November 1961

It was good to see your paper in the *Annals*[1] for that Journal needs the injection of a little sense and relevance. But I was sorry that you had let Tukey and Savage waste your time for those two able minds are themselves in such a mass of confusion and contradiction that they can scarcely fail to confuse and frustrate others. The last section of your paper seems to lack confidence. What to me needs clarification are such phrases as 'the frequency interpretation that customarily goes with confidence intervals'.

Do you mean, for example: — This interval calculable from the data will cover the true value in $(1 - \alpha)$ of repeated random trials?

The probability that the true value lies in this interval is $(1 - \alpha)$?

I gather the latter is unorthodox among the great herd of teachers in American mathematical departments, and it is certainly not a valid inference from a test of significance only. It is orthodox also to avoid questions of

Sufficiency, although their relevance was pointed out quite 25 years ago.

Still how can the former version above be called an 'interpretation'? It altogether avoids the specification of uncertainty, and makes no specific inference. It seems to ride on the shadow of the fiducial inference, which is often rigorously valid for the same interval (in very simple cases).

I am glad you see the sense of Behrens' test. I am sending a couple of offprints from *Sankhyā* but they will take time travelling about the world.

[1] Fraser, D.A.S. (1961). On fiducial inference. *Ann. Math. Stat.* **32**, 661–76.

D.A.S. Fraser to Fisher: 24 *November* 1961

Thank you very much for your letter and the kind comments it contained. . . .

The section on the correlation coefficient — you made reference to it — contains a regretful error. The fiducial distribution for ρ obtained by a group-theory analysis involving regression of y on x, as obtained similarly based on regression of x on y, and the original fiducial distribution as defined in your paper 'Inverse probability' [*CP* 84] are it seems identical and correspond to the equation

$$w = v \frac{r}{\sqrt{1 - r^2}} - u \frac{\rho}{\sqrt{1 - \rho^2}}$$

where w, u, v are independent and are distributed as a standard normal variable, a χ variable on $n - 1$ d.f., and a χ variable on $n - 2$ d.f. Solving for r yields the ordinary distribution of r given ρ; and solving for ρ yields the fiducial (as determined by any of the 3 methods) for ρ given r.

Also, for u, v given, the relationship of r to ρ is essentially of transformation parameter form. Thus the marginal for r with respect to ρ or ρ with respect to r is an average of transformation relationships . . .

Fisher to D.A.S Fraser: 29 *November* 1961

Thanks for your letter and for letting me know that you have now removed the cause of confusion which appeared in your discussion of the correlation coefficient.

The form you send me

$$w = v \frac{r}{\sqrt{1 - r^2}} - u \frac{\rho}{\sqrt{1 - \rho^2}}$$

is most interesting and suggestive. I wonder if it is capable of generalization to the trivariate case; for this I have found it possible to express the distribution of r_1, r_2, r_3 in terms of ρ_1, ρ_2, ρ_3 only. As you would expect the distribution involves the determinant of the r to one power $(N - 5)/2$ and that of ρ to another $(N-1)/2$ together with a well defined function of r and ρ, which for 3 variates can be written

$$\int_0^\infty\int_0^\infty\int_0^\infty (uvw)^{(N-2)/2} \exp - \tfrac{1}{2}(u^2 + v^2 + w^2 - 2\gamma_1 vw - 2\gamma_2 uw - 2\gamma_3 uv)\, dudvdw,$$

say $F_{N-2}(\gamma_1,\gamma_2,\gamma_3)$, where $\gamma_1 = r(\rho_1 - \rho_2\rho_3)/\sqrt{(1 - \rho_2^2)(1 - \rho_3^2)}$, writing from memory.[1] The distribution is easily generalized for t dimensions. I do not know why Sam Wilks could not do it.

You can imagine that an inversion would be of great interest.

It was not until about 1957 that I realized how completely dumb University teaching in Math Stat was in the U.S.. I could not believe it at first!

[1] cf. *CP* 288.

Fisher to D.A.S. Fraser: 18 *December* 1961

As I have not heard from you, it may be that my reply to your letter of Nov 24 has gone astray. I think I did include the Calcutta address as above.

With respect to the very elegant formula of your letter, on trying to reproduce it, I am led to put to you the possibility that it should be

$$w = v\sqrt{\frac{n-1}{n-2}}\,\frac{r}{\sqrt{1-r^2}} - u\frac{\rho}{\sqrt{1-\rho^2}}$$

where, as in your formula,

w is normally distributed about zero with unit variance,
v is a random variable χ for $n-2$ d.f.,
u is a random variable χ for $n-1$ d.f.,

and u, v, w all independent.

I put the question whether this same approach would not give the simultaneous distribution of ρ_{12}, ρ_{23}, ρ_{31} in terms of r_{12}, r_{23}, r_{31} observed.

I believe it should do so, but the approach is yours, and I think you should consider its extension.

D.A.S. Fraser to Fisher: 6 *January* 1962

Thank you for your letters of Nov 29 and Dec 18. . . .

Concerning the pivotal equation for the correlation coef.

$$z = \chi_{n-2}\frac{r}{\sqrt{1-r^2}} - \chi_{n-1}\frac{\rho}{\sqrt{1-\rho^2}},$$

this was derived by the group theory approach and made use of the canonical form for the Wishart distribution which is concealed in Mauldon *JRSS*, *B*17

(1955). For a sample of $n - 1$ from a bivariate normal with means equal to zero it yields:

$$\begin{pmatrix} \sqrt{\Sigma x^2} & \Sigma xy/\sqrt{\Sigma x^2} \\ 0 & \sqrt{\Sigma(y - bx)^2} \end{pmatrix} = \begin{pmatrix} \chi_{n-1} & z \\ 0 & \chi_{n-2} \end{pmatrix} \begin{pmatrix} \sigma_1 & \sigma_2\rho \\ 0 & \sigma_2\sqrt{1 - \rho^2} \end{pmatrix}.$$

This is based on positive upper triangular factoring of the positive definite matrices. The diagonal terms produce the equation:

$$\begin{pmatrix} \sqrt{\Sigma x^2} & 0 \\ 0 & \sqrt{\Sigma(y - bx)^2} \end{pmatrix} = \begin{pmatrix} \chi_{n-1}\sigma_1 & 0 \\ 0 & \chi_{n-2}\sigma_2\sqrt{1 - \rho^2} \end{pmatrix}.$$

Taking the inverses of these diagonal matrices and multiplying on the right in the first equation produces:

$$\begin{pmatrix} 1 & \Sigma xy/\{\sqrt{\Sigma x^2}\sqrt{\Sigma(y - bx)^2}\} \\ 0 & 1 \end{pmatrix} = \begin{pmatrix} \chi_{n-1} & z \\ 0 & \chi_{n-2} \end{pmatrix} \begin{pmatrix} \chi_{n-1}^{-1} & \chi_{n-2}^{-1}\,\rho/\sqrt{1 - \rho^2} \\ 0 & \chi_{n-2}^{-1} \end{pmatrix}$$

which yields

$$\frac{r}{\sqrt{1 - r^2}} = \chi_{n-1}\,\chi_{n-2}^{-1}\,\frac{\rho}{\sqrt{1 - \rho^2}} + z\,\chi_{n-2}^{-1}.$$

This rearranged is the pivotal equation.

A student and colleague of mine at Toronto is working on aspects of the higher dimensional case. The distribution of the r's can be expressed in terms of the ρ's in a pivotal form. This form can be inverted but certain symmetries may be lost because of the use of the partial correlations which are natural in the regression-group form. I write from memory and a little geometrical intuition.

$$\begin{pmatrix} S_1 & S_2r_{12} & S_3r_{13} \\ 0 & S_2\sqrt{1-r_{12}^2} & S_3\sqrt{1-r_{13}^2}.r_{23.1} \\ 0 & 0 & S_3\sqrt{1-r_{13}^2}\sqrt{1-r_{23.1}^2} \end{pmatrix}$$

$$= \begin{pmatrix} \chi_{n-1} & z_1 & z_2 \\ 0 & \chi_{n-2} & z_3 \\ 0 & 0 & \chi_{n-3} \end{pmatrix} \begin{pmatrix} \sigma_1 & \sigma_2\rho_{12} & \sigma_3\rho_{13} \\ 0 & \sigma_2\sqrt{1-\rho_{12}^2} & \sigma_3\sqrt{1-\rho_{13}^2}\cdot\rho_{23.1} \\ 0 & 0 & \sigma_3\sqrt{1-\rho_{13}^2}\sqrt{1-\rho_{23.1}^2} \end{pmatrix}$$

$S_1^2 = \Sigma x_1^2$, $S_2^2 = \Sigma x_2^2$. The diagonal terms can be used to form diagonal matrices. The inverse of these diagonal matrices multiplying into the above equation on the right will yield a matrix equation which contains implicitly the marginal fiducial (group method) distribution for the ρ's in terms of the pivotal variables. . . .

The general Wishart distribution is available from the equation above. The equation gives a parameter transform of the distribution in the standardized case. The standardized case has its distribution available by a simple geometrical-regression analysis argument: $(x_{11}, \ldots x_{1,n-1})$ has length with a χ_{n-1} distribution. $(x_{21}, \ldots x_{2,n-1})$ has a component in the direction $(x_{11}, \ldots x_{1,n-1})$ and this component has a standardized normal distribution; it has a component perpendicular to this with length having a χ_{n-2} distribution. And so on for trivariate and multivariate case — components in the orthogonal frame built up by earlier components have standardized normal distributions — the remainder has a χ distribution.

On November 28th I gave a fiducial talk at Berkeley and it was received with utter contempt by Neyman and bare tolerance by the others. . . .

D.A.S. Fraser to Fisher: 10 *January* 1962

I hope my letter of Jan 6 reached you in India. I have now heard from Keith Hastings at Toronto and shall quote several of his formulae — they tie in with those in my Jan 6 letter. [With] $g_{ii} = \chi_{n-i}$, $g_{ij} = N(0,1)$, $[i \neq j$, and all $g_{ij}]$ independent,

$$\begin{pmatrix} s_1 & s_2r_{12} & s_3r_{13} \\ 0 & s_{2.1} & s_{3.1}r_{23.1} \\ 0 & 0 & s_{3.12} \end{pmatrix} = \begin{pmatrix} g_{11} & g_{12} & g_{13} \\ 0 & g_{22} & g_{23} \\ 0 & 0 & g_{33} \end{pmatrix} \begin{pmatrix} \sigma_1 & \sigma_2\rho_{12} & \sigma_3\rho_{13} \\ 0 & \sigma_{2.1} & \sigma_{3.1}\rho_{23.1} \\ 0 & 0 & \sigma_{3.12} \end{pmatrix}$$

[i.e.] $\mathbf{S} = \mathbf{G}\Sigma$ say. Let $\mathbf{D}(\mathbf{S})$ for example be:

$$\begin{pmatrix} s_1 & 0 & 0 \\ 0 & s_{2.1} & 0 \\ 0 & 0 & s_{3.12} \end{pmatrix}$$

The pivotal equation is $\Sigma\mathbf{D}^{-1}(\Sigma) = \mathbf{G}^{-1}\mathbf{S}\mathbf{D}^{-1}(\mathbf{S})\,\mathbf{D}(\mathbf{G})$ which yields

$$\frac{s_2r_{12}}{s_{2.1}}g_{22} = \frac{\sigma_2\rho_{12}}{\sigma_{2.1}}g_{11} + g_{12},$$

$$\frac{s_3r_{13}}{s_{3.12}}g_{33} = \frac{\sigma_3\rho_{13}}{\sigma_{3.12}}g_{11} + \frac{\sigma_{3.1}\rho_{23.1}}{\sigma_{3.12}}g_{12} + g_{13},$$

$$\frac{s_{3.1}r_{23.1}}{s_{3.12}}g_{33} = \frac{\sigma_{3.1}\rho_{23.1}}{\sigma_{3.12}}g_{22} + g_{23}.$$

Hastings then quotes the following probability density function for r_{12}, r_{13}, r_{23}:

$$K\,D_r^{-(n+3)/2}\,D_\rho^{-(n-1)}\{(1-\rho_{12}^2)(1-\rho_{13}^2)(1-\rho_{23}^2)\}^{(n-1)/2}\times\int\int_0^\infty\int(uvw)^{n-2}\times$$

$$\exp\{-\tfrac{1}{2}(u^2+v^2+w^2-2uvr_{12}\rho_{12.3}-2vwr_{23}\rho_{23.1}-2uwr_{13}\rho_{13.2})\}\,du\,dv\,dw,$$

$$\text{where } D_r=\begin{vmatrix}1 & r_{12} & r_{13}\\ r_{12} & 1 & r_{23}\\ r_{13} & r_{23} & 1\end{vmatrix}$$

and this has the form in your Nov 29 letter with a difference in power for (uvw).

Solving the pivotal equation the other way for the ρ's yields acc. to his letter: p.d.f. for $\rho_{12}\,\rho_{13}\,\rho_{23}$

$$K\,D_r^{-(n-3)/2}\,(1-r_{12}^2)^{-1}\,\{(1-\rho_{12}^2)(1-\rho_{13}^2)(1-\rho_{23}^2)\}^{(n-2)/2}\times$$

$$D_\rho^{-(n+1)}\,(1-\rho_{12}^2)^{-1}(1-\rho_{13.2}^2)^{-1}\int\int_0^\infty\int u^n v^{n-2} w^{n-4}\exp\{\text{as above}\}\,du\,dv\,dw.$$

This seems asymmetric generally; definitely is for $r_{12}=r_{13}=r_{23}$. This distribution relates the ρ's to the r's by reference to a definite order in which regression is run on the variables 1,2,3. . . .

Fisher to D.A.S Fraser: 11 *January* 1962

I have just seen your letter of Jan 6. Do not forget to look up Walter Bodmer, who has also had some experience of being 'bawled down' by Neymanians. They intimidate Americans successfully enough, especially refugees anxious to get posts in American Universities. I do not think they need intimidate anyone else. For your encouragement I transcribe the first paragraph of a letter just received from a mathematical logician[1] at the Rockefeller Institute.

'I have just finished your book *Statistical Methods and Scientific Inference.* I wish I had seen it sooner. I am just delighted with it, for I have felt altogether alone in my dire suspicions of the logical inadequacy of the theory of statistical inference as expounded by Neyman, Pearson and nearly every other statistician I can think of . . .'. Later on, he says 'I am interested in tracking down everything that could possibly be of use to me concerning fiducial inference; for this seems to me to be the fundamental form of statistical inference — the decision theoretic approach being etc.'

I think I told you I had the simultaneous distribution of r_{ij} given ρ_{ij} for t variates. It can be written (if I have room)

$$\frac{\pi^{-t(t-1)/4}2^{-t(N-3)/2}}{\{(N-3)/2\}!\ldots\{(N-t-2)/2\}!}\,|\rho_{ij}^*|^{(N-1)/2}|r_{ij}|^{(N-t-2)/2}\,dr_{ij}F_{N-2}(\gamma_{ij})$$

ρ^*_{ij} is the cosine of the dihedral angle of the ρ figure, or of the side of the polar figure.

$$F_{N-2} = \int_0^\infty \ldots \int_0^\infty (u_1 \ldots u_t)^{N-2} \exp\{-\tfrac{1}{2}(u_i^2 + \ldots + 2\gamma_{ij}u_iu_j + \ldots)\}[du_1 \ldots du_t]$$

a useful function of γ_{ij} only [where] $\gamma_{ij} = r_{ij}\rho^*_{ij}$. All very compact and handy if put this way [cf. *CP* 288].

Starting with the distribution of ρ_{12} marginal, I should like you to get that of ρ_{13}, ρ_{23} conditional (a), then ρ_{14}, ρ_{24}, ρ_{34} conditional (b) and so on.

But of course we are thinking of the problem in rather different terms.

Rao here, who is quite first class, derives from your compact formula one for ρ which I like

$$\frac{1}{\pi(N-3)!}(1-\rho^2)^{(N-3)/2}(1-r^2)^{(N-2)/2}d\rho \int_0^\infty\int_0^\infty (uv)^{N-3}(1+\gamma uv) \times \exp\{-\tfrac{1}{2}(u^2+2\gamma uv + v^2)\}dudv$$

[where] $\gamma = -r\rho$.

Thanks for all the trouble you have taken. Would you consider a joint paper with Rao capping what I have done, with everything you two can do — for *Sankhyā*?

[1] H.E. Kyburg. For Fisher's reply to Kyburg's letter, see p. 187.

Fisher to D.A.S Fraser: 14 *January* 1962

You still tell me nothing about Walter Bodmer; won't you ring up the medical school and locate him? I am sure you will like him.

I feel sure your last two letters are a step back from the vantage point you had reached with your equation between r and ρ with $s_1, s_2, \sigma_1, \sigma_2$ eliminated. The general linear transformations in three dimensions do not transform spheres into spheres, or spherical triangles into spherical triangles unless they are rotations, and simple rotations do nothing.

I want to end my book [*SMSI*] (when it comes to a third edition) by saying that the right way of treating equation (231) had since been demonstrated by you, eliminating s and σ, in the form

$$w = \frac{r}{\sqrt{1-r^2}}\chi_{N-2} - \frac{\rho}{\sqrt{1-\rho^2}}\chi_{N-1}$$

from which for any chosen ρ the distribution of r is found as I gave it in 1915 and for any given r the distribution of ρ is what I gave in 1930.

That from it you and Rao had derived new forms of which one, parallel to 216, might be set out. Is it

$$\frac{\pi}{(N-3)!}\ (1-r^2)^{(N-2)/2}(1-\rho^2)^{(N-3)/2}d\rho \left\{\frac{\partial}{\partial(r\rho)}\right\}^{N-3}\left(N-2+\frac{\partial}{\partial(r\rho)}\right)\frac{\theta}{\sin\theta}$$

or what? I am scribbling from memory. Tell me what you make of it and I shall know what to put. Perhaps the last operand is $(\theta - \frac{1}{2}\sin 2\theta)/\sin^3\theta$.

It is not axiomatic that a triple or multiple distribution for ρ_{ij} exists. I guess it does, and that the only satisfactory demonstration is to take the variates in arbitrary order

$$\begin{matrix} & & \rho_{12} & & \\ & \rho_{13} & & \rho_{23} & \\ \rho_{14} & & \rho_{24} & & \rho_{34} \end{matrix}$$

and exhibit *symmetrical* formulae emerging. Will not the partial regression of z on x and y, conditional on a known distribution of ρ_{12}, give a symmetrical result?

Of course I should love to see the De Lurys again. I must consult in Adelaide before promising anything specific. Remember I am aged and pretty blind, but still *compos mentis*.

Do not forget Walter.

[P.S.] For your approval: proposed addition to *Stat. Meth. and Sci. Inf.*, p. 175.

The correct way of using the facts stated in (231) has been more recently demonstrated by Donald Fraser and David Sprott who eliminate s_1, s_2 and σ_1, σ_2 obtaining the equation in three random variables

$$w = \frac{r}{\sqrt{1-r^2}}\,\chi_{N-2} - \frac{\rho}{\sqrt{1-\rho^2}}\,\chi_{N-1} \qquad (233)$$

in which w is a normal variable with mean zero and unit variance, while the two others are χ-variables with $(N-2)$ and $(N-1)$ degrees of freedom respectively. Assigning any value to ρ the equation gives the distribution of r as it was given in my paper of 1915, while assigning any value to r, it gives the fiducial distribution of ρ as first given in 1930, and as used in this chapter.

Explicit forms for the distribution of ρ have been derived by C.R. Rao from Fraser's formula. For example the form corresponding with (216) is

$$\frac{1}{\pi(N-3)!}\ (1-\rho^2)^{(N-3)/2}(1-r^2)^{(N-2)/2}\ \frac{\partial^{N-3}}{\partial(\rho r)^{N-3}}\left\{\frac{\theta-\frac{1}{2}\sin 2\theta}{\sin^3\theta}\right\}d\rho \qquad (234)$$

D.A.S. Fraser to Fisher: 27 January 1962

In your January 11 letter you ask if I would consider a joint paper with you and Rao. I would be honoured and pleased to collaborate. . . .

The transformation approach to fiducial came to definite form with me two years ago and led to a talk in August 1960 here at Stanford to the Inst. of Math. Stat. That talk was written down to form the article in the *Annals*.[1] In preparing for that talk I only sketched the details of the application of transformation to the bivariate normal and in the hurry incorrectly inferred that the 3 marginal distributions for ρ were not the same. David Sprott at Waterloo Univ. took time to check the details and let me know in October last that the three seemed to be identical; he quoted a triple integral with a relationship between variables, which relationship becomes the equation

$$z = \frac{r}{\sqrt{1-r^2}} \chi_{n-2} - \frac{\rho}{\sqrt{1-\rho^2}} \chi_{n-1},$$

upon interpretation. My method gives the formula which I have interpreted, but in working with it I shall need to acknowledge Sprott's checking of the details on the derivation and picking up the incorrect statement in my *Annals* paper.

My transformation method using Mauldon's results gives a quick derivation of trivariate and k-variate correlation distributions and might be used in the proof of the trivariate and k-variate distributions you report in your letters of Nov 29 and Jan 11. The various approaches to fiducial distributions for correlation coefficients fascinate me and I feel that I can contribute to the analysis. Unfortunately I don't see time for this before the end of March. Is all this consonant with a joint paper with you and Prof. Rao?

You mention the changes for the third edition of your book [*SMSI*] and my transformation analysis which handles equation (231) and for which simple elimination in your equations yields the equation

$$z = \frac{r}{\sqrt{1-r^2}} \chi_{n-2} - \frac{\rho}{\sqrt{1-\rho^2}} \chi_{n-1}.$$

I would be pleased to receive acknowledgment for this. For noting that your marginal distribution for ρ is implicit in this, contrary to the statement in my *Annals* paper, Dave Sprott's name should be mentioned.

You record an interesting distribution for ρ that Rao has worked with — I have no preferences at present for one form over another — you have magnificent facility for handling such expressions.

The details of the successive analysis of the ρ's intrigues me — I at the moment don't see time to go into the details at this hectic and hurried place, Stanford, during this quarter. My colleague Hastings is working in this area and I shall certainly refer to you any results for this analysis. He claims in his last letter to have several other groups for the plane that yield symmetric fiducials. I think he is breaking with a certain aspect of my transformation approach — and getting interesting results — but at the moment I don't have clear feelings on their relationship to my work so far. . . .

[1] See p. 107.

Fisher to D.A.S. Fraser: 2 *February* 1962

I have your letter of Jan 27th. I am delighted that you think you see a straight path to the general inversion. May it not disappoint you.

The position about publication is this. *Sankhyā* has and is setting up my paper on the general r_{ij} distribution. In handing it in I said I did not much want to publish until you or someone else could follow it with the fiducial inversion. I had hoped that a paper by yourself and Rao might suit you both. He is an excellent collaborator, as well as first class in his own right.

You owe me an *amenda* for letting people think (as some are too eager to) that there was doubt as to the uniqueness of my solution of the bivariate case, which logically was elementary enough, and the most emphatic repudiation you can make of this misapprehension is to generalize it coherently with the multiple distribution I propose to publish.

It has also this importance: we know too little of the conditions in which fiducial distributions are to be expected. If exhaustive estimation is, as it may be, a sufficient condition, then $r_{ij} - \rho_{ij}$ is a test case. It is known not to be, strictly, a necessary condition, but it would be important to establish a group of sufficient conditions.

I expect to leave India before the middle of February. I believe you will be very welcome here if you can bear the hot weather expected by the end of March. No doubt Rao or Mahalanobis will write to make this clear.

Sprott has some difficulties about integrals based on my 5-parameter distribution, but I cannot think they are serious.

Send me a line on the strategy of the approach you think will work and give my love to Walter.

Fisher to D.A.S. Fraser: 13 *February* 1962

I thought I had made clear how pleased I was at the suggestion of my working for a year at Toronto. I expect De Lury will get around to making me a definite offer. I explained that I should have to discuss the matter in Adelaide. I should like to include the period of ISI meeting in Ottawa, which should be convenient.

I have heard from Sprott who also approves the wording I propose.

I shall be leaving for Adelaide on the 15th. I think you have my address there

Division of Math. Stat., CSIRO,
Adelaide University,
S. AUSTRALIA

I believe distributions, or equivalent general probability statements about functions of σ_1, σ_2, ρ can only be valid if based primarily on the marginal or unconditional distribution of ρ. Probably this has given Sprott and perhaps you some trouble. The individual statistics of an exhaustive set are not necessarily exhaustive for the corresponding parametric function.

However, you will have thought about all this.

Fisher to D.A.S. Fraser: 3 *May* 1962

I suppose you decided not to go to India after all. Have you had any useful correspondence with Rao lately or Sprott?

I heard from Dan and suggested coming about the end of March 1963 working (perhaps with advanced students chiefly) until the end of May, visiting Britain in June and July, and returning to Saskatchewan and later Ottawa in Aug and Sept; then serving a term until Christmas and perhaps spending that cheerful season with my daughter in Wisconsin. All this may not suit perfectly, but I hope it can be made to do.

You may remember my suggesting that the general inversion of the simultaneous correlation problem might be approached by steps. From the marginal bivariate, about which I hope and think there should be no further difficulties, one might graft on the formulation as in my 1928 paper on multiple correlation, the simultaneous distribution of ρ_{13}, ρ_{23} in an array with given r_{12}, ρ_{12}. You might well find the introductory section of the 1928 paper helpful, using ψ as pivotal instead of integrating it out as I did to find the distribution of R. I think if ρ is the true multiple correlation, $d\rho d\psi$ is

$$d\rho_{13}\, d\rho_{23} \;/\sqrt{1 - r_{23}^2}$$

or something like it.

Manipulation of the full explicit expression is rather tough and here it may be that your use of exact random variables may provide the key.

With good wishes to Sprott and Dan De Lury.

Fisher to D.A.S. Fraser: 7 *May* 1962

I have just seen yours of 1st May. I am indeed hoping to see a good deal of you after you are free from the fairly congested term's work Jan–March. I could come earlier than the end of March if that is desired, but I imagine that you and a few of your brighter men may be more free for concentrated thought after March.

Sprott also, I understand, is in the same Province, and I should be willing to take quite a bit of trouble to see something of him too this time (April, May).

. . .

Have you contacted Henry Kyburg of the Rockefeller Institute, New York? I have not yet seen his book, which is expensive even for an American book (Wesleyan Univ Press) and may be large. I suppose it will obliterate Keynes' book of 1928, which also was ambitious, but in retrospect not very good.

Have you heard at all further from Rao? Even if not this year you must visit the Stat. Inst. in Calcutta some time, but I should advise Dec and Jan. Rao is always worth talking to.

It is serious that so many young men, like you and George Box, have had to learn to disregard the indoctrination pressed upon them at their Universities. A live centre of teaching so near as Toronto would certainly make a number of U.S. centres sit up and take notice. I think they induce the belief that they already know everything by attending too many seminars.

Fisher to D.A.S. Fraser: 22 May 1962

I delayed answering your letter of May the 9th to give Alf Cornish a chance to consider it. Now he has returned in some triumph with a 3 million pound computer, as it were, in his pocket, I can say that it looks O.K., and I am prepared to arrange to fly to Toronto at the end of April. Dan had better arrange through the airline, presumably Qantas, for a credit at that time.

About the correlation problem, it does involve I fancy further intricate analysis, of a kind which the classical analysts have largely ignored. I mean something like this; the distribution of r_{23} depends only on ρ_{23}. Therefore there is a pivotal involving only these two, the distribution of which is independent of ρ_{12}, ρ_{13}, ρ_{23}. Equally if R is the multiple correlation of (1) on (2,3) the distribution of R has been expressed as a function of ρ only, where ρ is the corresponding parameter. I would swear, though I would not undertake to demonstrate, that these two distributions are statistically independent, so there is a pair of pivotals. I doubt if it is mathematically possible that there is not a third, involving say

$$\frac{\rho_{12} - \rho_{13}}{\rho_{12} + \rho_{13}}, \quad \frac{r_{12} - r_{13}}{r_{12} + r_{13}}$$

to complete the specification of statistics in terms of r_{12}, r_{13}, r_{23} and of the parameters in terms of ρ_{12}, ρ_{13}, ρ_{23}. It could be that this third distribution has a twist and fails to be monotonic. I will believe that when I see it. However, those pivotals with a joint distribution independent of the parameters, and each monotonic in the appropriate parametric function must give the fiducial inversion. Probably one needs something less evil than the probability integrals of the first two distributions, which is why I was excited by your use of random functions in

$$\frac{r}{\sqrt{1-r^2}}\,\chi_{N-2} - \frac{\rho}{\sqrt{1-\rho^2}}\,\chi_{N-1}$$

and the real reason why I threw the problem at you. Let me know.

M. Fréchet to Fisher: 8 January 1940

May I consult you on two points (even three). It happens that in two ways I am much more engaged this year in mathematical statistics than I was ever before (when I was more active in other fields).

My colleague Darmois being mobilized, I replace him in teaching of math. statistics and besides I have been appointed Director of a so called 'Laboratoire' of Mathem. Statis. and Calculus of Probability organized in view of researches interesting National Defence.

As the Directory of this 'Laboratory', I will probably soon be sent to England to establish contact between statisticians working [on] both sides of the channel for war purposes. This letter is then first intended to beg you to let me know whether you are fully engaged in your usual so valuable researches or whether you have to divert some of your labour for special researches for National Defence. In the last case, I should like to have from now your views concerning the different fields where statistical researches could be done for war purposes and during my visit in England we might précise [*sic*] some details which it is better not to précise in letters. Even in the case where you are not in touch with military departments, you might perhaps let me know the names of mathematical statisticians engaged in work for these departments. So much for this part of my new work.

Concerning the other part, that is, my course on mathematical statistics, I had to look more attentively than I had time to do before, on some new developments of statistical theory. And I would like to have your opinion on the following quotation of a paper by Deming and Birge[1] 'On the statistical theory of errors'. This is page 142, end of first column and beginning of the second. 'These particular values of σ and s are accordingly so related to each other that if σ were actually the S.D. of the parent population then there would be 19 chances in 20 that a sample drawn therefrom would have a S.D. as large as or larger than s; and conversely, since s has actually been observed, there is only 1 chance in 20 that the S.D. of the parent population is as large as or larger than σ.'

They have added a slip to introduce some corrections to their paper, but these quoted lines do not seem to have been corrected. I have underlined 'and conversely' because it seems to me that a logical confusion has there arisen. If I understand well, the probability mentioned there is computed from equation (30):

$$P_s = 1 - \Gamma_v\left(\frac{n-1}{2}\right) \Big/ \Gamma\left(\frac{n-1}{2}\right); \; v = ns^2/2\sigma^2.$$

This equation holds good when σ *being fixed*, s is a *random* variable computed from the observed values of the random variables $x_1, \ldots x_n$. Now in the second part of the quotation, the former value of this probability (computed *before the trials*) is considered also as valid when, s being observed, σ is considered as a random variable. As the whole proof of the formula is based on the first hypothesis and not on the second one, I do not see why the formula should be still valid in these very different circumstances.

It is so much important to clear up the validity of these quotations that it seems that this logical confusion is often met, though in sentences which are not clear enough, in many modern papers on statistics.

P.S. Vous avez obtenu la loi de probabilité de $(\bar{x}-\mu)/s$ (avec vos notations). A-t-on obtenu et publié (et dans ce cas, où?) la loi de probabilité de $u = \{\gamma_1(x_1-\mu) + \ldots + \gamma_n(x_n-\mu)\}/s$ où les γ sont des constantes $(\Sigma\gamma_i = 0)$, les x_i valeurs observées d'une variable gaussienne X, $s^2 = \Sigma(x_i-\mu)^2/(n-1)$, $\mu = \bar{X}$?

[1] Deming, W.E. and Birge, R.T. (1934). *Rev. Mod. Phys.* **6**, 119–61. See Fisher's letter of 25 September 1934 to Deming (p. 80).

Fisher to M. Fréchet: 17 *January* 1940

I am very glad to have your letter of January 8th with the quotation from Birge and Deming. I find these writers often very obscure, but as they are obviously influenced by the form of argument for which I am responsible, I should like to make my own position, at least, clear to you.

I enclose the paper on 'Inverse probability' [*CP* 84] in which I first introduced the fiducial argument, though, in fact, as my reference to M. Ezekiel shows, many people had been arguing in this way from the moment when theoretical distributions, such as χ^2, t and z, were first tabulated so as to show the values taken at different levels of significance (values of P) instead of showing the values of P for different values of χ^2, etc.

I should like you to keep the offprint so long as it is useful to you; but, as I have very few copies, perhaps you will send it back if, at any time, you find you no longer want it. A second paper (The concepts of inverse probability and fiducial probability referring to unknown parameters. *Proc. Roy. Soc. Lond. A, 139:* 343–348) [*CP* 102] also expresses well what is still my point of view. I have no copies of this, but the *Proceedings of the Royal Society* may be accessible to you.

With respect to your other mathematical question:—

If $x_1, \ldots x_n$ are values independently and normally distributed with standard deviation equal to σ, and if $S(Y) = 0$, then $S(Yx)/\sigma$ is normally distributed about zero with standard deviation equal to $\sqrt{S(Y^2)}$ and this distribution will be absolutely independent of that of

$$s^2 = S(x-\bar{x})^2/(n-1)$$

or of s/σ. So that the ratio of these two quantities

$$S(Yx)/s$$

will be distributed exactly as is $t\sqrt{S(Y^2)}$, where t is the value I tabulate for 'Student's' distribution.[1]

I emphasise this type of extension of 'Student's' original argument in a paper in *Metron* (Applications of 'Student's' distribution) 1925, Vol. V, Part 3: 90–104 [*CP* 43]. It is really the basis of the tests used in the Analysis of Variance.

On your mission to England I think you ought to see Professor Leonard Jones, at the Mathematical Laboratory, Cambridge, who will be able to put you in touch with war work involving mathematical statistics better than any one else. My laboratory has not yet been used for this purpose.

[1] See also Fisher's letter of 26 April 1940 (p. 134).

M. Fréchet to Fisher: 21 *January* 1940

I have read with great interest your paper 'Inverse probability'. I understand all the paper and agree with many of its statements, but I must confess that I find here precisely the *same difficulty* as the one I pointed out in Deming and Birge's paper. The difficulty arises in a very short sentence which looks as obvious as the other ones but of which the meaning has a capital importance (as the words 'and conversely' in Deming and Birge):

Page 533: 'we may express the relationship by saying that the true value of θ will be less than the fiducial 5 per cent value corresponding to the observed value of *T in exactly 5 trials in 100*'. Now the five per cent and the 5 trials in 100 refer to two probabilities of the same event, it is true, but the populations where these probabilities are computed are extremely different. In the first one θ is fixed whereas T is a random variable and the table has been computed on this assumption. In the second one T is fixed, θ is a random variable and the first table which is still used (or the corresponding formula) has *not* been computed under this second assumption.

Such is my difficulty and I have found the same difficulty in other statistical papers.

In fact there is a long time [*sic*] I had doubts on such a point but they had [arisen] in reading sentences which were not definite enough to lead to something more than suspicion. The 'and conversely' was the first sentence where I found clearly an identity admitted as obvious between probabilities of the same event in two very different populations.

As you kindly allow me, I will keep sometime your paper to think more of the matter. I feel that it may be possible that the identification which I do not admit as obvious may still be used with success frequently in practice, though perhaps not always.

I am very glad to have got the reference to your paper in *Metron*; it will save me the time of doing laboriously what has been already brightly done.

My thanks also for mentioning Prof. Jones' work which will simplify my quest in England—where I am supposed to land in London 28th Jan. morning.

Fisher to M. Fréchet: 26 *January* 1940

Many thanks for your letter on Inverse and Fiducial probability, which is so clearly expressed that there is no difficulty in specifying exactly the point where our thought diverges. You say, in your third paragraph: 'In the second case T is fixed, θ is a random variable, etc.' If this were so, then the probability statement under consideration would be precisely a statement of inverse probability and, as such, as we both agree, probably not true, and certainly not known to be true.

I tried to make clear the difference in logical content between statements of inverse probability and statements of fiducial probability in my last paragraph, in which I specify: 'Whereas, however, the fiducial values are expected to be different in every case, and our probability statements are relative to such variability, the inverse probability statement is absolute in form, etc.'. From this it is, I hope, quite clear, that in the fiducial statement we are not considering T to be fixed. If, on the contrary, we remember that T will vary from sample to sample, and with it the corresponding fiducial 5% value of θ, then it will be true that the true value of θ will be less than the fiducial 5% value in exactly five trials in 100.

Some people prefer to make the same statement in the form of a disjunction, namely, either something has occurred which is known to occur in only 5% of trials, or θ exceeds $\theta_{5\%}(T)$. The important point, however, is that statements of fiducial probability have a logical content different from the more familiar statements of inverse probability, and are not intended to be equivalent to such statements. They refer, as you clearly perceive, to a different conceptual population. I have done my best, though perhaps not very successfully, to make this clear from my first references to the subject.

When in England I hope you may find time to visit me here in Harpenden, where my home is, and where I should be happy to have an opportunity of introducing you to my family.

M. Fréchet to Fisher: 3 *February* 1940

Arrived today in Paris, I find here your letter. After answering it on one point, I would like to submit to you a complement to a theorem of yours, in a different Chapter.

I. On the question of fiducial probability I would not like to let you lose your time were it not that the question of principle, there, is important and also that I heard recently of similar objections. Some were presented to me by a very good young French mathematician and some have appeared in a recent paper by Gini reproducing a speech 'I pericoli della Statistica' (*Rivista di Politica Economica*, 1939) which—as far as I understand this speech in *Italian*—looks very important. We are then at least three to have come, as it appears—to the same doubts, quite independently.

Now, it may be understood that, as you say, T will vary from sample to sample and with it the corresponding fiducial 5% value of θ. Still I expect that everybody will understand that the probability 5 trials in 100 is computed in the population where T gets a fixed value. You can see by Deming's quotation that at least Deming and Birge understood that, and other instances could be quoted.

But, even if we leave that, it remains that identification is admitted between probabilities of the same event in 2 populations which are different and essentially different because *in one, θ is fixed* (and this is implicitly supposed in the computation of the corresponding first probability) whereas *in the second, θ is supposed to be a random variable*.

The second way to state the meaning of fiducial probability, which is found in the second page of your letter, is essentially different and I quite agree with it. In fact it coincides with a statement which by chance came recently to my attention by Jan Wiśniewski: A note on inverse probability, p. 417 of the *Journal of the Royal Statist. Soc.* 1937:—

> 'either p is within the limits' (defined by some inequalities) 'or an improbable event happened, "improbable" to mean one whose probability is less than ' (5% in our case).

May I add that since I met the Deming's converse proposition, I noted also a converse proposition which appeared to me as of the same kind, in your paper 'The fiducial arg . . ..' *Annals of Eugenics*, vol. VI, 1935, at pages 391–392 [*CP* 125]: 'Since . . ., we may state the probability that μ is less than any assigned value'.

In order to assign this value $\mu_1 = \bar{x} - st_1/\sqrt{n}$, it is not sufficient to give the value of t_1 corresponding to 5 trials over 100, but also to give $\bar{x}$ and s or at least the linear function $\bar{x} - st_1/\sqrt{n}$ of $\bar{x}$ and s so that here we have really a hypothesis similar to the fixed T, that is, this linear function which was random in the proof of the law of t, whereas μ was fixed, becomes here fixed whereas μ is random in the statement on the law of μ.

Furthermore, I cannot reconcile the quotation by Wiśniewski of a statement in your paper in *Phil. Trans*. 1922, p. 327 [*CP* 18] (according to which it is impossible to get anything about the confidence interval of p from the information supplied by the sample) and the statement of the other more recent paper according to which the observation of $\bar{x}$ and s might give you the law of prob. of μ.

II. In your important paper 'The logic of inductive inference', *Journ. Stat. Soc.* 1935 [*CP* 124], you find an upper bound i of $1/(nV)$ and you prove that this very bound is obtained when choosing θ by the max. likelihood. It seems to me that, to be complete, the second part should also prove that this method gives a statistic T which is distributed normally about θ.

Perhaps you have given that proof elsewhere and in such a case I would beg you to give me the corresponding reference.*

However, I have discovered that this complement is not necessary *if* the first part of your theorem is generalized so as to get rid of the hypothesis of the normality of T (either for any n, or even only for great values of n). Now I thought it might interest you to hear that I have obtained such a result. Of necessity, my proof is absolutely different from yours where the hypothesis of normality fundamentally intervenes in the demonstration.

As our time was very short and we had to limit ourselves to meeting those engaged in war work, I could not think of trying to see you, at least this time. But I have, all the same, very much appreciated the courtesy of your kind invitation.

[P.S.] *I just now notice that Dugué proved this in his thesis under convenient hypotheses.

M. Fréchet to Fisher: 5 February 1940

I believe that the method I communicated to you in my last letter enables me to obtain something similar to the second theorem of your paper on 'The logic of inductive inference'. But I find it difficult to guess what is the exact meaning of your statement

> 'that for certain estimates, notably that arrived at by choosing those values of the parameters which maximize the likelihood function, the limiting value of $1/(nV) = i$.'

For the method of max. likelihood gives a value θ_0 of θ *after* n trials have given a sample; and in the formula, V is the variance of a random variable, which is a function $H(X_1, \ldots X_n)$, H appearing as a function chosen *before* the trials. So that the word estimates appears to cover in the same statement two significations: the numerical estimate θ_0 after the trials; the function $H(X_1, \ldots X_n)$ defined before the trials, (though its numerical values are known only after the trials). The result of my method leads me to a statement where I have to distinguish sharply between these two meanings.

Allow me to come to a second question, a historical one, based on your answer (p. 77 of the same paper) to the discussion. Is it right to say that the *formula* $1/(nV) \leq i$ was found by Filon and Pearson, or Edgeworth, but that they applied it incorrectly, in some cases? Is it right to say that some authors found that the equality $1/(nV) = i$ holds (at the limit) when the estimate is obtained through different (right or wrong) methods, but that none of them proved the same thing when the method employed is that of maximum likelihood, before your paper was published?

Finally a last question. In the same printed reply of yours, I read

> 'I mean by mathematical probab. only that objective quality of the individual which corresponds to frequency in the population, of which the individual is spoken of as a typical member.'

I have given a definition of probability which is different but which appears to me as having the same purpose. I wonder whether you would consider your definition, though perhaps not equivalent to mine, still as being at least consistent with mine. My definition is given in my books but may be summarized as follows:

The probability p of an event E in a category C of trials (in a universe U) is a physical magnitude $P(E/C)$ of which the frequency r/n of E in n trials of C is an experimental measure. . . .[1]

[1] For Fisher's reply, see his letter dated 12 February 1940.

Fisher to M. Fréchet: 10 *February* 1940

I doubt if we shall be able to get to a clearer understanding of the problem of fiducial probability, unless you are willing to accept it as a fact which I demonstrated to you by quotation in my last letter, that for the population of cases relative to which a fiducial probability is defined, the value of any relevant statistic T is not regarded as fixed. This I have deliberately exerted myself to make clear since my first writings on the subject. You will understand, therefore, that I find it a little disappointing when, on the third page of your letter of February 3rd, I find you saying: 'this linear function which was random in the proof of the law of t, whereas μ was fixed, becomes here fixed whereas μ is random in the statement of the law of μ'.

I shall be glad to give you all possible support in dissuading mathematicians from thinking that they can obtain a true probability statement logically equivalent to one of the kind aimed at by Bayes' theorem, yet without using the approximate basis of this theorem. Believe me, I have never attempted anything so foolish. The inferences which can be drawn without the aid of Bayes' axiom seem to me of great importance, and quite precisely defined, but are certainly not statements of the distribution of a parameter θ over its possible values in a population defined by random samples selected to give a fixed estimate T.

. . .

P.S. You say you cannot reconcile the point of view I have been expressing since 1930 with a too sweeping statement I made in the *Phil. Trans.* in 1922. The explanation is that fiducial probability was discovered in the interim. My 1922 statement was simply intended as a rejection of probability statement[s] based on the Laplacian principle of insufficient reason, which were the only statements concerning the probability of parameters made, I believe, up to that date. I did not then realise that statements of a logically different kind were possible, and rigorously deducible from the data, It is sad, but true, that every advance in Natural Science, including mathematics, means that some previous work is erroneous, inadequate, or obsolete.

Fisher to M. Fréchet: 12 *February* 1940

In my paper of 1922, *Phil. Trans.*, which I think you have available, I comment on the use by Pearson and Filon in 1898 of the formula giving the minimal variance or standard deviation of an estimated parameter.

If I remember right, they arrive at the formula by a very obscure argument involving inverse probability; but probably what they have done is equivalent to proving that $\sqrt{n}$ times the error of an estimate, taken e.g. approximately at the mode of an inverse probability distribution, will in the limit when n is large tend to be normally distributed with variance $1/i$. One of the features which makes their treatment so obscure and indefinite is that they do not notice that different methods of estimation have, in the limit for large samples, different precisions. Consequently they apply the formula, without hesitation, to estimates found by fitting Pearsonian curves by moments, and obtained a number of erroneous formulae, which were, I believe, not corrected till early in the present century when Sheppard showed how the standard errors of moments could be calculated directly. The erroneous formulae seem thereupon to have been dropped by the Pearsonian school, but there seems to be no hint in *Biometrika* which would inform the reader that previous misleading formulae were being corrected.

Edgeworth wrote a series of papers about 1908 in the Statistical Society's *Journal*. His attitude seems to be rather over-cautious than over-confident. He refers his readers to the paper by Pearson and Filon in terms which leave little doubt that he regarded it as correct, although in numerous other passages in this series of papers I should have thought he must be taken as recognising that different methods of estimation possessed different precisions, that these precisions have an upper bound, and that an estimate having in the limit the highest possible precision can be found by the procedure which Edgeworth regarded as inverse probability, but which, as I was concerned to emphasise in the 1930 paper I sent you, can and indeed must be completely dissociated from inverse probability, and which in 1922 I called the method of maximum likelihood.

The confusion of associating this method with Bayes' theorem seems to have been due originally to Gauss, who certainly recognised its merits as a method of estimation, though I do not know whether he proved anything definite about it.

I do not know of any explicit statement of the properties, consistency, efficiency and sufficiency, which may characterise estimates prior to my 1922 paper. I had noted the functional peculiarity of sufficiency in an early paper, 1920, in the *Monthly Notices of the Astronomical Society* [*CP* 12].

Yes, I think your definition of probability is consistent with mine, though I should emphasise that the physical magnitude considered is a character of the universe (population sampled).

M. Fréchet to Fisher: 20 *February* 1940

When you pointed my attention on the lines of your first paper on fiducial statem. where you state that T is not fixed, I noted this fact. I have not seen a similar warning in the paper which I quoted from *Annals of Eugenics*, 'On the fiducial arg.'. But since you say that when you wrote that second paper, you intended to mean that in the fiducial probability, $\bar{x}$, s as well as μ were not fixed, I have to admit that you know better than I do what you intended to mean.

Thereupon my question remains: on which ground are identified (or rather equalized) the two concerned probabilities referring to the same event (one particular inequality) but two different populations:

	Paper on inverse probability	Paper on the fiducial argument
1st population	T random, θ fixed	$x_1, \ldots x_n$ random, μ fixed
2nd population	T and θ random	$x_1, \ldots x_n$ and μ random

I mean, are those two probabilities equalized:

because it is obvious to you that they are equal?
or because it is a logical deduction of classical principles of Cal. of Prob., and then how?,
or because of the admission of a new principle and then which one?
or etc?

I thank you for the historical information which you gave me.

P.S. Trying to extend to small samples the second theorem of your paper: the logic of inductive inference, I find that in order that the estimate $T = H(X_1, \ldots X_n)$ which gives $1/(nV) = i$, should come from a function $H(x_1, \ldots x_n)$ independent of θ, it is necessary that f should be of a special form

$$f(x,\theta) = \exp\{\mu_\theta'(h(x)-\theta) + \mu(\theta) + g(x)\}$$

[with] μ, g, h arbitrary as long as they are such that $\int_{-\infty}^{+\infty} f\, dx = 1$. And then $T = \Sigma h(X_i)/n$. Among these functions f is for instance:

$$f(x,\theta) = (1/\sigma\sqrt{2\pi}) \exp\{-(x-\theta)^2/(2\sigma^2)\},$$

more generally, $f(x,\theta) = (1/\sigma\sqrt{2\pi})\, h'(x) \exp\{-(h(x)-\theta)^2/(2\sigma^2)\}$ and so on. (It had seemed necessary to admit that $H(x_1, \ldots x_n)$ be independent of θ, to get a rational theory of estimates).

Fisher to M. Fréchet: 26 February 1940

Thanks for your note of February 20th. I think logically the fiducial argument proceeds in three stages, setting aside for the moment my usual cautions about using the whole of the information.

1) A continuous distribution is found for T for samples of a given size drawn from a population having parameter θ. θ is then also a parameter of this distribution of T.

2) A relation is established between the true value and any percentile point T_p of the distribution of T. We shall suppose that this also establishes a univalent inverse relationship from which, given T_p, θ may be found. It is then true for all samples of the given size (or otherwise specified by ancillary statistics) that the inequality T exceeds θ will occur with given frequency when T and θ are mutually related as defined above.

3) In these circumstances *I think it proper* to refer to p as the fiducial probability that θ is less than T. This as it stands is a definition of the phrase *fiducial probability*. I believe it is, properly speaking, a probability, measuring as it does the relative frequency of one out of two or more well defined outcomes of a well defined procedure. I think it may be described properly as the admission of a new principle, if this phrase means, as I suppose it does, the thinking of a given situation in an unfamiliar way. Alternatively, I have no objection to regarding stages 1) and 2) as logical deductions, and 3) as an arbitrary definition. The definition is, however, a matter of choice and not a matter of chance.

The outstanding difference from inverse probability lies in the population of events of which the particular one, to which the probability refers, is regarded as a member. In the case of inverse probability this population is that of all samples of a given size, selected to have a given value for the estimate T, drawn by chance from a population which has itself been drawn by chance from a super-population having a given specification in respect of the distribution of the parameter θ.

The population of events referred to in fiducial probability consists of all samples of a given size drawn from any population defined by some value or other, θ. It is obvious that the frequency of a given event in members of this last population may be unequal to the frequency of the same event in the population considered in the theory of inverse probability.

I hope this will do something to clear up this knotty problem.

M. Fréchet to Fisher: 5 March 1940

Thanks for your interesting letter of Feb 26. We are near a point where you will understand in what consists my difficulty and I will be able to see exactly how you come to your conclusion. But not yet quite.

I fully appreciate the difference which you establish between inverse and fiducial probabilities—or at least I appreciate that there is between both an outstanding difference. What remains obscure for me is why you equalize, as a matter of course, two probabilities related to the same event (a given inequality) but computed in 2 different populations. You would be ready to admit this equality as a new principle. This would, in a way, get rid of my difficulty if this new principle would mean an assertion which might be admitted or denied by people who all accept the usual axioms of Calcul. of Prob. But as you consider this new principle as 'the thinking of a given situation in an unfamiliar way', I understand that this is not at all what I meant. Is it right to say that according to your explanation, the considered equalization is an unfamiliar but real consequence of the ordinary rules of Cal. of Probab.? If it were right I think it would be worth the trouble to show how it is a consequence.

You consider also an alternative. The fiducial probability as you have defined it would be an arbitrary definition and I quite admit that it is an 'arbitrariness' (?) limited by rational grounds. But does it mean that after the definition has been given we are not entitled to treat this f. probability as the other probab. (to say that it is defined in a specified population, that it obeys the theorems of total and compound probabilities and so on ?). If it means that, it would be necessary to complete the definition so as to be able to derive something of it in practice. If it does not mean that, if it is a probability like the other ones, a probab. of a given inequality in a given population and if we say that it is equal to the prob. of the same ineq. in a second given population, there is no more a new definition, but a new theorem which has to be proved starting from the ordinary rules of Cal. of Prob. or which has to be admitted as independent of these rules.

You are right in qualifying this problem as rather knotty but I am afraid that many statisticians will use wrongly fiducial probab. as long as all statisticians have not succeeded in agreeing about its real meaning.

Fisher to M. Fréchet: 8 March 1940

It was a pleasure to read your interesting letter of the 5th March. You there open up problems, the discussion of which would carry us far, and which I cannot easily attempt by way of correspondence. If someone would supply me with a list of the 'usual axioms of the calculus of probability', or the 'ordinary rules of the calculus of probabilities', I think you or I could, without much difficulty, ascertain whether the probability statements of the kind I call fiducial can be derived from them by a rigorously deductive process.

Now, I do not possess any such list, for an interesting reason which applies to much more of mathematics than our particular problem. Mathematics is the oldest discipline of the human mind, and mathematical truths seem to be the most enduring kind which the human mind can discover. They are not only as solid, but as precious as adamant, but the formulation of the axiomatic bases from which these truths might be derived is beset with difficulties which we cannot ignore, if only for the reason that the axioms have to be reformulated so frequently. Our palace of adamant rests upon foundations of gossamer which have to be renewed two or three times a week by the indefatigible labours of mathematical logicians, and yet the superstructure seems to be secure and quite habitable.

Certainly one can show, and easily, that fiducial probability, defined as I previously explained, satisfies the laws of multiplication and addition, in fact that it represents, in a well defined population, the proportion of events which belong to a well defined class.

M. Fréchet to Fisher: 11 *March* 1940

We are quite in agreement concerning the difficulty of ascertaining which are the true bases of each science. Though this last problem interests me*, I was simply thinking in my last letter of naive proofs as are given in the many papers on probability. Such a naive proof would quite satisfy me, if using the same principles as those which enable [one] to prove that

$$\{\text{Probability of } (\bar{x}-\mu)/s < t/\surd n \text{ when } \mu \text{ has a given value}\}$$
$$= A \int_{-\infty}^{t} (1+z^2)^{-n/2} dz,$$

it were similarly proved the different assertion that

$$\{\text{Prob. } (\bar{x}-\mu)/s < t\,[/\surd n] \text{ when } \mu \text{ is not given}\} = A \int_{-\infty}^{t} (1+z^2)^{-n/2} dz.$$

As you write that this last equality enables [one] to find the probability distribution of μ, I expect that perhaps the sentence 'μ is not given'means that μ is (like $x_1, \ldots x_n$) a random variable in the second equality.

[P.S.] *I will send you a reprint of my lecture in English at one Cambridge congress 2 years ago on the foundations of the Calc. of Prob.

Fisher to M. Fréchet: 18 *March* 1940

I think the naive proof that you want could run as follows: what we choose to call an 'event' consists of n values x drawn at random from a normal population. The mean of the population we shall designate by μ, the mean of the sample by $\bar{x}$, and the estimated variance by

$$s^2 = S(x-\bar{x})^2/(n-1).$$

These events may be divided into two classes, (a) the successful events for which $\bar{x}-\mu$ exceeds $st/\surd n$, and [(b)] the unsuccessful events in which it is equal to or less than $st/\surd n$. Considering the population of all events that occur, without selection, the probability of success is the 'Student' integral from t to infinity, e.g. t can be chosen so the probability of success is 5%, 2%, 1%, etc. So far this is pure mathematics, and, I believe, unexceptionable, Now, suppose we are confronted with a concrete sample for which we accept the belief that it has been drawn from a normal population of which nothing is known except what the sample tells us of the parameters μ or σ; we say 'here is an event which *may be legitimately regarded* as one chosen at random from the population of which the theory was investigated above'. The probability that it is a success is 2%. If this is so, the value of μ is less than a quantity which I can calculate from the sample. In this sense, therefore, the probability that μ is less than this quantity is 2%. The same argument applied to other percentile values gives a consistent series of values of μ, i.e. one in which μ increases when the percentage is increased. The statement that the event with which we are confronted is one chosen at random from all such events is therefore made a basis from which we deduce a frequency distribution for μ, and which may be called its fiducial frequency distribution. In relation to its frequency distribution μ is, properly speaking, a random variable whatever may be the physical origin of the sample considered.

The stipulation that nothing beyond the sample is known of the population sampled is relevant to our decision to regard the concrete sample as one chosen at random from the population of all such samples. It is, of course, only in the sense implied by that decision that the distribution of μ exists.

M. Fréchet to Fisher: 13 *April* 1940

Please excuse my delay in answering your letter of March 18th. . . .

I read your letter with great interest. However, I expected some detail not given in your publications and which would answer the question which I had put out at the very beginning. This question was and remains unanswered: how is it that in your proof, you equalize without comment, as a matter of course, the probabil. of one same inequality in two different populations?

I will try in §I to point out the point in your letter where this arises. In §II and III, I will raise two other questions which were always in my mind, but which I had formerly not stated or at least not clearly stated.

I. Your proof has 2 parts. The first one is, as you say, unexceptionable; but to prepare what I have to say on the second part, I would précise [*sic*] that your probability of success refers to a population where μ (though arbitrary) is given and fixed.

I should think that we both agree,—since you mention that the first part is unexceptionable,—that it is the second part which is difficult.

As a secondary remark, I would say first, that when you write

'here is an event which *may be legitimately regarded* as one chosen at random from the population of which the theory was investigated above'

I would replace 'from the population of which' by: 'from one of the populations, the theory of each of which'. But the capital point is in the following 3 sentences. There, to be quite clear, it is necessary to explicit [*sic*] each time the population which is spoken of, though it would be pedantic to do so in one of the usual reasonings on other matters. I insert in your text my addition: 'The probability that it is a success is 2%' (it refers here to the case where μ is *fixed*, though perhaps unknown, if it can be considered as proved). 'If this is so, the (fixed) value of μ is less than a quantity which I can calculate from the sample (which is random, so that the probability is computed before the sample has been drawn, otherwise it would be 0 or 1). In this sense therefore the probability that μ is less than this quantity is 2%' (provided μ is still fixed (though unknown) and the quantity be still random). 'The same argument . . . from which we deduce a frequency distribution for μ . . . .' Then we suddenly pass from the case where μ was fixed to the case where μ is random and we speak of *the* probability of $\bar{x}-\mu > st_0/\sqrt{n}$ (t_0 corresponding in 'Student's' distribution to 2%) as being 2% as well when μ was fixed—the case when 'Student's' dist. is proved—as when μ is random: a different case where 'Student's' distribution *has to be* proved.

II. The conception of the new population which is introduced, formed by all possible normal populations with arbitrary μ and σ, (the drawing of all elements of one sample being done in one (random) of these populations) is not an easy one; and the fact that, nothing being said about the way of choosing at random one of these populations (before drawing a sample in it), there is, however, a definite distribution of μ, may look strange.

III. When a proposition becomes wrong when one of its hypotheses is abandoned, it is always possible to find exactly the point of the proof where this hypothesis had to be assumed.

Now, I understand that you consider that your proof would be wrong if the sample would enter in the proof through inefficient statistics. In fact, you explain in your Paris lecture at the Soc. de Biotypologie [*CP* 156], that the *result* of the proof could not safely hold in that case; but you do not show where, *in the reasoning* itself, the efficiency of $\bar{x}$ and s enters. In no place of the proof quoted above, efficiency or information were mentioned, so that the proof applies or appears to apply to both cases.

The reader may get the impression that you formed your proof first, that you recognized after, that the result of the proof was not valid when inefficient statistics are utilized and that instead of revising the proof, you introduced a new hypothesis. I cannot say that this was your procedure, but that it is how your procedure looks to have been. Perhaps precisely in making explicit the point where this hypothesis appears you may let disappear the gap, or what appears to me the gap, mentioned in §I of this letter.

I wonder whether I have mentioned that by informing me of the existence of Prof. L. Jones' laboratory in Cambridge you increased considerably the value of my enquiry in England. Prof. L. Jones gave me very useful information.

Fisher to M. Fréchet: 18 *April* 1940

I have your letter of April 13th and mine of March 18th before me. As I stated the case in my letter, I do not think it is correct to say that in my proof I 'equalise without comment, as a matter of course, the probability of one same inequality in two different populations'. The only population which I consider in that proof is that of events as defined in the first paragraph.

The question I proceed to discuss is whether, when confronted with a concrete sample, we may legitimately regard it as an event chosen at random from that population. In that population I ought to insist that it is indifferent whether μ is the same from sample to sample, or varies from sample to sample, that is, from event to event, for samples from populations having any value μ are, in fact, members of the population of events defined. I therefore demur at your comment that the value of μ, though arbitrary, is given and fixed. I go on to suggest that, when we have a concrete sample for which we accept the belief that it has been drawn from a normal population of which nothing is known except what the sample tells us of the parameters μ and σ we may legitimately regard it as an event drawn at random from the population investigated. This step seems to me strictly analogous to that which is made in all applications of the theory of probability; for example, when a man whose personal and family history in no way differentiates him from the actuarial population accepted as healthy lives is accepted by a Life Assurance company at its standard rates. This is certainly an act of judgment, and not a deduction from any axiomatic basis, or at least from none that I should be prepared to put forward. Your amendment involving the words 'from one of the populations, the theory of each of which has been investigated' does not seem to me necessary, if the full extent of the original population is once grasped.

If I am not mistaken as to your meaning, this reservation applies to the other difficulties which you feel, which turn, I think, always on the idea that the population of events defined refers to drawings from normal populations having the same mean. The distribution of t is certainly independent of any possible variation in the parameters μ and σ.

Under section III of your letter you raise a distinct point, which I have discussed in some of my addresses on the subject, namely, why I reject analogous arguments based on inefficient statistics. This point is fundamental, and I think quite easily explained. It is that, whereas in deductive reasoning we may make any selection we please from our axiomatic basis, and, reasoning from the selected axioms only, it may be possible to derive certain rigorously justifiable consequences, it is a characteristic of inductive reasoning that the whole of the information available must be utilised. Every statistician is aware that by an arbitrary selection from his data, and the subsequent use of this selected portion as though it were the whole of the information available, he could make a show of justifying any number of false conclusions. The use of inefficient statistics appears to be indistinguishable from a selection of part out of the whole of the data available (see The logic of inductive inference. *J.R.S.S. 98*, 39–82 [*CP* 124]). Of course there is also a sense, elaborated by Neyman and his colleagues, in which formal inferences may be drawn from inefficient statistics, but such formal inferences when properly understood, or when fully stated, are seen to be irrelevant to the objects of scientific research.

M. Fréchet to Fisher: 24 *April* 1940

I will first reply to your last letter.

But there is no hurry to read this reply; whereas I would be much obliged if you could look at my P.S. appended on a *separate sheet* and either reply to this P.S. or send to me the *Metron* paper mentioned there, the soonest being the best.

Now to the thorny question where it seems to me that we are marking time both sides.

I. Concerning my observation about inefficient statistics, you repeat what I read in your printed papers, that is, a valuable *explanation* of how the *result* could not be right with ineff. statistics. But in my letter I warned that I had read this explanation and that what I would like to get was totally different, that was: in which part of the *proof* (composed of 2 parts) of the distribution of μ is introduced the hypothesis that the statistics are efficient?

II. I cannot agree with the statement that you only consider one population; and perhaps our difference comes from the fact that you think that I do not accept what I call your second population. Though, I had not first understood exactly what was this second population, when I thought that μ was still given, you have explained to me your view, substantiated by your paper on inverse probab.

Therefore, I do *not* reject the population mentioned in your last letter, where μ is random, when you come to the *second* part of your proof concluding with the distribution of μ. But, I understand that in the first part of the proof, the one which is unexceptionable, μ is given and fixed and this is where I find a first population $\neq$ [*sic*] from the second though the 2 proba. refer to the same event (the same inequality).

Perhaps one day we may clear the matter, and then I know very well that it will be useful to several statisticians and mathematicians beside myself.

P.S. You were kind enough to indicate to me the solution of the following question (avoiding to me the necessity of computing some multiple integrals already computed):

Let $x_1, \ldots x_n$ [denote] random values independently and normally distributed with some law, $s^2 = S(x_i - \bar{x})^2/(n-1)$, $\bar{x} = Sx_i/n$.

To find the distribution of

$$w = S\gamma_i(x_i - \bar{x})/s = S\gamma_i\, x_i/s,$$

where γ_i are constants such that $S\gamma_i = 0$.

You answered on January 17th that w has the distribution of $t\sqrt{S\gamma_i^2}$ where t has the 'Student's' distribution. Applying this result my assistant found some unexpected consequences and trying to find where was the difficulty, he noticed that the 'Student' t may vary from $-\infty$ to $+\infty$, whereas by Lagrange Inequality:

$$|S\gamma_i(x_i-\bar{x})/s|^2 \leqq S\gamma_i{}^2 S(x_i-\bar{x})^2/s^2 = (n-1)S\gamma_i^2$$

so that $|w|$ is always $\leqq \sqrt{n-1}\cdot\sqrt{S\gamma_i^2}$, from which it appears that there is something wrong somewhere.

I expect that it is due probably to some misunderstanding somewhere and easy to settle.

As you mentioned that your result can be found in one of your papers in *Metron* 1926, Vol. V, pp. 90–104, and as *Metron* is not easily accessible here, I should be much obliged if you were kind enough to send me (or lend to me) your *Metron* reprint.

(I have not forgotten that you expect me to send you back your paper on Inverse Prob., when I have no more to utilize it. Unless you require it sooner, I will keep it for some time).

Fisher to M. Fréchet: 26th *April* 1940

Looking at the P.S. of your letter of April 24th in order to give you a quick reply, I see that the paradox is partly my own fault. I had sent[1] the solution of the distribution of:

$$S(Yx)/s$$

where x is distributed normally about zero, without noticing that your problem with $S(Y) = 0$ introduces a restriction which diminishes the degrees of freedom by one. For the common form of the analysis of variance we then have

	Sum of Squares	Mean Square
1	$S^2(Yx)/S(Y^2)$	s'^2t^2
$n-2$	$S(x-\bar{x})^2-S^2(Yx)/S(Y^2)$	s'^2
$n-1$	$S(x-\bar{x})^2$	s^2

so that the ratio you enquire about may be equated to

$$S(Yx)/\{s\sqrt{S(Y^2)}\} = t\sqrt{n-1}/\sqrt{n-2+t^2}$$

the distribution of which is easily derived from that of t, being that of the sine instead of the tangent of an arbitrary angle. t has now, of course, $n-2$ degrees of freedom.

[1] See Fisher's letter of 17 January 1940 (p. 119).

M. Fréchet to Fisher: 14 *June* 1947

I send you separately a typewritten copy of my report on the 'Estimation of parameters' about the enquiry organized on that subject by the Intern. Inst. of Statistics. This report shall be presented at the September Session in Washington of this Institute and followed by discussion.

As I tried, in pages 35–36, to explain your position, I think best to send this copy to you, because if your answer would change my mind, I might still introduce corrections when I receive the proof sheets.

Furthermore, I wonder whether you are aware of criticisms of fiducial probability which have been published by such eminent scientists as Gini and Serge Bernstein and von Mises. I am quite convinced that their reasonings are rigorous and that the difficulties lie in what they have understood that fiducial probability is.

As even those who think they closely follow you have been mistaken (you remember perhaps those few lines of Deming and Birge which you repudiated[1]), I think it would be most important if you would scrutinize the papers by these three colleagues (of so high standing) and show where the discrepancy happens. . . .

[1] See Fréchet's letter of 8 January 1940 (p. 118).

Fisher to M. Fréchet: 21 *June* 1947

Thank you for your letter and courtesy in sending me your critical commentary. I shall not be inclined to argue the matter, and no doubt the great statisticians whom you mention will discover in time how the matter should properly be expressed. It seems to me obvious that serious mental obstacles must always have existed to the apprehension of a form of reasoning which seems to me new and valuable, and that they should exist in illustrious minds trained in earlier ideas is not at all to me surprising. I should be glad if anyone reading my works should take what good they may find in them and make good use of it, and trouble themselves little about what they think to be false or defective.

M. Fréchet to Fisher: 18 *October* 1951

Vous savez que j'ai la plus grande admiration pour l'ensemble de votre oeuvres. Vous savez aussi que je ne suis pas toujours d'accord avec vous sur les détails.

Ce sont sur quelques-uns de ces points que j'ai attiré l'attention dans le Rapport au Congrès de Washington de 1947.

Je préfère vous les indiquer moi-même au moment où ce rapport vient de paraître, plutôt que de vous laisser les apprendre indirectement et beaucoup plus tard.

Vous avez sans doute reçu comme moi ces jours-ci le volume contenant ces rapports 'Vol. III, Part A, *Proc. Intern. Stat. Conf.*, 1947, Washington'. Les passages où des observations vous concernent sont aux pages 373, 4e ligne; p. 380, dernière ligne; p. 381; p. 378.

Certains auteurs consultés dans mon enquête, font des objections qui vous concernent, en particulier, von Mises: p. 390.

J'ai tenu à vous mettre au courant, mais bien entendu, il ne m'appartient pas de dire s'il convient ou non que vous répondiez à ces observations.

Fisher to M. Fréchet: 23 *October* 1951

Thank you for your considerate letter of October 18th.

I am not much tempted to enter into controversy, since I have been much impressed now for many years with the number of highly intelligent men in mathematical departments, masters for the most part of an exquisite mathematical facility, who have blundered quite dreadfully in trying to rehandle the mathematical methods which have a purpose to serve in Statistics. Where I have been consulted in time, I have been able on many occasions to dissuade authors from making extravagant claims under a misapprehension of the nature of the problem, but this has not always been possible, and I have come to the conclusion that, given time, and zealous emulation, most of the errors committed in ignorance will clear themselves up, though in a language difficult for the applied statistician to understand.

M. Fréchet to Fisher: 3 *November* 1951

Vous admirez Laplace et pourtant vous n'êtes pas toujours d'accord avec ses écrits.

C'est dans le même esprit que tout en admirant l'ensemble de votre oeuvre et constatant combien vous avez fait progresser la statistique mathématique, je ne puis *parfois* vous suivre dans vos conclusions.

Voici un point que je soulève dans mon rapport et sur lequel je *crois que vous me donnerez raison*. C'est un point *d'histoire*.

Si j'ai bien compris la page 528 de votre mémoire: Inverse Probab. (*Proc. Cambridge Phil. Soc.*, vol. XXVI), vous avez été mal informé. Il vous suffira de lire la citation de Laplace ci-jointe pour voir que Laplace n'a pas ignoré, tout au contraire, le cas des probabilités a priori *inégales*. J'ai en outre, dans mon rapport, donné d'autres références sur ce sujet, p. 372–373.

Il reste un point sur cette question, qui peut être débattu. D'après von Mises et d'après Molina, Bayes lui-même a ignoré ce cas. Il reste possible que Bayes l'ait écrit dans un passage qui leur aurait échappé. Peut-être pourriez-vous le faire connaître? Jusqu'à ce que cette preuve soit donnée, je propose donc d'appeler la formule complète, formule de Bayes–Laplace.

Dans votre dernierè lettre, vous dites que des mathématiciens ont commis des 'blunders' en s'occupant de statistique mathématique. *Je suis tout prêt à l'admettre*, et je crois qu' un mathématicien qui commence à s'occuper de statistique mathématique a tout intérêt à consulter des 'applied statisticians'. Toutefois, *il y a réciprocité* et les erreurs, même non mathématiques, ne sont pas toujours du même côté. Vous vous souvenez peut-être que je vous avais signalé un passage d'un mémoire, d'ailleurs, en général, bien fait, par Birge et Deming et que vous aviez été *d'accord avec moi* pour reconnaître que ce passage renfermait une dangereuse erreur d'interprétation. Pourtant Deming est un 'foremost' 'applied statistician'.

En résumé, et je crois que sur ce point aussi vous serez d'accord avec moi, statisticiens et mathématiciens ont intérêt à se contrôler et à s'aider mutuellement.

[Enclosure]

On a parfois écrit que Laplace avait réduit la formule de Bayes au cas où les probabilités sont égales.

Monsieur Itard signale dans *l'Essai Philosophique sur les Probabilités* de Laplace, 6ème édition, 1840, page 264, le passage suivant, où nous soulignons une phrase, qui suggère une conclusion exactement opposée: 'Bayes dans les *Transactions philosophiques* de l'année 1763, a cherché directement la probabilité que les possibilités indiquées par des expériences déjà faites sont comprises dans des limites données; et il y est parvenu, d'une manière fine et très ingénieuse, quoique, un peu embarrassée. Cet objet se rattache à la théorie de la probabilité des causes et des évènements futurs, conclue des évènements observés; théorie dont j'exposai, quelques années après, les principes, *avec la remarque de l'influence des inégalités qui peuvent exister entre les chances que l'on suppose égales.*'

Fisher to M. Fréchet: 19 *November* 1951

. . . I still feel somewhat strongly that, as I said at that time, 'it is not to be lightly supposed that men of the mental calibre of Laplace and Gauss . . . could fall into error on a question of prime theoretical importance without an uncommonly good reason' [*CP* 84]. The reason, to which later in this paper I ascribe the contradictions which are historically unmistakeable, is the assumption that uncertain inference of all kinds, irrespective of the logical situation in which it is attempted, can be adequately expressed in terms of the single concept of mathematical probability.

I should submit for your consideration now, that if it were indeed true that this single concept were adequate for all purposes we should confidently expect that definitions of probability should have become more exact and better understood with the progress and study of this subject. In a recent and not unintelligent book, however, (*Probability and the Weighing of Evidence*, by I.J. Good) five very distinct meanings of the word 'probability' are found necessary for the discussion.

The procedure which I have preferred in face of a situation of this kind is to choose the oldest clear and useful definition that I could find, namely that of

Bayes, based on, and appropriate to, expectations in games of chance, and in the case of other concepts distinct from this, yet which may seem sufficiently analogous to cause confusion or which are equally relevant in processes of inductive inference, I have preferred to seek for distinct appropriate names, of which 'likelihood' and 'quantity of information' may serve as examples.

The passage of Laplace to which you kindly draw my attention is one which I had occasion to quote some years ago in addressing the Tercentenary Conference at Harvard [*CP* 137]. I think it does show that Laplace was not prepared to appreciate the caution and scepticism which actually prevented Bayes from publishing his treatise during his lifetime. It was, as you know, published posthumously at the instance of his friends.

H. Gray to Fisher: 15 *June* 1951

The bearer of 'news', and the candid friend, make themselves unliked; but I am willing to take the risk. The *Scientific Monthly* may not come under your notice, being aimed at laymen, *Science* at professionals; both publ. by Am. Assn. Advancement of Science. Also you may not care to notice Neyman's notions.[1] If it annoys, forgive me.

[1] Neyman's review of *CMS* in *Sci. Month.* **72**, 406–8, (1951).

Fisher to H. Gray: 2 *July* 1951

The second word of the phrase 'candid friend' is sometimes not taken very seriously. If I mistake not, however, you mean it to be in this case. Hence what follows:—

Neyman is, judging by my own experience, a malicious mischief-maker. Probably by now this is sufficiently realised in California. I would not suggest to anyonc to engage in scientific controversy with him, for I think that scientific discussion is only profitable when good faith can be assumed in the common aim of getting at the truth.

There are, however, in the review two points in which he seeks to impugn my own good faith, and in such a case it may have seemed to you a friendly office to show that some of my friends in California do not share his opinion.

(a) The use of Euclidean hyperspace of N dimensions to represent a random sample of any size N. Before ascribing this innovation to me, Mahalanobis[1] had looked into the matter. His claim, of course, is not that I invented Euclidean hyperspace, but that this particular use of it enabled me, starting in 1915, to give the exact solutions of a number of problems of distribution, which had been unsuccessfully sought by other means. In 1900 Pearson had used hyperspace to represent the correlational properties of a number of different variates. Each variate was given one dimension. This is not the same as discussing a finite sample and assigning a dimension to each observation. If the latter process had occurred to Pearson fifteen years earlier, it would be curious that neither he, nor the abler mathematicians (e.g. Sheppard and Soper) associated with him at the time, had found the solutions they were seeking; which indeed are nearly obvious, and come out at once, when my method is used.

I can see no source for Neyman's imputation save a desire to be spiteful. His whole exposition of tests of significance is based on my 'sample space', but such indebtedness he is not inclined to admit. Likewise his 'power function' is, in disguise, a special application of my Likelihood.

(b) (see *J.R.S.S. 98*, p. 77)[2] Edgeworth's paper[3] of 1908 has, of course, been long familiar to me, and to other English statisticians. No one could now read it without realising that the author was profoundly confused. I should say, for my own part, that he certainly had an inkling of what I later demonstrated. The view that, in any proper sense, he anticipated me is made difficult by a number of verifiable facts. (i) He based his argument on Bayesian inverse probability; my results are free from this assumption and represent an entirely different approach. (ii) He ends his paper by explaining that he has been writing only of measures of central tendency, and not of 'the fluctuation', nor, it may be presumed, of measures of correlation, etc. (iii) The formula common to his work and mine, that for the variance of an efficient statistic, he obtained from Pearson and Filon (1898),[4] who believed it to give the sampling variance of statistics obtained by the method of moments. Already in 1903, using a direct method due to Sheppard, these erroneous values had been corrected, but Edgeworth does not refer to this, and may not have known of Sheppard's work. Obviously Sheppard's calculations had raised the questions:— Had Pearson and Filon's variances any validity at all? Does any class of estimate actually have these variances? If so, how can such an estimate be obtained in general? But Edgeworth would have been far ahead of his time had he asked them.

Setting aside the disputed postulate of inverse probability, and having regard to parameters of all kinds, irrespective of what quality of the population they measured, I showed in 1922 that statistics could in general be found with variances given by this formula, and that it was the least limiting variance possible.

Anyone wishing to ascribe my results to Edgeworth should at least have ascertained that he accepted them, which so far as I know he never did.

[1] Mahalanobis, P.C. (1938). Professor Ronald Aylmer Fisher. *Sankhyà* **4**, 265–72.

[2] Fisher's reply to discussion of his 1935 paper, 'The logic of inductive inference' (*CP* 124). See *Collected papers of R.A. Fisher*, Volume III, p. 310.

[3] Edgeworth, F.Y. (1908). On the probable errors of frequency constants. *J. R. Statist. Soc.* **71**, 651–78.

[4] Pearson, K. and Filon, L.N.G. (1898). Mathematical contributions to the theory of evolution. IV. On the probable errors of frequency constants and on the influence of random selection on variation and correlation. *Phil. Trans.* **A 191**, 229–311.

Fisher to R.F. Harrod: 3 *December* 1956

My new book this year, *Statistical Methods and Scientific Inference*, coincides in time so closely with yours[1] that I was more than a little anxious to ascertain what point of view it was that you were expressing, and as I think I indicated, I found from such soundings as I have made that I was, allowing for differences in the problems being discussed and occasionally in the meanings attached to the words used, very much indeed in agreement with your point of view. When you care to glance at mine I hope you will be amused at the indications which I try to give, from several different points of view, of the reasons why mathematicians have cut such a comparatively poor figure in this business.

The two books seem to me to be frankly complementary. You are developing the subject in the philosophic tradition, and I almost exclusively in the mathematical; namely, I am concerned with the exact quantitative treatment of particular cases in which uncertain inference seems to be possible, with the specification for such cases for both the nature and the extent of the uncertainty. The book is sprinkled with warnings against hasty generalizations.

Quite naturally we use the word 'probability' rather differently, I with my eye fixed on those noble and ardent gamblers for whom Montmort and de Moivre were writing, and I am therefore concerned with, *senso strictu*, Mathematical Probability as developed by these mathematicians. Keynes's phrase 'measure of the degree of rational belief' obviously covers a much wider field which would shatter the old bottle of mathematical probability. It is, however, an admirable phrase, and I show that in many instances objective quantitative measures of this kind occur in situations where any statement of Mathematical Probability would be invalid.

Somewhere I noticed you turn aside to stick a pin into the word 'likelihood'. I should like to leave it to you to judge, after reading what you care to of my book, whether the use of this term is really unsound.

[1] Harrod, R.F. (1956). *Foundations of inductive logic*. Macmillan, London.

R.F. Harrod to Fisher: 8 *January* 1957

I have read through your book once. I curse myself for not having insisted on being taught more mathematics at school. It is difficult to pick these things up late in life. I cannot pretend that I have understood more than about half the book at most. But there are matters on which I hope I may still gain an understanding through better acquaintance with some of the forms of art. I am ordering *Statistical Methods*. Also I want to understand more about the 'fiducial argument' which is central to so much in your book. According to the refs. I think I ought to read your 1930 article.

I think I was in sympathy with the greater part of your thought—and much enjoyed your controversial passages. And I naturally rejoiced when I found, as I sometimes did, passages which almost echoed certain things in my book.

If you have read that, I fear you will be deeply shocked at my crudity and sloppiness whenever I approach the fringe of mathematical problems. My impulse, as you say, was a philosophical one. You may not realize how completely the philosophical world has given up induction in despair. Philosophers teach their pupils that the validity of inductive reasoning can in no wise be established; we are bidden to admire the *de facto* success of scientists and to take their methods on trust. They are to be justified solely by their pragmatic success. That would incidentally put your work out of court completely! You are under an illusion, they must suppose, if you think that by studying the logic inherent in certain methods, you can actually improve those methods. For according to them there is *no* logic inherent in them.

At Oxford the majority of clever boys not reading Maths or Natural Science do philosophy, whether in Greats (Classics leading on to modern philosophy) or in Modern Greats (Philosophy, Politics and Economics). All these are sent out into the world fully convinced that induction is a bit of fraudulent trickery, which just happens to work. I think this bad. It means that this large group, who have some influence on our affairs, are basically anti-rational.

Philosophy is in the same state everywhere. The only philosopher I can find who believes that there is any validity in induction is the man, Williams of Harvard, to whom I have referred. Unfortunately his book is not a very good one, containing much rather loose controversializing.

I believe that your fiducial argument, if only I could grasp its essence aright, will be found to require something like my fair sampling postulate. The philosophers are up in arms saying that we have no right to suppose with any degree of probability, however low, that our successive experiences do not consist of events which, given the real condition of nature, are excessively improbable. That is why I think my 'principle of experience' to be of real logical value, *viz.* the principle that one is not likely to be on the edge of a great surface. For this has the happy consequence that it does not matter whether the fair sampling postulate that we assume in making inductions is true or not.

I am sorry about my footnote on likelihood! What I had in mind was philosophical writing in which the distinction was between objective and subjective, or, if not subjective, then at least purely intuitional probability. If I have understood aright, your likelihood is fully as objective as probability.

The passage where I felt least happy was on pp. 41–2.[1] I wholly agree with your remarks on p. 110[2] about confusion in regard to infinity and the limit (and have said something similar). But I do want to define probability as arising from data of a certain *logical* character (e.g. a sample of a certain size and sort) that ensures that beliefs arising by inference from the data shall be more often right than wrong. This seems to imply a set of occasions on which one is confronted with data of that logical character. And I would *prefer* not to assent to your indifference (p. 109)[3] as between object, event and proposition as the predicand of probability statements; the objects or events you have in mind can presumably always be converted into contents of propositions. I do want to keep probability as a logical relation, *viz.*, between data and the inference from them. Incidentally I have studiously avoided any reference to 'propositions', always writing in terms of 'beliefs', so as to avoid being bogged down by the symbolic logicians and their tortured paradoxes. . . .

[1,2,3.] Pages 44-5, 114, and 113 respectively, in *SMSI* .

Fisher to R.F. Harrod: 15 *January* 1957

It was nice of you to write me such a long letter. For my part I am quite as lost in philosophical terms as you can be in mathematical. In different languages we do seem to be doing very much the same thing, you being primarily concerned at a much deeper level, and in more qualitative terms, with what I am concerned with in trying to give an exact mathematical formulation of the steps from data to inference in cases where, *more mathematicorum*, I choose carefully what sorts of data I am prepared to discuss, and so am working at a more accessible level.

Both of us are building, I believe, fundamentally, on the logical force of what I call a test of significance, an inferential force which I describe as more elemental than that of a statement of probability; and here I am using 'probability' in a stricter, i.e. less inclusive, sense than Keynes' 'degree of rational belief', because I am concerned to distinguish the historical Mathematical Probability from various other measures of rational belief appropriate to circumstances in which no mathematical probability can be asserted. This seems to be a rather remarkable conformity of intention, scarcely at all frustrated by the differences attached to the meanings of the words used.

May I take now some of the interesting points you raise, not that any of them seem to involve a fundamental difference but because their clarification may well lead to a better understanding.

Thank you for your corrections; I am returning a few that I have found, including some improvements in the wording that have occurred to me. The only one of yours needing discussion is where you refer to the statement on page 19 [*SMSI*], ' The expectation from a mating between two heterozygotes is 1 homozygous black, to 2 heterozygotes, to 1 brown'. My intention in saying this is that the numbers enumerated should be taken out of their total of four, so that 1/4 are said to be homozygous black, 1/2 heterozygous black, and 1/4 brown. I expect the way I put it is elliptical and should be expanded. Its substance, however, is merely to lead to the justification of the prior probability ratio of 1 homozygote to 2 heterozygotes for a black mouse from such a mating. I agree entirely that probability, either in the narrow sense of the eighteenth century gamblers or in any wider philosophical sense you may prefer to use, refers to strength of rational belief properly to be inferred from given data. On page 41–42,[1] however, I am trying to say that we also have rational grounds for belief in many cases in which probability (or frequency) statements cannot properly be made, and that this is typical of simple tests of significance, from which, in my narrow sense, no probability statements can properly be inferred, although I do point to a somewhat peculiar class of cases in which, not by a simple test of significance but by a more sophisticated fiducial argument, true probability statements are available. They do not in my opinion make the inference much stronger, though they do make it somewhat stronger in respect to the specification of precision as elaborated in Section 4, as contrasted with, for example, that of the likelihood graphs on page 72.[2]

I agree with you and with Keynes that the predicand of a probability statement can always be taken to be a proposition, but I would plead that the popular use which applies the term to objects or events is also correct and sufficiently vivid to supply a warning of the kind that I think Keynes needed in the remark I quote on page 44.[3]

I do not think I have referred much to propositions, but I have often used the term 'assertions' where you use 'beliefs', because I am not concerned with beliefs in themselves and as such, but with their communication to other rational minds.

[1,2,3.] Pages 44–5, 76, and 48, respectively, in *SMSI*.

W.E. Hick to Fisher: 6 *October* 1951

Several years ago I ventured to ask you some elementary questions about statistical inference, and your kindness then has encouraged me to approach you again, on a matter which, though elementary, I feel is of some practical importance.

The question has to do with the theory of significance tests. Most experimenters, of course, have neither the time nor the inclination to go deeply into the theory of critical regions, power functions, etc. Yet many (of those who think at all!) are puzzled about when to use 'one tail' or 'two tails' and why one should compute the probability of a sample 'as unlikely as, *or more unlikely than. . .*'. Involved in all this is our *bête noire*—chi-squared, with its deceptive simplicity.

The question the experimenter thinks he is asking himself is surely, 'What is the probability, on the chance hypothesis, that I should have got the result I did get?' It is difficult to tell him straightforwardly why he should care about the probabilities of all the other, less likely, results that he did *not* get. . . .

Fisher to W.E. Hick: 8 *October* 1951

I am a little sorry that you have been worrying yourself at all with that unnecessarily portentous approach to tests of significance represented by the Neyman and Pearson critical regions, etc. In fact, I and my pupils throughout the world would never think of using them. If I am asked to give an explicit reason for this I should say that they approach the problem entirely from the wrong end, i.e. not from the point of view of a research worker, with a basis of well grounded knowledge on which a very fluctuating population of conjectures and incoherent observations is continually under examination. In these circumstances the experimenter does know what observation it is that attracts his attention. What he needs is a confident answer to the question 'Ought I to take any notice of that?'. This question can, of course, and for refinement of thought should, be framed as 'Is this particular hypothesis overthrown, and if so at what level of significance, by this particular body of observations?'. It can be put in this form unequivocally only because the genuine experimenter already has the answers to all the questions that the followers of Neyman and Pearson attempt, I think vainly, to answer by merely mathematical consideration. In fact the practical experimenter does not often put up a damn-fool test of significance but it is a labour of many years and much art for the 'Theory of Testing Hypotheses' to avoid such tests.

Fisher to H. Hotelling: 3 *October* 1941

I have just read your letter of September 5th with pleasure, for it is always a pleasure to have a letter from you. I am very glad you noticed E.B. Wilson's article,[1] for he is a man of such generally sound judgment that one could not have expected that the test of significance he advocates for the four-fold table would have appealed to him. The point is, however, really rather a knotty one, for he might have adduced as the ground for choosing $p = \frac{1}{2}$, not that it was the value observed for the total of experimental and controlled animals, but that this value maximises the aggregate probability of the six possible outcomes of the experiment in which the surviving controls exceed the surviving experimental mice by four or more. Such a criterion would, of course, work out differently in other cases, and would be extremely troublesome to apply, but it would, I think, also be extremely troublesome to argue against.

With respect to Behrens' test, I should agree with you that Behrens' own statement seems extremely confused and unconvincing. I should not rely on it without developing an independent argument leading to the same result. I should like to be clear, however, that, far from ignoring the connection between t and s in samples from populations having a given value of σ, I regard this connection as essential to the development either of Behrens' or of 'Student's' test. Given σ, in fact, the distribution of t is not independent of s, and what is wanted is the product ts, the distribution of which depends on σ only. It is only because the Gaussian test, using the probability integral of the normal curve, is not available for rigorous purposes when σ is not known, that the observable quantities t and s come into consideration.

The logical development of the rest of the argument depends, so far as I can see, on three points, two of which are mathematically demonstrable, while the third, which justifies their relevance, is a matter of judgment.

(a) In random samples of n' from normal populations the ratio s/σ has a distribution dependent only on n', and therefore known.

(b) All other information supplied by the data is irrelevant to the value of σ. This is the sufficiency property.

(c) (which is, I think, undoubtedly incapable of mathematical proof). The sample under examination is unbiassed in respect of the values $\bar{x}$ and s obtained, so that for the population of possibilities to which our probability statement shall refer, the ratio s/σ takes all its possible values with the theoretical frequencies, independently of $\bar{x}$, or the fact that we can use the mathematical short cut of calculating the distribution of $\bar{x}/s$, and using in probability statements probabilities derived from its distribution. This act of judgment in the logical basis of 'Student's' test has, I think, obscured the issue to mathematicians much more than to practical experimenters. I think, however, that an analogous act of judgment enters into all applications of probability theory.

Of course this is, as Behrens' test shows, incompatible with any theory of tests of significance, such as that of Neyman, which makes it axiomatic that a given percentage of the observations in repeated trials shall be found to be significant. It allows for the fact that some of the unknown parameters may have the effect of making the test more or less sensitive, so that on fixing one of these we should find that the percentage of cases recognised as significant differed from the percentage with which the test of significance was to be entered. As I understand it, however, a test of significance is a logical statement to the effect that either a well defined event having a rarity definable in terms of the observations has occurred, or else that a certain hypothesis is untrue. Events the rarity of which can only be defined in terms of unknown parameters cannot in this sense provide tests of significance.

I hope you will write further about all this.

[1] Wilson, E.B.(1941). The controlled experiment and the fourfold table. *Science* **93**, 557–60.

Fisher to J.O. Irwin: 8 *November* 1935

I am glad you have given me an opportunity of reconsidering the arguments in the neighbourhood of page 721 of my paper on Statistical Estimation [*CP* 42]. The main point, which I think only becomes clear when the paper is read as a whole, is that we are concerned with the difference between two systems of equi-statistical surfaces or regions, and usually not with the difference between different regions of the same system. I am returning your letter for reference, together with a few sheets which I hope clear up the main difficulty, though I don't discuss the analytic handling of the variance of the second degree deviations between corresponding surfaces of the different systems. I will go into this with pleasure if you would like me to.

[Enclosure]

1. The origin of the loss of information incurred by using an estimate from a sample, in conjunction with the sampling distribution of all such estimates from samples of a given size, has been traced to the fact that the equistatistical surfaces on which $\partial L/\partial\theta = 0$ for different values of θ do not coincide with the system of surfaces on which $\partial L/\partial\theta = C$ for any one value of θ. The systems clearly have one surface in common, and the neighbouring surfaces of the two systems will generally be inclined to each other at only small angles; e.g. if $\hat\theta - \theta$ is small, the surface of the first system $\partial L/\partial\theta = 0$ may be chosen, with a certain harmless arbitrariness, to correspond with one of the second system, such as that which meets it on the locus of expectation

$$\frac{\partial L}{\partial\theta} = S\left\{\frac{\hat m}{m}\,\frac{\partial m}{\partial\theta}\right\}.$$

Then my assertion is that the angle between these corresponding surfaces of the two systems will, in the neighbourhood where $\hat\theta$ is near to θ, be proportional to the difference $\hat\theta - \theta$. The equations of the two surfaces are in fact

$$S\left\{\frac{x}{\hat m}\,\frac{\partial \hat m}{\partial\theta}\right\} = 0,$$

and

$$S\left\{\frac{x}{m}\,\frac{\partial m}{\partial\theta}\right\} = S\left\{\frac{\hat m}{m}\,\frac{\partial m}{\partial\theta}\right\},$$

and the coefficient of x in the first may, if the differential coefficients of log m are finite, be expanded as

$$\frac{1}{m}\,\frac{\partial m}{\partial\theta} + (\hat\theta - \theta)\frac{\partial^2}{\partial\theta^2}\log m + \frac{(\hat\theta-\theta)^2}{2}\,\frac{\partial^3}{\partial\theta^3}\log m + \dots$$

Using the convention that the square of the cosine of the angle between surfaces

$$S(px) = C,$$

and

$$S(p'x) = C',$$

is

$$\cos^2\alpha = \frac{S^2(mpp')}{S(mp^2)S(mp'^2)}$$

one can obtain

$$\sin^2 \alpha = \frac{(\hat{\theta} - \theta)^2}{I} S \left\{ \frac{1}{m} \left(m'' - \frac{m'^2}{m} \right)^2 \right\} + \text{smaller terms.}$$

2. It being agreed that we have to evaluate the variance of $\partial^2 L/\partial\theta^2$ over regions for which $\partial L/\partial\hat{\theta} = 0$, the question is whether it is sufficient to evaluate it within the regions for which $\partial L/\partial\theta = C$. Might one not argue as follows? The variance within any one of the regions $\partial L/\partial\hat{\theta} = 0$ will be the same as that evaluated, together with a corrective term proportional to $\hat{\theta} - \theta$, and other terms of higher order. Averaging all these according to their frequency of occurrence in the limit for large samples, we obtain a corrective term proportional to the variance of $\hat{\theta} - \theta$, or of order $1/n$ compared with the value obtained.

I think this also covers the logical though not the analytical difficulty on your p. 3.

With respect to p. 4 one might, perhaps, put the position in this way:— By basing our notion of precision only on a single estimate and the size of the sample, we are not neglecting such information about precision as our estimate provides, but we are neglecting other information provided by the sample respecting the precision of an estimate of given magnitude derived from it. Common formulae for standard errors, for example, are of the form

$$\sigma = f(\theta, n).$$

Consequently, variations in precision associated with variations in θ are taken into account through the magnitude of our estimate, but variations in precision for samples of the same size and giving the same estimate are neglected.

Fisher to A.T. James: 4 *October* 1961

It was nice to get your letter which I found waiting for me when I got back here yesterday.

I shall soon be sending out two papers from *Sankhyā* which indeed you may already have seen, extending to other problems the method by which Behrens' test was justified in 1935. I think you will be interested in the derivation of the simultaneous distribution of the variances, when the means are known to be equal. I do not know how, otherwise, this distribution could have been obtained. Of course, all three problems are inaccessible to Neyman and Pearson's approach to testing hypotheses.

I have not done anything further with the problem of the simultaneous set of correlations among t normal variates. The simultaneous distribution is not difficult, and I ought to publish it, but I have not fished it out since coming back, and it may not be invertible, though I feel it ought to be. The example I take up first shows that Exhaustive Estimation is not strictly necessary, and it may not indeed be sufficient, though it would be surprising if it were not. The example with two variates shows that it is not too slick analytically.

I think consultant work is immensely important, but on condition that the consultant is concerned to learn, and not merely to pontificate.

A.T. James to Fisher: 14 *January* 1962

. . . Your discussion in 'Sampling the reference set' [*CP* 284] made it clear to me that in the test of the difference of means with unknown variance ratio, the distribution of

$$d = (\bar{x}_1 - \bar{x}_2) \Big/ \sqrt{\frac{s_1^2}{n_1 + 1} + \frac{s_2^2}{n_2 + 1}}$$

in the conditional distribution given s_1/s_2 averaged over the fiducial distributon of σ_1/σ_2 is the same as the distribution of

$$\left(\frac{s_1 t_1}{\sqrt{n_1 + 1}} - \frac{s_2 t_2}{\sqrt{n_2 + 1}} \right) \Big/ \sqrt{\frac{s_1^2}{n_1 + 1} + \frac{s_2^2}{n_2 + 1}}$$

. . .

Fisher to A.T. James: 13 *March* 1962

Your letter of Jan 14 has got around to me at last.

I am glad you liked the paper on Sampling the Reference Set. The business you call averaging out the distribution of σ_1/σ_2 was only an experimental way of putting the fact that the distribution of σ_1/σ_2 is exhaustively estimated by s_1/s_2 independently of $\bar{x}_1$, $\bar{x}_2$ and on this marginal distribution is superposed the array distributions of μ_1 and μ_2 for each σ_1/σ_2 that can be encountered. What I liked best and what rather surprised me is that the same experimental mechanism should give the simultaneous distribution of σ_1 and σ_2. I wish one could always set up the reference set so explicitly. . . .

I am not so sure that the tide of real feeling in the U.S. is so backward as it was 15 years ago, when so many leading posts fell to Neymanians. I think a good many are dissatisfied with mere 'decisions', which essentially evade the problem of specifying the nature of uncertainty, or the true grounds for belief. . . .

Fisher to G.S. James: 13 *August* 1955

Thank you for your letter.[1] It has been interesting to me to see the recent development of interest in Behrens' solution to this problem, due, I suppose, to recognition that the grounds on which it was at first attacked, principally by Bartlett under the influence of J. Neyman, are no longer tenable, for I suppose few except Egon Pearson now believe that the strength of the significance under which a composite hypothesis can be rejected is to be identified with the frequency with which such evidence is obtainable, or indeed that the various simple hypotheses comprising the composite will necessarily yield the same frequency. As a by-product, however, of this misunderstanding, a number of efforts have been made to discredit the mode of argument which I have termed 'fiducial'. I do not believe that practical men have ever had any difficulty in understanding, or doubts about using, Behrens' test, and during the current year I have been notified of new tabulations to facilitate its use. There is a good deal of material for this in tables 3 and 4.

[1] James had written asking for a copy of Fisher's 1941 paper, 'The asymptotic approach to Behrens' integral, with tables for the *d*-test of significance' (*CP* 181).

Fisher to G.S. James: 16 *August* 1955

Thank you for your long letter, which, however, broaches many topics beyond those of which I wrote. Returning to the content of my letter of August 13th, I gather you do believe that the strength of the significance under which a composite hypothesis can be rejected is to be identified with the frequency with which such evidence is obtainable.

I had thought that few other than Egon Pearson accepted this view, to which there are a great many counter-examples. It would therefore be useless to argue the matter. However, with respect to what you say on fiducial probability, I see that you have been misled partly by early misapprehensions of my own, because for a time I imagined that the logical content of a statement of fiducial probability could be distinguished from that of a classical probability, in the sense of Bayes and the early writers. I had also not sufficiently emphasized that fiducial probabilities can be demonstrated by an entirely rigorous step-by-step argument which excludes all possible ambiguities. However, I am averse to controversy in print or in letters.

Fisher to H. Jeffreys: 23 *February* 1934

I understand that the Royal Society are prepared to publish two articles from yourself and from me, on the Principles of Inductive Inference, on condition that we see each other's contributions before the articles are accepted. I enclose a short note[1] which gave rise to this decision, which embodies the arguments that weigh with me on some of the topics that may be raised. I should be glad to have it back for a revision with the first draft of your contribution.

[1] Eventually published as *CP* 109.

H. Jeffreys to Fisher: 24 *February* 1934

Thanks for loan of your paper. I have glanced through it but of course have not yet had time to make up my mind about all the points. However my main reaction is that the difference between you and me on any questions that concern you directly is not great. Wrinch and I showed in our original *Phil.*

Mag. paper[1] that in sampling problems, provided the sample is large, that for a wide range of distributions of the prior probability the posterior probability is practically proportional to the p. of the sample given the whole class, and therefore to your 'likelihood'. Consequently allowance for prior probability in these matters makes little difference except in freak cases, as when the sample consists of five white balls in a bag we knew to contain 100 in all of which just 5 were known to be white. Your expressions in ordinary cases are good approximations to the posterior probability; if I claim in one case to have got a slightly better one I don't see that you need worry. If in your original *Phil. Trans.* [*CP* 18] you had referred to this result of ours as providing an alternative justification of your results you would have been on strong ground. I notice that in your paper you don't defend your statement that the ratio of two infinite numbers has a definite value; and this is my crucial objection to your point of view.

My point is that we need a theory that will allow us to draw inferences in cases where sampling methods do not apply, and it must have two properties. It must make it possible to assimilate new experimental knowledge; and it must give us ground for believing the laws that we do believe or else say definitely that our inferences are fallacious. You would accept the first; but for the second you must introduce *a priori* considerations of some kind. A quantitative physical law is on an entirely different footing from any type of sampling, for reasons given in the Quantitative Laws chapter of my *Scientific Inference*.[2] It would be a quite logical position to maintain that all quantitative laws are merely interpolations and that all values of y for other than the observed values of x are equally likely; e.g. that the predictions in the *Nautical Almanac* are meaningless; but if you really mean that, I think you ought to say so.

My quarrel is not with you, but with Eddington and similar people. E. would say that the law of gravitation is determined by purely *a priori* considerations; I say that it is partly *a priori* and partly experimental. You would deny the *a priori* considerations altogether. So it is rather anomalous that in a recent Phys. Soc. paper on measurement E. throws as much mud as he can at probability (while introducing probability considerations at every turn without noticing them) and at the same time quotes some of your work with approval. Such an attitude means a complete *a priori* theory and an experimental one, with no possible point of contact. I prefer to have just enough *a priori* material to give experiment a chance to decide what is true. You give too little; Eddington gives too much.

I cannot follow your objection to the generalization to all probabilities of the laws of probability obtained for samples. You do not deal with my argument for the need for an *a priori* postulate, and your argument would say equally well that we cannot determine the distance of a star from its parallax.

[1] Wrinch, D. and Jeffreys, H. (1919). On some aspects of the theory of probability. *Phil. Mag.* **38**, 715–31.
[2] Cambridge University Press, 1931.

Fisher to H. Jeffreys: 26 *February* 1934

Thanks for your letter. I am glad you think the differences in our points of view do not go so deep as one might judge, as it would be a pity if we occupied the space offered by the Royal in imitating the Kilkenny cats.

I myself feel no difficulty about the ratio of two quantities, both of which increase without limit, tending to a finite value, and think personally that this limiting ratio may be properly spoken of as the ratio of two infinite values when their mode of tending to infinity has been properly defined. I gave a rigorous statement on this point in the *Proc. Camb. Phil. Soc.*, at the beginning of a paper entitled 'Theory of Statistical Estimation' [*CP* 42]. The proposition there stated is not difficult to prove, and I cannot see that it leaves any ambiguity as to the meaning of frequency ratios in infinite populations. The question has been discussed in other terms by von Mises, but his definitions applied, I believe, only when the populations are denumerable—an unimportant but necessary restriction, seeing that we often use probabilities proportional to lengths or areas.

I do not object to the generalisation to all *probabilities* of the laws appropriate to the games of chance, but I do think, and indeed claim to have shown, that there are also logical situations in which a rigorous statement of the nature of uncertainty in our uncertain inferences is expressible not in terms of probability, but in terms of likelihood, a quantity which does not obey these laws. The derivation of probability statements from statements involving likelihood, in the special cases where such derivation is possible, interests me greatly, and seems the right starting point for exploring the almost unknown field of the relations between probability and likelihood.

Here is an example of the kind of problem in this connection which puzzles me. A man makes genetical tests on a number of plants from a wild population. He finds three of them may be called a, two more b, two more c, and one each of d, e, and f. His data thus consists of the partition $(32^2 1^3)$ of the number 10.

Let him know or be willing to assume that different types occur in the population with frequencies in the ratio $1{:}r{:}r^2{:}r^3{:}\ \ldots$ For any value of r he can calculate the probability of getting his observed partition, supposing his sample has been chosen really at random. He thus knows the likelihood of all values of r, but unless he has prior knowledge as to the distribution of r, I think you will agree with me that he does not know the probability of r exceeding any assigned value. For each value of r there is, I think, a calculable probability that the next plant to be tested will be of a type not hitherto found, and this probability will, I suppose, increase from 0 to 1 as r increases from 0 to 1. What sort of information has he then about this probability?

You would, I think, approach the problem introducing some sort of prior knowledge, which would make this probability definite though it would depend on the prior knowledge introduced, but you would not be unwilling to conceive that a rational being might happen to lack the prior knowledge of the kind introduced, and yet you would, I suppose, be unwilling to assert that no amount of experience without such prior knowledge could give him any guidance as to whether to expect new types or not. It looks as though some sorts of rational inference require both the concepts of probability and likelihood in a rigorous statement of the nature of their uncertainty.

H. Jeffreys to Fisher: 1 *March* 1934

Thanks for your letter. I think that even if we can't agree and have to leave a lot for other people to argue about, we ought to be able to agree about what our points of difference are. At present it seems that the relations between us are of the form 'A thinks B has done all the things that B has been at special trouble to avoid'; where if A = J, B = F and conversely. With me might I think be included F.P. Ramsey and C.D. Broad; there are two very important papers by the latter in *Mind*, vols. 27 and 29,[1] which I read when they came out, and they seem (especially the first) to have influenced me to the extent that I assimilated the ideas and forgot where they came from. I'll have to make a belated acknowledgement somewhere.

On the question of ratios of infinities first. In your *Phil. Trans.* [*CP* 18] you quite explicitly excluded the notion of a limit and stated the definition of a probability as the ratio of two infinite numbers. The limit was Venn's dodge, and you dropped it. But when it comes to the point neither you nor Venn use your own definitions. You have never taken an infinite class and counted the number of a sub-class within it, nor has Venn ever found a probability as the limit of a ratio when the number of trials tends to infinity. (By the way, you use the word 'probability' in your definition; but later on you generally use 'frequency'. I suppose you mean the same thing by both, but am not sure.) When you want actual numerical estimates you assess directly the frequency of a sample in the form ${}^{r}C_{\ell}\ {}^{n-r}C_{m-\ell}/{}^{n}C_{r}$. So far as I can see this is got by counting cases in a perfectly correct way and then saying that all cases are equally probable. On the *a priori* view of probability there is no more to be said; on the frequency view it may be right or not, but the point I want to make is that you don't avoid making an *a priori* assumption. You have to assume that if you take an indefinitely large number of classes of number n, each of which contains r ϕ's, and sample them, the ratios of compositions of the samples will occur in just the frequency given by your estimated probability. You do not know this by experience and therefore it is an *a priori* postulate. You may think that it is more plausible than the direct use of non-sufficient reason; but that is not a reason for condemning the latter on the ground that it is not known by experience. So far as I can see your practical results (in the *Phil. Trans.* I mean; I have not followed your later work in much detail) can be taken over into the *a priori* theory of probability without change of their quantitative statement, though some of them may need a change of language.

To take another case; suppose that by your methods you compare two methods of growing potatoes and show that in 90 per cent of cases method A will give a greater yield than method B. Suppose a farmer asks you 'What reason is there to suppose that *I* will get a greater yield by method A?' It seems to me that your only answer is in terms of the view that probability is intelligible without definition.

There is a criticism of the *a priori* view that I often meet and have never succeeded in understanding, and you give me some clue to it at last. It is the fundamental objection to the idea that $P(p|q)$ has a definite value whatever p and q may be. I am getting inclined to think that behind this objection is an idea that I think that q is *relevant* to the truth of p in all cases, which is another matter altogether. In the majority of cases we have things like $P(p|qr) = P(p|q)$ whatever r may be over a very wide range, i.e. in words, r is irrelevant to the truth of p given q. E.g. on my present knowledge of your movements it is as likely as not that you will be in Cambridge before you are at Rothamsted; the probability is $\frac{1}{2}$. If I find that you still have a house at Rothamsted the probability sinks nearly to 0; but it is only such quite special additions to information that will materially alter the probability.

You also seem to suggest in your paper that when I assess a prior probability I am expressing an opinion on the ratio of the numbers of cases in the world. My attitude, on the contrary, is that it is only on my view that it is possible to make any progress without expressing such an opinion. I started from the pure phenomenalist position, but found that phenomenalism needs a good deal of amplification before it can deal with the problem of inference. We don't know all about the world to start with; our knowledge by experience consists simply of a rather scattered lot of sensations, and we cannot get any further without some *a priori* postulates. My problem is to get these stated as clearly as possible. The tests available cannot be experimental; the conditions required, in my view, are that we want just enough *a priori* hypothesis to make it possible to settle the rest by experience. We cannot exclude logically the possibility that when a law looks well established the next attempt at verification may jigger it up altogether, so that general laws are never certain at any stage but only have a certain degree of probability. I got as far as I could with the principle of non-sufficient reason, but it turns out in some cases to give answers quite contrary to general belief. E.g. as Broad pointed out, it will never give a reasonable probability to a general law of the form 'all crows are black'. He tried to get a workable alternative but got no answer very convincing either to him or to me; I discussed the point very shortly in *Scientific Inference* 191–197, but I think the attempt in my *Camb. Phil. Soc.* paper on sampling[2] is on the right lines. The case of quantitative laws is even more to the point. E.g. in the case of uniform acceleration discussed in *Scientific Inference* 37–41, we could choose an infinite number of laws that would fit the data perfectly, but nevertheless do choose one, on *a priori* grounds, that only fits them approximately. If likelihood was the only thing that mattered we would choose any of the exact solutions in preference to the one we actually do choose.

The principle of non-sufficient reason is intended to serve simply as an expression of lack of prejudice; in a sampling problem we want to give all constitutions of the whole class an equal chance of acquiring a high probability by experiment. But in these cases of general laws there seems to be prejudice; I cannot help it, but there is a general belief in the possibility of establishing quantitative laws by experience, and I am not prepared to say that the general belief is wrong. I think I have stated a postulate that expresses it sufficiently clearly for practical purposes, but the postulate cannot be proved experimentally. But since such a postulate must from its nature be believed independently of experience that is a recommendation. What I want is, since an *a priori* postulate is needed anyhow, to choose it in such a way that the maximum number of alternatives are left for experience to select from. Eddington, e.g., chooses it too drastically by saying that the law of gravitation *must* be generally covariant and that space *must* have a finite curvature, deliberately excluding 0 from the admissible values of a certain coefficient.

Your paper on Inverse Probability in the *C.P.S. Proc.* [*CP* 95] involves a subtle point. Suppose we have a huge population and from it make up classes each of 10000 members, containing respectively 0,1,2 . . . with the property ϕ. We choose one of these at random and select a sample of 100, of which 30 have the property ϕ. You say, and I agree, that this establishes a high probability that the class chosen contains about 3000 with the property. If on the other hand we choose 10000 at random, sampled them as before with the same result, we should still estimate by inverse probability, using n.s.r., that the class contained about 3000 ϕ's. You think it absurd that the two should give the same result. Actually they do when you only sample one class; but if you take two classes there is a difference. In the first case the fact that one class has been estimated to contain about 3000 ϕ's slightly reduces the probability that the second will contain a number in this range, because one of them is excluded from the possible compositions of the second. But in the second case the first sample of 100 is effectively a sample of the whole super-population, and establishes that the ratio in this is probably about 3 to 7. Hence there is an *increased* prior probability that the second 10000 sampled will be in about this ratio. If we again get 30, we shall feel increased confidence that the number is about 3000 (and reduce the probable error); but in the first case what change there is, is in the other direction.

I don't understand your remark about a verbal point in your criticism of my remarks on Keynes. Mises said something similar in reviewing my book. My immediate reaction is that I cannot take it as obvious that people's linguistic habits are meaningless; but I think there is something else and can't see what it is.

I keep on getting other jobs pushed on to me and have not yet managed to read your paper carefully, though I have re-read a good deal of previous work. . . . In your discussion of my 1/3 business [*CP* 102] you seemed to assume that the third observation was equally distributed in some way; this is contrary to the postulate that the probability follows the normal law. I don't think there is much to add to Bartlett's discussion. My own got rather muddled through getting statements in the wrong order. The point is that if we know anything about h to begin with, and the first two observations are separated by a big multiple of $1/h$, there will be an extra probability that the third will lie between them; if they are separated by a small multiple, it will be probable that the third will lie outside them by a distance of order $1/h$. When the first two are fixed, the probability that the third will be between them is 1/3 only if we have no previous knowledge about h.

I will think about your problem of the trees.

I think that as a result of this you may feel inclined to alter your paper somewhat; if I answer it as it stands I think I shall miss the important points and say too much about others. Should I send it back for any modifications you think desirable and then I can have another try?

[1] Broad, C.D. (1918). On the relation between induction and probability—(part I). *Mind* **27**, 389–404. Broad, C.D. (1920). The relation between induction and probability—(part II). *Mind* **29**, 11–45.
[2] Jeffreys, H. (1932). On the prior probability in the theory of sampling. *Proc. Camb. Phil. Soc.* **29**, 83–87.

H. Jeffreys to Fisher: 21 *March* 1934

I have got this more or less done at last; sorry to be so long. I have not typed it as there may be further revision.

I think it may help readers if we could put things the other way round. I have arranged my paper so that it will be an explanation of my own point of view; how far the opinions I have dealt with are yours I don't know, but as they are some people's I think they should be discussed. I think your paper is not relevant to my real arguments and that as it stands it will tend to obscure things; so I suggest that mine might come first and you might add comments at the end. That of course is as you like. I could put a note at the beginning to day that this is the result of discussion between us.

I have thought a bit more about your question about the wild plants, but am hazy about what you want. If it is a sampling problem, total number of individuals n, with $r_1, r_2 \ldots$ of the various species, I think that
$P(r_1, r_2, \ldots | n)$ should be constant for all values subject to the sum being n. This seems the obvious generalization of Laplace. W.E. Johnson had a paper in *Mind*[1] about a year ago (posthumous, edited by Braithwaite) in which he did something of the sort; the remarkable thing about it was that he did the whole thing on the posterior probability, but he had an undetermined

constant left over. That's the trouble about philosophers; they talk sense up to a point (some of them I mean) but run away when they're on the verge of saying something useful. On the other hand you may be thinking of a Mendelian problem where there is one species and the question is about the incidence of one factor. Then I think my *C.P.S.* paper is relevant. But I may have missed the point altogether because I couldn't see where your ratios $1, r, r^2$ . . . could have come from.

Frank Ramsey seems to have had the gift of thinking clearly about these things more than anybody I know, though I have hopes of Broad. If somebody of that type would get on to it I think we could make some progress. I'm not desperately interested myself, being willing to take approximations and even orders of magnitude when it is too much trouble to get anything better; but if e.g. the simplicity postulate could be stated more precisely than I have done it would serve as a better guide than we have yet as to when we ought to introduce a complication. The sort of thing that bothers me is this. In seismology we get times of transmission to various distances, and fit a polynomial of degree 3, say, to them. The significance of the last term really involves the prior probability that such a term will be present. The usual thing is to keep it if it is some arbitrary multiple of its standard error, but I think it ought to be possible to frame a rule with *some* sort of argument behind it. . . .

[1] Johnson, W.E. (1932). Probability: the deductive and inductive problems. *Mind* **31**, 409–23.

Fisher to H. Jeffreys: 29 March 1934

Thanks for your letter. We seem to have a deal of ground to cover in our ten pages apiece. I think your draft shows very well what your line of approach is; at least, I find it much more reasonable in this form than I did in the *Proceedings*.[1] I want rather to modify the tone of my note, without much altering its substance, e.g. instead of saying that in your definition notions analogous to probability and those analogous to amount of information are 'confused', I have altered the word to 'admissible' [*CP* 109, p. 6.].

From the point of view of interesting the general scientific public, which really ought to be much more interested than it is in the problem of inductive inference, probably the most useful thing we could do would be to take one or more specific puzzles and show what our respective methods made of them. This might mean more work than you want to undertake and will certainly require that we should agree quite exactly as to what the data were and what questions we want to elucidate from them. I had this in mind in mentioning the plant problem, as it was the first case I had come across in which the notions of probability and likelihood appeared to be harnessed tandem, instead of only one or the other of them.

In case you feel it is a point worth discussing I will state it again in a form which, apart from complications, which I think are logically unessential, is that in which it might arise in practice.

A man collects a number of plants of the same species from a wild population, grows them in experimental cultures, and after some years of much labour, has succeeded in classifying a small number. Three of those tested he finds to be identical in respect to his method of classification and these he calls type *A*. Two more differ from *A*, but agree with each other. He calls these *B*. Two more differ from *A* and *B*, but being alike, constitute a new type *C*. The remaining three which he has tested differ from *A*, *B* and *C*, and from each other, and so constitute solitary representatives of types *D*, *E* and *F*. He has then tested ten plants and found them to belong to six different types with frequencies of occurrence represented by the partition $(32^2 1^3)$ of the number 10.

He knows, or thinks he knows, that the frequencies with which the different sorts of plants distinguishable by his tests occur in the wild population constitute a geometric progression like the frequencies of Planck's oscillators containing 0,1,2,3,. . . quanta of energy. He is not concerned to test this hypothesis, but using it as part of his data to estimate the probable outcome, in respect of the discovery of new types, of continuing his labours by testing more specimens from the wild population, the kind of question he might properly ask is, 'What is the probability that the eleventh plant tested will turn out to belong to one of the six groups already found, or alternatively to add a new type to the list?'

If he knew the value of r, the constant ratio of his geometric progression, this probability would, I think, be determinate and for different values of r from 0 to 1, the probability of getting a new type next time must also, I suppose, run from 0 to 1. He does not, in fact, know r, but his observed partition gives him some information about r, e.g. if the observed partition had been (1^{10}) he would think r was higher than he would if the observed partition had been (10). He can in fact assign to each value of r a certain likelihood, which will be higher if his partition is made up of a lot of little parts and lower if it has a few big parts, and for any observed partition he can specify the particular value of r which is most likely and the likelihood relative to it of any other value of r.

But this takes me no further than to tell the botanist that I know which of the possible values of the probability he is seeking has the highest likelihood, and I know how much proportionately the likelihood is lower for any other value of the probability. The real question is whether he has any right to expect from me more definite information than this and this may be considered under the heads (a) are any further deductions to be drawn rigorously from the bad data as stated? (b) can we *properly* supplement these data by axiomatic truths relevant to our problem, which will enable further conclusions to be drawn?

So much for the case that puzzles me. Without suggesting that you should modify your draft, it may be worthwhile for me to make a few comments. You say I object to the introduction of an *a priori* element, [so] I should like to get this notion a little more precise. What I object to is the assertion that we must, in considering the possible values of an unknown parameter used to specify the population sampled, introduce a frequency distribution or a probability distribution of this parameter, supposingly known *a priori*. If we really have knowledge of this kind, as in some problems, which can be reconstructed with dice or urns, I do not deny that it should be introduced, but I say that we often do not possess this knowledge, i.e. either that we are in absolute ignorance or that our knowledge is so vague or unsatisfactory that we may properly prefer not to introduce it into the basis of a mathematical discussion, but rather prefer to keep it in reserve and see what the observations can prove without it and whether it, for what it is worth, is confirmed or contradicted by the observations. I claim, in fact, that it is at least a legitimate question to ask: 'What can the observations tell us when we know the form of our population, but know nothing of the *a priori* probability of the different values that its parameters may have?' This is the situation which I treat as the typical one in the Theory of Estimation, but it would be quite legitimate to say that our assumed knowledge of the form of the distribution, which is needed before there can be anything to estimate, is in the nature of *a priori* knowledge. In fact, such knowledge seems to me essential in what I call problems of specification, but out of place in the next stage, when problems of estimation arise.

I think this may explain your difficulty about fitting a smooth curve to a number of discrepant observations. If the form of the curve is given, e.g. a polynomial of the fourth degree, then you will probably agree that the best curve of this form will be one specified by maximum likelihood. This would not prevent one from concluding, if after much experience it were found that some simplification of the formula were possible, e.g. that the coefficient of the fourth power seldom differed significantly from the difference between the coefficients of the second and third powers, that our specification could be revised with advantage, but while any specification is in use the problem of estimation can be made perfectly definite and it is primarily from the advantage of separating all the complications of the theory of estimation from the logical difficulties of specification, that I prefer to think of the thing in these two stages. In practice, as we know, simplicity is sought in specification, but whether it is to be justified primarily by convenience, subject to the satisfaction of tests of goodness of fit or whether, as you have proposed, it should more properly be justified in terms of probability, I have not any strong opinion. If our object were to explain the habits of a group of organisms, known as scientific men, I do not see that convenience is a less effective explanation, though certainly a less distinguished one than probability.

This is enough and more for one letter. I am returning the drafts.

[1] See Fisher's letter of 26 September 1933 to Bartlett (p.46).

H. Jeffreys to Fisher: 10 *April* 1934

I find your genetics problem rather hard. I think it might be tackled like this. Suppose you have a population of one type and breed from them; and that the probability of any new one showing a mutation is r. Then I should say that the prior probability of r is uniformly distributed if you know nothing about r to begin with; i.e. if $f(r)\,\mathrm{d}r$ is the p.p. [prior probability] that r is in [the] range dr, $f(r)=1$. More accurately perhaps, if you are testing the occurrence of mutation in a particular factor, the prior p. that $r=0$ is 1/3, $r=1$ is 1/3, and $f(r)=1/3$ otherwise. For another factor the same values; and as far as we know to begin with the two are independent so that $f(r_1, r_2)\mathrm{d}r_1\mathrm{d}r_2 = f(r_1)\mathrm{d}r_1 \times f(r_2)\mathrm{d}r_2$. I can't see what would make your biologist think that r is the same for all factors, but if he is sure of it then the same $f(r)$ holds for all. I think that if he knows this he must know a lot of other things too and this form may be quite wrong. We might have to allow for our knowledge that in fact most things breed true, which would concentrate $f(r)$ towards the lower values; but I think the above is right if he really knows nothing.

Newman once asked me the following: a man arrives at a railway junction in a town in a foreign country, which he has never heard of before. The first thing he sees is a tramcar numbered 100. Can he infer anything about the number of tramcars in the town? Newman thought the question was significant and so did I, and we both had a feeling that there were probably about 200. I tried it on M.S. Bartlett, who thought it was meaningless but had the same feeling about 200. This may have some analogy with the question of the existence of a further factor liable to mutation. My very doubtful solution is: let $f(n)$ be the prior probability that there are n cars; the probability that number 100 would be observed, given n, is $1/n$ for $n \geqslant 100$, 0 for $n<100$. Hence the posterior probability of n is $\propto f(n)/n$. If $f(n)$ is constant $\Sigma f(n)/n$ diverges and we can infer nothing (or rather, that it is practically certain that $n>200$!); but if $f(n) \propto 1/n$, $f(n)/n \propto 1/n^2$ and the probability that n exceeds m (>100) is

$$\sum_{m}^{\infty} \frac{1}{n^2} \bigg/ \sum_{100}^{\infty} \frac{1}{n^2} \approx \frac{1}{m} \div \frac{1}{100}$$

and it is about as likely as not that $m>200$.

I have just read Bayes's paper again for the first time since 1919. I see he defines probability in terms of expectation of value, as Ramsey does; R. has really given only conditions for consistency. Bayes gets $P(pq|h) = P(p|h) \times P(q|ph)$ quite correctly; I can understand him rather better than Ramsey but that is probably my fault. What bothered me about R. was that I did not see that people would agree in their ordering of values any better than in their ordering of probabilities; it really involves an ideal man who would always act on the alternative that gives him the greatest expectation of value. But it's really no worse than mathematics, which really assumes an ideal man that always gets his arithmetic right.

Towards the end of your letter you speak again of unknown *a priori* probabilities as opposed to those arising with dice. My attitude is that you are asking the wrong question. My question is, what distribution of probability corresponds to absence of previous knowledge? I think you are regarding a probability as a statement about the composition of the world as a whole, which it is not and on a scientific procedure could not be until there was nothing more to do.

I have just been away cycling for ten days.

Fisher to H. Jeffreys: 13 *April* 1934

Thanks for your letter of April 10th. You have not got hold of my botanical problem yet, for it has nothing to do with mutations or mutation rates, but solely with the proportions in which different types, which can be recognised by experimental breeding as distinct, exist in a wild population, i.e. for any two plants the experimenter can, in time, ascertain whether they are alike or not alike, and so can find that he has 6 different sorts in the 10 plants tested. But he knows nothing about which is the mutant type and which the parent type.

The information which I supposed him to have, that the frequencies of different types in nature were in geometric progression, might have been different without altering the logical form of the problem, e.g. an alternative problem might have been to interpret the same data on the supposition that there was in nature a finite but unknown number of types, all equally frequent. One could then infer the likelihood of any proposed number as great as or greater than the number so far found and so the likelihood of each of the possible values of the probability of getting a new type at the next trial. Either item of supplementary information leads to essentially the same logical situation though with a different mathematical content.

I like Bayes' approach to the notion of probability, because it is historically true that expectations, as of cargoes at sea or contingent rights in property, must have been bought and sold long before the value placed upon them was formally analysed into the two components, probability of possession and value if possessed. Indeed it is curious, though I believe it is a fact, that this analysis was not carried out by the Greek or Arabian mathematicians or indeed by anyone until the sixteenth century, if then. It seems to me the basis for what we may call the frequency aspect of the probability theory.

The tram problem is a good one. If instead of a whole number, you have a continuous variate, e.g. if the traveller arrives by parachute near a stone marked 'City of X, 1 km. from city centre', he has quite a nice basis for estimating the radius of the city supposing, of course, that he knows that all the cities are circular and that distances are marked from their geometrical centres. He then has an unambiguous fiducial argument as follows. 'If the radius of the city exceeds R kms., the probability of falling on or within the 1 km. circle is [less than] $1/R^2$. If this event has happened the fiducial probability that the radius exceeds R kilometres is therefore $1/R^2$ and the fiducial probability of it lying in the range dR is $2dR/R^3$. The fiducial median city has a radius $\sqrt{2}$ kilometres and an area 2π. The fiducial mean radius is 2km. and the fiducial mean area is infinite. The most likely [radius] for the city however, is 1km., for this maximises the probability of his observation.

I don't think this quite applies to the tram-cars. If No. A has been observed one can argue that if n is the total number the probability of meeting with the number A or less is A/n and if n exceeds any value N_0 this probability is less than A/N_0. But one clearly cannot differentiate and at first sight I do not see how to handle the inequalities. In fact I fancy that the reason why the fiducial type of argument was for so long unrecognised is just that the earlier writers always took frequencies, that is discontinuous observations, rather than measurements as the typical data in the theory of probabilities. Continuous variates with their standard deviations and regression coefficients are, in this matter, susceptible to a much simpler rigorous treatment.

H. Jeffreys to Fisher: 20 *April* 1934

I've got this typed at last. I have rearranged and altered it a bit. Would you mind filling in the reference to a paper of yours on p.9? . . .

Fisher to H. Jeffreys: 26 *April* 1934

I am sending you back the type-scripts, so that you can see my last alterations, at least I hope they may be the last. If you think they will do, will you send them both in to the Royal, as they are agreed documents. I think, on the whole, the suggestion of the Mathematical Committee[1] has been justified and that our notes, within the limits of space assigned, will be of more general interest than if we had written independently.

[1] See Fisher's letter of 23 February 1934 (p.149).

H. Jeffreys to Fisher: 9 *May* 1937

A question has just arisen about the excess motion of the node of Venus. It is 3.5 times the standard error, the probability of a random deviation exceeding which is 0.00041. Eddington says that as it is one of 15 it can be accepted as normal. The p. that one of 15 would exceed 3.5σ is 0.006. What I should like to know from you is whether there is another case on record where a statistician has accepted at sight a deviation beyond your 1% limit as random? (The other 14 give a χ^2 of 15).

By my test the thing is probably random on account of the large number of observations combined, but there's not much to spare, and the situation would be altered if some *specific* systematic error was before the House.

I should be grateful if you would send me some of your papers. I'm not much interested in agriculture but I should particularly like (1) the one where you got the distribution of χ^2, (2) the one where you introduced z, (3) one that Wishart mentioned to me the other day about χ being distributed nearly normally.

Fisher to H.Jeffreys: 11 *May* 1937

I am sending a few mathematical offprints that may, I hope, be of interest to you. The problem you set me is clearly one of judgement, and not of calculation, and indeed you seem to have done all the calculations needed. I should be inclined, naturally, to accept Eddington's judgement on an astronomical point, especially as your own test seems to confirm it. On the other hand, *prima facie*, i.e. on an assumption ordinarily made, the probability 0.006 is amply small enough to claim significance, and would be used for this purpose with complete confidence, I have no doubt, if anyone had a theory which required such a deviation.

H. Jeffreys to Fisher: 18 *May* 1937

Thank you very much for your letter and papers. Your letter confirms my previous impression that it would only be once in a blue moon that we would disagree about the inference to be drawn in any particular case, and that in the exceptional cases we would both be a bit doubtful. I am particularly glad to note your treatment of an unknown systematic error as a motion not before the House, while a specified one can be treated on its merits. This is what I have been saying about the 'excluded middle' for a long time.

Your letter encourages me to ask whether you also agree with me about the ideas of Eddington and Milne that all valid observations can be deduced by pure thought, without reference to other observations. As far as I am concerned this opinion does not agree with experience, and I prefer the alternative explanations that E. and M. are either very skilful in selecting their observations or very lucky. I agree completely on this with Dingle's recent *Nature* article.[1]

It was my disagreement with Eddington over this matter that first brought me into the subject in 1919, and I have been trying at intervals ever since to do for induction what Whitehead and Russell did for deduction in the introduction to *Principia Mathematica*: to get the postulates down to a minimum. You are doing the same. My only objection to your treatment of estimation is in the infinite population, which you introduce only in order to give a meaning to probability, while recognizing that it is hypothetical. I recognize that probability is hypothetical, so that the number of hypotheses is the same. It only happens that I find my way easier to understand. Any theory of induction is bound to be a bit fuzzy at the edges, but apparently you recognize that too, though you put the fuzziness at the fiducial limits, whereas I put it into the practical difficulty of assessing the prior probability when the previous data are abundant but inadequately classified.

I am writing this because there is a tendency about to attribute what I believe to be an entirely exaggerated idea of our disagreement to us, for which we are both possibly partly responsible, and I think an occasional mention of cases where we agree would be for the good of the subject. I have a paper coming out soon in the *Proc. Roy. Soc.*,[2] partly on what had seemed to me the entirely mysterious fact that 'Student's' *z* distribution and mine obtained by the inverse theory are identical; there turns out to be a perfectly genuine reason why they should be the same in form, and with a trifling extra assumption 'Student's result can be made into an alternative proof of mine.

[1] Dingle, H. (1937). Modern Aristotelianism. *Nature* **139**, 784–6.
[2] Jeffreys, H. (1937). On the relation between direct and inverse methods in statistics. *Proc. R. Soc.* **A 160**, 325–48.

Fisher to H. Jeffreys: 19 *May* 1937

I have been a great deal interested in Milne's recent writings, and have been a little puzzled by, I think, the same point as you, namely, his taste for developing the subject as much *a priori* as possible. I do not think he always felt like this about it; indeed his first reaction to Eddington's *a priori* approach was, if I remember right, rather that the interesting thing about the universe was that we had to look at it and see how it worked, instead of inventing it by introspection. It may be now that he has been working on some general and reasonable postulate which he has not yet made fully explicit, such as, that the observer has a right to assume that he is not too exceptionally placed in respect of the observations he can make, or that the laws of motion must be expressible in terms of relative only. I do think that his dynamical work is a marvellous achievement, and it seems at present to be in the way of clearing up the whole tangle for which Einstein and his following are responsible; but, much as I am impressed by his work, I am afraid I have not followed it closely enough to give an opinion of real value. . . .

H. Jeffreys to Fisher: 20 *May* 1937

Milne's general principle is something to be found in a part of Mach's work that I possibly never read, and from his earlier work I should be surprised if Mach thought it important: that as mass can be inferred only from relative motion, the mass of any body is caused by the motions of other bodies. It looks to me like a lapse into a metaphysical idea of causality that Mach had already disposed of. But unfortunately in order to get a frame of reference Milne has to assume a substratum with a finite density, tending to infinity at an outer boundary, which is itself spreading out with the velocity of light; and then he assumes that motions with respect to this substratum are small. As the substratum is unobservable I don't see that he has any justification for the second assumption. I should begin to be interested if he explained the things where the velocity of light and gravitation appear to be mixed—the three Einstein crucial tests. Judging simply by capacity for predicting observations I should prefer Eddington to Milne. On philosophical outlook my order of preference would be Einstein first, Eddington and Milne nowhere. Einstein has, I think, always made it clear that he is making suggestions that look hopeful and working out consequences that may lead to a test.

I am sending you a few early papers of mine that you won't find mentioned in any of the official relativity books; but they constitute my reason for taking the whole business seriously. . . .

H. Jeffreys to Fisher: 5 June 1937

. . . I have asked the Press to send you a copy of the new *Sci. Inf.* with the addenda. I should really have liked to scrap the lot and do it again, but at the present rate it looked as if the thing would take about 60 years to sell out. Looking at it again I don't seem to have made enough of the places where I agree with you, but I think I have remedied that in an article I have sent to *Nature* for the Dingle controversy.[1]

When I have time I should like to write a note showing how some of your methods fit into my theory. Maximum likelihood is obvious (I can't understand why K.P. was sticky about this, since m.l. for a large number of observations is an immediate consequence of inverse probability, and K.P. always accepted inverse probability) but sufficient and efficient statistics also have a definite place. The situation in my method indeed is that it goes straight for the efficient statistics and the mathematical difficulty comes in when we have to decide what to do when somebody has provided only inefficient ones.

[1] Jeffreys, H. (1937). Physical science and philosophy. *Nature* **139**, 1004–5.

H. Jeffreys to Fisher: 30 *July* 1937

I have read your paper [*CP* 149] on the methods (plural intentional) of moments with interest. A bit on the severe side, perhaps, but probably called for. However you tempt me to ask whether you would consider a short paper for your journal on the relation of maximum likelihood to inverse probability. An idea seems to have grown up that they are opposed, and I think this is very much in need of correction. It is most odd that K.P. always accepted i.p. and yet was sticky about likelihood, and adopted a method of fitting that was inconsistent with i.p. . Wrinch and I gave the essence of the relation, for sampling, in our 1919 paper,[1] but it would stand a bit more generalization. We might have gone further at the time, but for one thing we thought it was obvious, and for another we thought that statisticians already used it; as indeed they did for sampling and fitting normal distributions.

I am interested in your doubts about Pearson curves as such. They have not come my way much; I once got a distribution of the x^p $e^{-ax}dx$ form, in radioactivities of rocks, but have not had anything more complicated. In seismology the law of error is practically a normal law with a uniform distribution superposed; the infinite Pearson curves don't seem to fit the outlying observations very well, but it seems possible to locate the normal distribution very well without anything more complicated than the dodge I introduced in my Strasbourg paper, which is quite complicated enough when there are some hundreds of groups to fit. However I should like to know whether there is any evidence (1) that the K.P. function with infinite tails both ways is better than the exponential of a quartic, which would make the method of moments right, (2) whether the one with finite limits $-a$ and $+b$ is any better than is got by taking a new variable $y = \log\{(x+a)/(b-x)\}$ and fitting a normal distribution? The functions would no longer look like a single family, but I don't think that is any great objection. The K.P. ones aren't really because an imaginary tail is rejected to keep the results real.

In astronomy the question of correcting an observed frequency distribution for a known standard error is continually coming up. Eddington gave a method a long time ago (reproduced in Brunt) which is mathematically equivalent to solving the heat conduction equation for negative time. If he had the true probabilities presumably it would be all right, but it is applied to the observed numbers and if applied completely I think would necessarily diverge. Do you know any way of doing it otherwise than for the normal law? The kind of thing it is applied to is the distribution of absolute magnitudes of stars.

[1] See Jeffreys' letter of 24 February 1934 (p. 149).

Fisher to H. Jeffreys: 16 *August* 1937

I had to be rather down on Pearson, as for a time it seemed likely that his attack would wreck Koshal's statistical career, and I particularly wanted to make clear to our Indian brethren that unfair attacks on foreigners were as much resented in this country as they must be by their victims.

Like you, I have fairly often run into $x^p e^{-ax}dx$, but with the terminus at zero, and not at an arbitrary point as in the Pearson form. This makes the curve altogether more manageable.

I have long thought it a pity that the form with a quartic exponent had not been more fully studied. One or two attempts have been made, as by O'Toole in America, but it involves a transcendental function as interesting, I should think, as the hypergeometric which seems never to have been sorted out. I have very often found translation, i.e. functional transformation of the variate, extremely useful to make a distribution sufficiently or indeed nearly usually normal. I think, myself, it would be unreasonable to expect cases from all sources to conform to a single family.

The question you raise of removing a known amount of variance from a distribution is really a most intricate and intriguing one. I suppose the logical approach is by correcting the cumulants, but that, of course, only gives a finite series of the first few cumulants of the corrected curve, and to infer the true distribution from these not only sometimes involves approximations that fail to converge ultimately, but too often have a bad practical convergence of the terms available.

[P.S.] Do not hesitate to send along anything you would like to publish, but do not be offended if I think it unsuitable for the journal.

H. Jeffreys to Fisher: 25 *August* 1937

I enclose the paper[1] in question. My own feeling about it is that it rather labours the obvious, but as K.P. never saw the point, and as it came as a surprise to Yule when I mentioned it to him the other day, I think it will be unfamiliar to 90 per cent of statisticians. So I should be glad if you can find room for it.

I am not sure if I have got the right reference on p.4 about order less than $n^{-\frac{1}{2}}$; I have an impression that I have seen something by you that refers to the point more definitely, but a search has not revealed it.

[1] Eventually published as Jeffreys, H. (1938). Maximum likelihood, inverse probability and the method of moments. *Ann. Eugen.* **8**, 146–51.

Fisher to H. Jeffreys: 8 *September* 1937

As I do not like the practice of editorial comments being made without the author's knowledge I am sending you the enclosed which I propose as an editorial note[1] on your paper. I hope and think that you will find it unobjectionable.

[1] See *Ann. Eugen.* **8**, 151 (1938).

H. Jeffreys to Fisher: 5 *March* 1938

I heard yesterday that you were back; I hope you found India enjoyable. Meanwhile I have got out a number of things that may interest you. . . .

I read your *Design of Experiments* recently with much interest. It leads me to query your disparagement of your intuition in the note at the end of my paper in the *Annals*. At any rate wherever my methods, which are really based on the *Grammar of Science*,[1] lead to criteria differing from K.P.'s, you seem to have recommended something either identical with mine or so similar that hardly anybody would notice the difference. . . .

By the way where did you get the identification of a prior probability with a frequency in the external world? K.P.'s habit of calling his tables of probabilities frequencies suggests that he identified them; but I haven't found anything analogous in the work of anybody else that has used inverse probability. It is however always hard to be sure that a particular statement is not in Laplace somewhere.

[1] Pearson, K. (1892). *The grammar of science*. Black, London.

Fisher to H. Jeffreys: 7 March 1938

Many thanks for your interesting letter. I am very glad you like *The Design of Experiments*, as I had been concerned for some years previously with the kinds of logical difficulties that experimenters run into, and found it interesting and enjoyable to try and make an orderly statement of what can be done about them, putting the various dodges which have proved successful into their proper relations with each other. . . .

On your last point about identification of a prior probability with a frequency in the external world, I think I was quite a little influenced by Bayes, who, you may remember, defined probability as 'the ratio of the expectation to the value expected', and clearly regards expectation as a value averaged over objective frequencies of occurrence. I have always thought that this point of view must have been fairly widely spread in the 18th century, as it is the natural one for gamesters to take for such problems as evaluating expectations on unfinished games, but I do not know anyone who has brought it out so explicitly as Bayes.

H. Jeffreys to Fisher: 14 *March* 1938

I have just looked through Bayes's paper again, and I cannot see anything that suggests a frequency interpretation to me. He unfortunately does not give any examples of what he means by the value of an expectation, but I see nothing in his remarks inconsistent with Frank Ramsey's idea of comparing the desirability of different courses of action. (Ramsey does not appear to mention Bayes — or me — and he is as naive about causality as Russell; but he makes some good points). There would be little point in defining probability in terms of expectation if the latter was then going to be defined in terms of frequency — it would be simpler to define probability in terms of frequency straight away, and expectation would become a superfluous notion. Note too the remark on p. 410: 'It should be carefully remembered that these deductions suppose a previous total ignorance of nature'. This is, I think, clear enough that Bayes was not interpreting a prior probability as a statement of frequency.

The chief difficulty of Bayes's presentation is, I think, in his Prop. 1, where he assumes expectations additive. This needs a good deal more analysis, which Ramsey has supplied. I mean that the value of two dinners on successive nights may be additive, but that of two on the same night may be less than that of one, at any rate if the morning after is taken into account. So further restrictions are needed before the definition will work — it must be clear that the expectations added do not interfere with one another. I am not sure when the Weber–Fechner law was invented; if Bayes had known of it I think that he would have taken precautions; but personally I think it much simpler to take probability as a primitive idea and define expectation in terms of it when occasion arises.

The first definite identification of probability and frequency that I know is by R. Leslie Ellis, in *Camb. Phil. Trans.* 1843; Keynes [1] refers also to Cournot in the same year. Laplace in using the *m*/*n* ratio always sticks in the proviso that all the events must be equally probable, thus making probability a primitive idea, and this is also in Todhunter's *Algebra*, from which I got my first ideas. In a lot of later works this has been left out — e.g. by Neyman and Levy. Thus the Weldon–Pearson experiments on the bias of dice and the usefulness of Tippett's numbers as against personal choice to get a random sample become pointless — matters on which we should agree perfectly.

I notice that the same volume of the *Phil. Trans.* as contains Bayes's paper also contains a proof by him of the divergence of Stirling's series; I suppose this was the first.

I have been reading or rereading a good deal of mathematical logic during the last few weeks and am beginning to wonder whether a good deal of pure mathematics is not inductive. Whitehead and Russell, Ramsey and Wittgenstein all seem to agree that a number is a class of classes of distinguishable individuals, and therefore, one would think, that the existence of classes up to a certain number is an empirical proposition depending on the number of individuals in the world, which may well be finite. In that case there would be no such real number as $\sqrt{2}$, since the notion of a limit breaks down. The only possibilities seem to be that mathematics is either wholly idealist or that it is developed as science indicates that it may be wanted. Apparently even consistency cannot be proved. I sounded Littlewood about this and he shares my doubts. Hitherto I have regarded pure mathematics as fundamental and probability as using its rules, but essentially as a separate idea; but the alternative that pure mathematics is merely a special case of probability is going to need serious examination. At any rate on Wittgenstein's arguments it seems perfectly possible that the whole logic of infinity and therefore of limits and continuity is just bunk, and nobody can apparently see anything wrong with them. I don't see much prospect of solving the puzzle myself but I think that I may manage to limit the questions to a few definite alternatives.

By the way you said somewhere that the variation of observations from a formula is just as essential a part of the law as the formula itself, but I cannot trace the reference when I want it. It is a most important point from the logical and philosophical point of view.

[1] Keynes, J.M. (1921) *A treatise on probability*. Macmillan, London.

Fisher to H. Jeffreys: 24 *March* 1938

I have tried circularising your letter in my Laboratory with a view to discovering your reference to where I said that the variation of observations from a formula is just as essential a part of the law as the formula itself. Like you, I am quite sure that I have said so and constantly impressed it in lectures, but cannot lay my finger on it in print.

H. Jeffreys to Fisher: 17 *September* 1938

. . . I am wondering whether more attention shouldn't be given to the median. Its uncertainty seems much less sensitive to the form of the law than that of any other simple statistic, and significance tests for it are easy. The kind of case where it may arise is where there are only 8 or 10 observations and nothing much to indicate the character of the law. The mean of all or the mean of the extremes may have little value, but there's always something to be said for the median. I have some cases where the law is pretty certainly not normal, but whether it is $e^{-x/a}(x>0)$ or rectangular or $\{1-(x/a)\}$ $(0<x<a)$ there is no evidence whatever.

Fisher to H. Jeffreys: 19 *September* 1938

Thanks for your letter. I agree very much with your point about the median. It has the great advantage of having always an intelligible meaning, whatever the form of the curve. In writing Gosset's obituary for the forthcoming volume of the *Annals* I was interested to find that one of the first applications he made of his distribution was to apply it to a sample of two, and thereby get out an inference respecting the median irrespective of the form of curve. In writing the obituary [*CP* 165] I took the opportunity of developing this idea somewhat further, i.e. to any percentile point and to the simultaneous distribution of different percentile points. [1]. . .

[1] For Jeffreys' reply with comments on 'Student' and randomization, see his letter of 22 September 1938 (p. 270).

H. Jeffreys to Fisher: 20 *September* 1938

. . . The median of the *law* has an important position in my theory, because with any law, symmetrical or not, involving parameters of location and scale that are unknown initially, I find that the posterior probability, given the first two observations, that the median of the law lies between them, is always $\frac{1}{2}$. This is not true for any other parameter of location. For any other it is $<\frac{1}{2}$. (*Camb. Phil. Soc.* paper[1] of 1936, 423–5). 'Student' presumably got the analogous thing by his way of looking at it, which could be turned into mine by the argument I used in 'The relation between direct and inverse methods'.[2]

. . .

[1] Jeffreys, H. (1936). Further significance tests. *Proc. Camb. Phil. Soc.* **32**, 416–45.

[2] See Jeffreys' letter of 18 May 1937 (p. 162).

H. Jeffreys to Fisher: 30 *September* 1938

. . . If you are bringing out a new edition of *Statistical Methods* I wonder if you would modify the passages about inverse probability in it. I didn't agree with them when the book first appeared, but at that time i.p. fell between two stools. The logicians like W.E. Johnson, Frank Ramsey and Keynes accepted it, but did not develop it to any point where it was much use for actual application; a lot of others that accepted it also seem to have thought that the uniform prior p. of Bayes and Laplace was an essential part of the principle, and if this was so it would say that any estimated difference must be accepted — the null hypothesis would always be rejected. The practical men, in the circumstances, had a good deal of sense in not trying to use it in the state it was in. But it is now in such a state that it can be used. Either method would get the right answer some time or other; I think i.p. would give it, on the whole, more quickly, but as I still haven't found an actual case where our decisions would differ I haven't any strong views. The chief difference is about what order we should say the same things in.

I have been wishing for some time that a public benefactor would subsidize the C.U.P. to scrap *Scientific Inference* and give me a chance of writing something up to date. I should have written it in 1923, when Einstein was still news . . . .

Fisher to H. Jeffreys: 11 *October* 1938

I find I have left your letter of September 30 unanswered. My position about inverse probability is as follows: the term has been used, not always for quite the same thing, during the last 150 years, but fairly consistently to designate Laplace's extension of Bayes' method based on the concept of 'equal possibility', which I submit was used by Laplace merely to introduce a postulate of equal probability by a verbal subterfuge. A series of writers, notably in England, Boole, Venn and Chrystal, protested at different times against this hidden postulate, and even orthodox exponents, such as de Morgan, were evidently shaken in their faith by the arbitrary way in which the postulate was sometimes applied.

During the 17th and 18th centuries I think that mathematical probability was used consistently as a means of expressing expected frequency, but in the course of the controversy aroused by Laplace's doctrine a psychological view, well expressed in Keynes' phrase 'degree of rational belief', came to be advocated.

The opinion I express in *Statistical Methods* is, I think, quite clearly intended to refer to this old controversy, and not to the recent Cambridge work you refer to. I could, of course, in later editions — though I am afraid not in the 7th — add a sentence to say that I was not in particular discussing your theory of scientific inference, or your later work; but I really do not think that anyone yet has taken it to be so, but has only drawn the obvious inference that you were attempting a resolution of the old enigma of a different kind from mine.

H. Jeffreys to Fisher: 12 *October* 1938

It is liable to be hard to follow old birds like Bayes and Laplace. People of that period often spent a chapter on a piece of mathematics that we should do in a page, and then slid over fundamental points with hardly a mention. Laplace's casualness about references in this case makes it harder. I think the real question is, did the critics of B. and L. from 1840 to 1900 or so understand them, or were they just attacking misinterpretations of them? In any case the misinterpretations have a long history, but they arose when B. and L. were not in a position to defend themselves. So far as I know K.P. was the only person that ever tried to *use* inverse probability and a frequency definition at the same time. I got my own first ideas from Todhunter's *Algebra*, the first relevant passage in it being practically a translation of Laplace, and it never occurred to me that any other interpretation than the one I make now could possibly be intended. I think it is relevant that I should regard a frequency definition of the prior probability as even more nonsensical than you do; you at least think it worth criticizing, whereas I should not think it worth mentioning if people didn't keep inflicting it on me. But if there was such a great difference between Bayes and Laplace and me as you suggest I should expect that reading them would make me feel sick, and it doesn't. I can't see anything in them that is not perfectly intelligible in terms of my own ideas, and I think the reasonable explanation is that they had the same fundamentals.

I don't agree that Laplace's 'provided all the cases are equally possible' is a verbal subterfuge; I should call it an indication that he had thought a bit more carefully than his predecessors. You strengthened my belief in this recently by referring me to de Moivre about something else. He starts off with a simple definition in terms of numbers of possible cases, as innocently as Neyman does. Laplace must have known of this, and the extra clause must mean that he saw that de M.'s definition didn't do what is needed. On the latter, so long as no face of a die is absolutely incapable of being thrown, the probability of any face is exactly 1/6 whatever other information we have. Laplace's allows for the possibility that if we get 50 sixes in 100 throws the p. of a 6 may be more than 1/6, even if all the others have actually occurred. Again, suppose we have two boxes, one containing a white ball and a black one, the other a white one and two black ones. Choose a box at random and then pick a ball at random out of it. What is the probability that the ball will be white? Answer, $\frac{1}{2}\cdot\frac{1}{2}+\frac{1}{2}\cdot\frac{1}{3} = \frac{5}{12}$. But on de Moivre's definition the assessment would be 2/5 — which I don't for a moment suppose he would actually say, but the only way of avoiding it is to recognize that the two white balls are not 'equally possible'. That is, even within the range of direct methods, a probability is not a mere matter of counting possible alternatives, and Laplace knew it.

Bayes's expectation theory is a bit muddled, because his rules correspond to what Laplace called mathematical expectation, but he is speaking of expectation of benefit, which is L's moral expectation. The distinction when the benefits to be added are in terms of money was actually brought out in Daniel Bernoulli's discussion of the Petersburg problem about 30 years before Bayes. Bayes's argument will still work, however, if the benefits compared or compounded are such as not to interfere with each other; e.g. the benefits to me of two dinners the same night would be far from independent, but those of a letter from you and a bike ride might be nearly so. Ramsey's theory attends to this point, but it is a bit cumbrous.

I'm afraid that more people than you think have the idea that what you say in *Stat. Meth.* is meant for me; and I should guess that about 20 times as many people have read *S.M.* as have tried any of my stuff. If you could find room for a remark that it isn't I should be very grateful. The sixth edition has about half a page blank at the end of the first chapter, which could hold a warning sentence easily!

H. Jeffreys to Fisher: 14 *November* 1938

. . . By the way who introduced the hypothetical infinite population? Your *Phil. Trans.* paper [*CP* 18] was the first place where I saw it mentioned, but it seems to be implicit in K.P.'s habit of calling chances frequencies in his tables, and it may have a longer history. There is a curious point about Laplace's $(p+1)/(p+q+2)$ rule of succession. It is given in the *Grammar*, and I proved it myself about 1915 for a population of any size — it is exact even for a finite population. I took it for granted that this was as old as the hills, but apparently it was new, at least I can't find it anywhere. Laplace's constant chance at each trial would mean either drawing with replacement or a population so large that the extraction of the sample produces a negligible change in the proportions in it. So perhaps Laplace may be responsible after all!

[P.S.] It also follows, if the sample is all of one type and of number p, the probability that the next s will all be of the type is $(p+1)/(p+s+1)$ whatever the number in the population. There are further extensions.

Fisher to H. Jeffreys: 15 *November* 1938

. . . I believe the hypothetical infinite population was latent in the earliest notion of sampling error, e.g. in Bernoulli's treatment of the binomial, but it seems to have lain latent for an unduly long time; at least I was conscious of wanting to make it explicit when I coined the slightly provocative phrase you quote. It is just possible, however, that it was previously used by Venn, who, in some passages, was wrestling with the same idea.

No, I did not recognise that the rule of succession was unchanged in form in sampling from a finite population.

H. Jeffreys to Fisher: 9 *January* 1939

About the question of $ns^2 = \Sigma(x-\bar{x})^2$ or $(n-1)s^2 = \Sigma(x-\bar{x})^2$. In response to wails from my class last term, I have decided that a probability book from me is due, and have started it. I don't want to make unnecessary departures from the usual practice, and in particular I am adopting the practice of using italics for estimates and Greek letters for quantities to be found — it was a matter of random selection after getting fed up with σ and σ' that made me adopt the opposite practice. However I'm bothered about n or $n-1$. The argument for $n-1$ is, I think, that it brings the mode of the z distribution to $z=0$. With n there is a bias, though the formulae are similar enough for one to have to look pretty carefully to see the difference. The z distribution goes over into my theory in an estimation problem: given one set of deviations, what is the probability of a given scatter in a new series of observations or means where there is reason to expect σ to be the same?

On the other hand in the method of least squares what comes in is $\{1+n(\alpha-a)^2/\Sigma c^2\}^{-(n-m+1)}$, and in significance tests for one new parameter the same thing with index $-(n-m+2)$ where in the former m is the total number of parameters found, in the latter the number of old ones; c is a residual. So it is really $\Sigma c^2/n$ that comes in, the m appearing only in the index; and taking $\Sigma c^2/(n-m)$ will complicate the algebra badly. Also it's the sort of algebra that I usually get wrong.

So it rather looks as if I shall have to stick to n. . . .

Fisher to H. Jeffreys: 12 *January* 1939

I always think that the best interests of mathematical notation are served by each man using the convention which suits him best. Consequently, I am not inclined to argue about using n or $n - m$. I don't quite follow what your letter says about least squares. However, since I always find it convenient to develop the theory in terms of undivided sums, e.g. in the coefficients and right-hand sides of the equations for regression coefficients and in the formulae for obtaining the sums of squares of residuals, as I see it the only question arises when one wants a mean square derived from this sum of squares, and then I am glad you agree that the old procedure of dividing by $n - m$ is the one to use. Of course, there is no great harm in doing what the biometricians do, and dividing all sums of squares by n automatically at any other point in the work, for such divisors cancel out in estimating the regressions. In fact, I regard regression work from another point of view as a good example of ancillary information, in that the precision of the regression does not really depend on the number in the sample, but only on the sum of squares of the independent variate, or, in general, on the dispersion sums of squares and products of the *set* of independent variates. Since the actual values of these are provided by the sample, there is no need to consider estimation in respect of the variances and covariances of these independent variates. In fact the whole work is completely independent of how they may be distributed in the population sampled. . . .

H. Jeffreys to Fisher: 5 August 1939

Bartlett has just sent me another paper on the Behrens–Fisher test, and I see you are having another go too [*CP* 164]. I sent Yates's paper on but have heard nothing further.[1] I have never seen the full form of your function and should be interested to know whether it is the same as my methods give for the *estimation* problem. The significance one is in a *Proc. Roy. Soc.* paper on the comparison of means.

I should state the problem thus. x and y are two unknown location parameters, σ and τ two unknown standard errors, θ and θ' two sets of observations intended to estimate x and y. What, given θ and θ', is the probability distribution of $y-x$? We have (the four unknowns being supposed irrelevant)

$$P(\mathrm{d}x, \mathrm{d}y, \mathrm{d}\theta, \mathrm{d}\tau|H) \propto \mathrm{d}x\, \mathrm{d}y\, \mathrm{d}\theta\, \mathrm{d}\tau / \sigma\tau$$

$$P(\theta, \theta'|x, y, \theta, \tau, H)$$

$$\propto \sigma^{-m}\tau^{-n} \exp\left(-m\{(x-\bar{x})^2 + s^2\} / 2\sigma^2 - n\{(y-\bar{y})^2 + t^2\} / 2\tau^2\right)$$

$$\therefore P(\mathrm{d}x, \mathrm{d}y, \mathrm{d}\sigma, \mathrm{d}\tau \mid \theta, \theta', H) \propto \sigma^{-m-1}\tau^{-n-1} \exp(\_\_\_\_)\, \mathrm{d}x\, \mathrm{d}y\, \mathrm{d}\sigma\, \mathrm{d}\tau.$$

Integrating with respect to σ and τ

$$P(\mathrm{d}x, \mathrm{d}y \mid \theta, \theta', H) \propto \{1+ (x-\bar{x})^2/s^2\}^{-\frac{1}{2}m} \{1+ (y-\bar{y})^2/t^2\}^{-\frac{1}{2}n}\, \mathrm{d}x\, \mathrm{d}y$$

and then putting $y = x + z$ and integrating for x

$$P(\mathrm{d}z \mid \theta, \theta', H) \propto \mathrm{d}z \int_{-\infty}^{\infty} \{1+ (x-\bar{x})^2/s^2\}^{-\frac{1}{2}m} \{1+ (x+z-\bar{y})^2/t^2\}^{-\frac{1}{2}n}\, \mathrm{d}x$$

and it only remains to do the arithmetic. At all events it is short and no information has gone down the drain; and it is perfectly clear what has happened to σ and τ.

This is what I call an estimation, not a significance, problem. The distinction is in the form of the question, but also affects the answer. Suppose you get an estimated difference z within the acceptance limit; do you say that the true value is as likely to be $2z$ as 0, or that it is more likely to be 0? If the former, it is what I call an estimation problem; if the latter, one of significance. It seems to me that most direct comparisons in agricultural experiments are estimation problems, but if you are testing whether a

Mendelian ratio agrees with 3:1 it is a question of significance.

It seems to me that in your original fiducial argument you were speaking of the probability of the true value, given the data, in precisely the same sense as I should, and you and 'Student' both make the point clearer in various places by emphasizing the uniqueness of the sample, thus saying what I say by using the prior probability that expresses previous ignorance. The trouble is that your language is not adequate to get into symbols what you are trying to say, and consequently you don't state all the steps that I filled in in my paper[2] about 'Student'. . . .

[1] See Fisher's letter of 20 February 1939 (p. 352).
[2] See Jeffreys' letter of 18 May 1937 (p. 162).

Fisher to H. Jeffreys: 8 August 1939

Many thanks for your interesting letter. I think your analysis is identically equivalent to mine and would lead to the numerical values given by Sukhatme. I hope you know his table, as it is a very usefully arranged one. I see, and like, your distinction between *estimation* and *tests of significance*, but am not quite sure of its application in this case. You are probably right that the language used by 'Student' and myself is not adequate to get into symbols what we are trying to say; but, of course, the effort to fill in all the steps is full of the danger of saying something quite different.

I am beginning to realise that we know much more about the methods we practise in reasoning than about the systems of logical postulates necessary to justify these methods. Consequently, such a question — which sounds urgent and important to purely deductive minds — as: 'Is there any new principle or postulate required or assumed in this method?' cuts very little ice with me because I do not know that the principles or postulates required for previous methods have ever been satisfactorily taped. Unless that has been done, and I respect Whitehead and Russell, Keynes, etc. as I respect the leaders of forlorn hopes, the first question asked is really meaningless. After all, may not the recognition of logical rigour be as empirical as the recognition of the three-dimensionality of space? . . .

Fisher to H. Jeffreys: 4 November 1939

I have just re-read your note on the Behrens-Fisher formula, which you sent me some time ago, and I forget whether you were inclined to publish it in the *Annals of Eugenics* or in the *Camb. Phil. Soc. Journal.* It would seem perfectly suitable for either.

I find I can follow your arguments perfectly, and should disagree, if at all, only on terminology, for you use the distinction between tests of significance and estimation differently from the way I do, including in the latter cases in which the answer is not, properly speaking, an estimate. However, apart from this and the propriety of using the *a priori* factor $\frac{1}{2}$ when a precise null hypothesis is specifically in view, I think your paper enables me to appreciate your point of view a great deal better than I have previously done.

I have already printed in the *Annals* rather more than I should have liked on what is really a technical point in theoretical statistics, with no genetic or human applications immediately in view; but, if you prefer the *Annals* to the *Cambridge Philosophical Journal*, I should be glad to put it in the first number of Vol. X.

H. Jeffreys to Fisher: 6 *November* 1939

Glad to see 'The Galton Laboratory' is still — or again — a recognized address.

I should prefer the paper to go to the *Annals*. . . .

I am not sure what cases you have in mind when you say that I include in estimation problems cases where the answer is not an estimate. In those I had in mind it seemed to me that there was always an estimate not very far away. E.g. suppose you have m sets of n observations; one standard error is estimated from the means and another from the scatter within the groups. If the former is too large it suggests a further variation, possibly normal, affecting the whole of a group, and the comparison leads to an estimate of this. But this is a typical z problem. I have extended this to unequal groups, as this trouble is continually turning up in astronomy and geophysics; there is no neat answer, but an approximate one can be got. . . .

Fisher to H. Jeffreys: 8 *November* 1939

Thanks for your letter. I will put your paper into *Annals*, Vol. X, No. 1.[1] . . .

[1] Jeffreys, H (1940). Note on the Behrens–Fisher formula. *Ann. Eugen.* **10**, 48–51.

H. Jeffreys to Fisher: 29 *March* 1942

Do you know anything about the following problem? It turns up pretty often, but I have just had it again in a rediscussion of H.S. Jones's determination of the mass of the moon. We have two sets of estimates of x, with different standard errors, estimated from n_1 and n_2 observations. There is no doubt about whether the two estimates of x refer to the same thing, so it is an estimation problem. The question is, how to allow for the difference of the estimates in finding the uncertainty of the combined estimate of x? The exact solution is

$$P(\mathrm{d}x|\theta, H) \propto \mathrm{d}x\,\{1 + (x-\bar{x}_1)^2/s_1^2\}^{-\frac{1}{2}n_1}\,\{1 + (x-\bar{x}_2)^2/s_2^2\}^{-\frac{1}{2}n_2}$$

so that it is the Behrens–Fisher rule with the two x's the same and no integration therefore to get a distribution for the probability of their difference. But can we get a useful approximation to this anyhow? The point is that in my problem $\bar{x}_1$ and $\bar{x}_2$ are based on $n_1 = n_2 = 13$ but differ by a shade under twice the standard error of their difference. This is really sure evidence that s_1 and s_2 are estimated a little too low and we could get a better idea of the uncertainty of x by allowing for the other degree of freedom expressed by $\bar{x}_2 - \bar{x}_1$. If only the standard errors in the two sets were equal it would be easy, but they aren't.

I see you have a new paper on this sort of thing but it can't be borrowed yet and I don't know exactly what you've done.

Fisher to H. Jeffreys: 30 *March* 1942

You are quite right that the problem you mention is very near to that of Behrens, but owing to the additional degree of freedom interpreted now as additional evidence on the values of s_1 and s_2, the significance test tabulated for Behrens' test is not exactly applicable. The analysis is, however, very similar in form, and I was tempted to try to tabulate values for your problem, and only deterred by there being one more (4 instead of 3) parameter with which the table must be entered. Tabulation would therefore be rather extravagantly elaborate, and the use of effective approximations based on the asymptotic approach of 'Student's' to the Normal distribution is all the more important.

I have discussed this approximation rather fully in the enclosed paper [*CP* 181]. . . .

The fundamental relations used are set out in Section 2, and the asymptotic approach in Section 3. Your case differs from Behrens in that, though both depend on a well determined line in the plane of t_1 and t_2, Behrens is concerned with the total frequency integrated on one side of the line, whereas you have to deal with the section of the frequency surface cut by the line itself. The maximal likelihood solution is the mode of this section, and to assign fiducial limits to this estimate one is concerned to determine the points on the lines which cut off $\frac{1}{2}$%, or $2\frac{1}{2}$%, or whatever you want of the sectional area at each tail, i.e. at p. 152 you would put $y = d$ in the expression for t_1 and t_2 and integrate for x only. I imagine the analysis will be about as difficult as what I have done on this and the next page, but you may not need so many approximative terms as I have introduced. At least, using definite values from the start, you are in a good position to see where to stop, whereas with algebraic investigation the relative importance of different terms is different in different parts of the table.

In the new edition of *Statistical Tables* also I am putting table 6 of this paper, together with Sukhatme's tables and an illustration of the sort of application I should think appropriate in physical or astronomical work. I should be very glad to hear what you think of these two parallel attempts, especially from the point of view of logical exposition, since such a disturbance has been raised about this aspect of the problem.

H. Jeffreys to Fisher: 9 *April* 1942

Thanks for your letter. The extra parameter finishes the thing off! . . .

I haven't anything to say about Sukhatme beyond what I said in my *Ann. Eugen.* paper — that it's right as a method of estimation but it isn't what I should call a significance test.

By the way you are inclined to blame omissions in the older methods on the teaching of inverse probability. I should rather blame it on the fact that it took a long time for people to see that getting the right answer depends on stating the question properly. With inverse probability it can be stated rightly or wrongly, but at least it has to be stated. Without it there is a continual risk of muddle through people not seeing what the problem is. As a matter of fact in 1921 or so I did the 'Student' problem using $P(\mathrm{d}h \mid H) \propto \mathrm{d}h$, but wasn't happy about it and didn't see how to put it right till *Scientific Inference*. But the difference between n and $n+1$ in 'Student's' formula is far less than the difference between you and Bartlett!

The awful problem in physics etc. is to get people interested at all. They seem to like multiplying by 0.6745 if they can be bothered to work out an uncertainty at all, and any attempt to tidy up the theory rather annoys them because it restricts their liberty of guessing. So for your tables I should say that it is no earthly use putting in physical applications that involve any appreciable amount of arithmetic. . . .

C. Jordan to Fisher: 14 *May* 1934

I thank you very much for your letter and the highly interesting paper you kindly sent me. I know that you establish your theory of maximum of likelihood independently of the Theory of Probability. I have shown that your method is equivalent to the following; an approximation of y_i by $f(x_i)$ is considered as being so much better as the geometrical mean of the geometrical deviations $y_i/f(x_i)$ is less (for $i = 0, 1, 2, \ldots n$).

This is quite different from the older methods of approximation, in which the minimum of a function of arithmetical deviations (standard deviation) is required. The geometrical mean and deviations seem to be according to Weber's and Fechner's law of psycho-physics much more important to us than the arithmetical mean and deviations.

Your method, applied to Laplace's and Poisson's formula of repeated trials, conducts, as does the method of moments (but not the other methods) to the usual results.

As you reject Bayes' theorem, it is all the more remarkable that the method of approximation according to the principle of maximum of likelihood is a necessary consequence of Bayes' theorem applied to Bernoulli's rigorous formula of repeated trials.

This is interesting, since there are many (I am among them) who think that it is impossible to form a rule for choosing between two hypotheses without admitting some postulate, and who think that the postulate of equal probability of the different possible hypotheses is at least as good as any other; for them Weber and Fechner's law and Bayes' theorem are the justifications of your method which show that it is preferable to any other method of approximation. . . .

Fisher to C. Jordan: 18 *May* 1934

. . . The parallelisms you mentioned are exceedingly interesting. I cannot help feeling, however, as indeed you recognise, that the logical basis of the use of likelihood must be really independent of Weber and Fechner's Law; i.e. of the mechanism by which differences in the intensities of a sensation are appreciated, especially since this law can only hold over a definite range of the intensity of the stimulus.

The connection with Bayes' Theorem, or rather with the postulate, which Bayes proposed, after proving his theorem quite rigorously on the data of the problem, has a more intricate history. Supposing that the postulate were true of a particular unknown parameter θ, i.e. it is known with certainty *a priori* that the probability that θ lies in the range $d\theta$ is proportional to $d\theta$ for all values of θ, then Bayes' method gives us an inverse probability distribution of θ having, if θ is a continuous variable, a determinate form, and therefore determinate values for its mean or mode or median or any other property of the distribution in which we might be interested. If I believed in this inverse probability distribution, I should be most interested in those of its properties, such as the median, which are invariant at least for a wide class of functional transformations of the variable θ. I should attach less importance to features such as the mean and mode, which may be brought to any value at will, by an appropriate functional transformation. The method of inverse probability, therefore, based on Bayes' postulate, does not point with any special emphasis to the mode of this distribution which, as is well-known, is in fact the solution of the equations of maximum likelihood. In arriving at the mode by this method two arbitrary steps have been taken, neither of them, in my opinion, justifiable, which in effect cancel each other out: (i) the choice of the particular parametric function θ, to which to apply Bayes' postulate, and (ii) the choice of this particular function in relation to which to determine the mode of the inverse distribution. If the same function is used for both processes, the maximum likelihood value is reached and this is in fact invariant for all functional transformations, and though Gauss justified his choice of the maximum likelihood by the inverse probability argument, I feel very sure that he would not have in fact adopted it if it had not been invariant.

I think it is well worth while to get to the bottom of the confused history of this intricate subject, since at the present time it seems possible at last to put at least some useful processes of inductive reasoning on an entirely rigorous basis.

Fisher to M.G. Kendall: 2 *November* 1942

I was much interested in your paper 'On the future of statistics',[1] and should have made a point of being present if any notice had reached me.

On p. 73 you ascribe to Neyman and Pearson a certain advance in the theory of inference. I could refer you to a good many things I had published on the subject before Neyman and Pearson knew anything of what I was doing; but in view of what you say I really think you ought to read a short note published in 1930, [*Proc.*] *Camb. Phil. Soc.* Vol. 26: 528–535, which I wrote under the title 'Inverse probability' [*CP* 84].

P.S. When you have read the reference I think you will understand why I should look to you positively to combat the fable of Neyman and Pearson introducing fiducial inference, for others beside yourself have been deceived by the attitude they have taken up.

[1] *J.R. Statistic. Soc.* **105**, 69–80 (1942).

M.G. Kendall to Fisher: 7 *November* 1942

Thank you very much for your letter of 2nd November. I am glad you found some interest in my speculations and am sorry you were unable to be present. I assumed at the time, of course, that you would receive a notice on a card in the usual way, but apparently yours must have gone astray. Had I thought there was any danger of such a thing I would have made a point of letting you know specially, as I am sure the discussion would have benefitted from a contribution by you.

As regards inference I am sorry if you feel I have not given credit in the right quarter, but I am sure you know me well enough to accept my assurance that anything of the kind was quite unintentional. The question of confidence intervals versus fiducial probability still puzzles me a lot. Like several other people, I had supposed that they were rather different ways of looking at the same thing; but from the argument between yourself and Bartlett and the recent paper by Neyman in *Biometrika*[1] it appears that this is not so, since the two methods can give different answers to the same question. In more peaceful times I would have tried to get over to Harpenden to discuss it with you, but what used to be half-an-hour's trip in the car has now become a full day's journey.

An opportunity to set out the whole position fully and fairly will, I think, present itself when I get down to writing the second volume of my book on the Advanced Theory of Statistics. The first volume is now in the galley proof stage and ought to appear some time next spring; but it does not deal with inference systematically, although I have introduced likelihood on more or less intuitive grounds. The second volume will take about another two years to write, even if the war comes to an unexpected end, and I very much hope that during that time we may have an opportunity of a talk about the whole position. It is part of my programme to give an exposition of fiducial inference, but there are several points on which I should like to be sure that I have grasped exactly what you mean. I suppose you couldn't spare the time to lunch with me the next time you are in town? I can usually be found at the Chamber of Shipping, Avenue 5381.

P.S. You will, of course, receive a copy of my book from the publishers.

[1] Neyman, J. (1941). Fiducial argument and the theory of confidence intervals. *Biometrika* **32**, 128–50.

Fisher to M.G. Kendall: 9 November 1942

Thanks for your letter. I am sure there are plenty of differences, not only in exposition, but also, very probably, in results, between my approaches and Neyman's. However, the particular innovation of reasoning which you specified to the Statistical Society in the words[1]:

'The essential concept of the theory of confidence intervals, if I have understood it correctly, is the replacement of these specified limits by . . . functions of the sample numbers.'

is to be found quite explicitly in the paper to which I referred you. I may say it took a lot of private coaching by me to get this simple modification into Neyman's head. When he had grasped it, he proceeded, with Pearson's help, to expound it to the world with the minimum of reference to its origin.

I should love to meet you for lunch some time, and am fairly often in town on A.R.C. business, but their meetings usually make one too frightfully late. Is the Statistical Society likely to make an occasion?

[1] *J.R. Statist. Soc.* **105**, p. 73 (1942).

M.G. Kendall to Fisher: 23 October 1943

The publishers have just sent me a copy of your review of my book[1] in *Nature*, and I should like you to know how much I appreciate some of the kind things you have said and your understanding of the difficulties of writing it.

Perhaps the second volume will be more after your own heart. It has certainly been more interesting to work on, though more difficult to pull into line. I have nearly finished a first draft, although a further six months will be necessary to put it into final shape for the printers.

I had been intending asking your help on volume two in the following respect. In dealing with estimation I decided to have four chapters: Estimation by Likelihood; Estimation by other 'point' methods (e.g. minimum chi-squared, least-squares, the Aitken–Silverstone approach and so forth); Estimation by Confidence Intervals; and Estimation by Fiducial Inference. It is the last two which present the difficulty. Until the controversy over the Behrens–Fisher test I had supposed that confidence intervals and fiducial intervals were essentially two different ways of looking at the same thing. It now appears that this is not so, and to avoid any possibility of mixing up the two I decided, as I say, to have them in separate chapters.

I hardly think it would be proper for me to intrude my own personal views as to the relative merits of the two methods. In a book such as this I am anxious to give an objective presentation. The two chapters, therefore, will attempt to give an account of each theory and to explain the differences between them, but will not reject one in favour of the other.

I think you will agree that it is very important that the presentation of each theory shall be fair and accurate; and I wonder therefore whether you would be so kind as to read the draft of the chapter on fiducial probability and let me know whether you approve? It will not be ready in a readable form until some time after Christmas so there is no hurry. To safeguard myself on the confidence interval side I am asking Egon Pearson to read the chapter on confidence intervals. . . .

[1] *The advanced theory of statistics*, Vol. 1. Griffin, London (1943). For Fisher's review, see *Nature* **152**, 431.

Fisher to M.G. Kendall: 3(?) *November* 1943

Thanks for your letter. I shall be certainly be glad to do all I can to help, but you have, even in your letter to me, already adopted part of the language and terminology which Pearson rather maliciously as I thought introduced (after ignoring my work for fifteen years) with a view to misleading the student as to its scope and nature. The so-called 'point' or 'single-value' methods of estimation constitute the whole problem or subject of estimation in the sense in which this phrase was used for at least fifteen years. When later I introduced the fiducial argument and showed how fiducial limits could be calculated, this did not in my opinion constitute a new method of estimation but rather a new method of constructing probability statements valid in respect of the unknown parameter, obviously quite a closely related topic, but one applicable to a smaller range of cases.

The criteria of estimation originally developed were based in respect of efficiency on the variance of the sampling estimate, a fact which alone shows how misleading is the phrase, 'single-value' or 'point' estimation, applied to such methods. In respect of sufficiency the exact form of the whole sampling distribution is taken into account or, looked at otherwise, the entire course of the likelihood function.

For several years it is certain that Neyman thought, in speaking of 'confidence-intervals', that he was only systematising and developing a a new exposition appropriate to the fiducial argument I had put forward and pressed upon his notice. . . .

The diversity of opinion as to Behrens' problem arises, I think, by an historical accident. In our statements of probability, e.g. tests of significance, neither 'Student' nor I took the problem to relate to a population of samples all of the same size and drawn from an identical population. In 'Student's' problem we were both thinking of a population of samples of the same size drawn from populations having their variances fiducially distributed, i.e. that unique distribution of populations for which alone the observed estimate s^2 would be a typical estimate. It so happens, however, that when there is only one unknown parameter, and in a few other cases of special simplicity, the probability which defines the level of significance can be reinterpreted as a probability related to a population of samples drawn from the same population. So to interpret them is to my mind always to wander from the point, but Neyman and Pearson have made this convention the headcornerstone of their exposition and, as the Behrens' case shows, are much aggrieved when what is frankly the only possible solution of a practical problem does not conform with their convention. This however will generally be the case whenever tests have to be made in situations involving multiple estimation. It is indeed surprising that Neyman and his followers, who so often discuss a comparison of two samples, should be unable to provide any solution to this rather fundamental and primary problem. I presume, of course, that the method, following Bartlett, of choosing one out of $n!$ equally eligible tests of significance and using its results without reference to that of any of the others, is not to be regarded as a solution of the problem. Moreover, the hope, which I suppose Bartlett originally entertained, that there was an appropriate test which would reject a fixed proportion of samples for all values of the variable parameters when the samples are taken to be drawn from a fixed population, seems now to have been finally abandoned.

One fundamental objection to basing the theory on samples of a fixed size from a fixed population is the obstacle it introduces to the rational use of ancillary information. Properly speaking, the sample-number or set of such numbers used to specify the sample are ancillary statistics, and in particular cases are found to be supplanted by more appropriate ancillary statistics, though of course in other cases such statistics are additional to the sample-number. When, in such cases, estimation is exhaustive, the likelihood function is completely specified by the set of ancillaries, and it can be shown that all other statistical calculations add precisely no information to that already in hand; it seems absurd to pretend that there is a fundamental distinction between such cases and those which I originally called 'sufficient', which differ from them only in that in these the sample-number alone supplies all the ancillary information required. Yet in the method of exposition adopted by Neyman and Pearson these cases are separated from the first, and indeed the use of ancillary information seems to be almost ignored.

M.G. Kendall to Fisher: 7 November 1943

Thank you for your very interesting letter. I will send a chapter along when it is ready and hope you will be able to approve. In the meantime, a few comments on your letter:

I haven't used the 'point-estimation' phraseology at all up to the present. It must have slipped into my last letter because I had just been reading Wilks' latest effort on Mathematical Statistics (have you seen it? — I wasn't very impressed). As I understand it you conceive of the estimation problem as that of determining a value surrounded, so to speak, by an unavoidable band of doubt; whereas in Neyman's treatment (as he now states it) the interval is everything and there is no pretension of locating the estimate at one point rather than another.

I don't know what you thought of Jeffreys' comments on the matter in the *Ann. Eugen. 10*, 48,[1] but he seemed to me to have made a point when he compared your attitude with his own towards posterior probability. Yates also speaks of the fiducial distribution as a sort of distribution of doubt, with a mode at the estimate and tailing off on either side.

Would you agree if it were put this way? If one is not prepared to make any additional assumptions or postulates, or to restrict the nature of the inference, one can get no further than Neyman and the Behrens' problem is insoluble. It becomes so if one is prepared to [do] one of three things: (a) follow Jeffreys and use probability as a measure of rational doubt, not necessarily related to frequency; (b) restrict the inference to the family of parents for which the sample variance is distributed in the specified Type III form: (c) adopt a new postulate to the effect that differentials like ds can be replaced by those of type-$d\sigma$. Probably (b) and (c) are the same thing in the ultimate analysis and no doubt Jeffreys would maintain that they are both equivalent to (a) — though I am always nervous about saying what he might think because I never seem to express his views in the way he would like.

[1] See Fisher's letter of 8 November 1939 to Jeffreys (p. 176).

Fisher to M.G. Kendall: 10 *November* 1943

Thanks for your note. I suppose that all science, including all but the most highly purified mathematics, requires some such postulate as that observations are not generally made by especially privileged or exceptional observers. For most human purposes, though not for cosmology, something of the kind is needed merely on the ground that what is going to be called science is to be public knowledge, true for all (or nearly all) mankind, whereas the observations of the exceptionally privileged observer are by definition available and repeatable only by an equally privileged one. In cosmology the argument seems to take a different form but to be equally necessary, as is well illustrated by Milne's statement of the same principle.

The form the postulate takes when faced with what 'Student' called a 'unique sample' is that the only basis upon which a unique body of information subsumed in such an estimate as s^2 can be utilised is that s^2 is typical or representative, or not selected, biassed or sophisticated. On the other hand, for refined purposes, we shall never want to assume that s^2 is equal to the true unknown variance of the population sampled.

The known sampling distribution for s^2, given σ^2, then provides us with a well-defined random variable σ^2 such that for the population of parent populations having such a random variance, s^2 is a typical or representative estimate. This, I imagine, was clearly implicit in 'Student's' argument in 1908 and made quite explicit in my paper introducing Fiducial Limits in the Cambridge Philosophical Society's proceedings [*CP* 84].

I doubt if axiomatics go to the root of the difference of opinion, for any rational approach to the interpretation of data must contain some axiom justifying the propriety of using the data at all, or basing any inferences upon them. So far as I can see, the difference is due wholly to Neyman having seized upon an unessential feature of the simpler tests of significance in wide use when he first became acquainted with the subject, namely that the level of significance was in fact, in these simpler cases, the proportion of samples obtained by repeated sampling of a fixed population which would be rejected by the proposed test. This, I think, only *happens* to be true; the tests are good not because it is true but because of the different proposition that the proportion of the population of samples, of which that presented is a representative, which are rejected by the test is in fact the fraction specifying the level of significance.

As to axiomatics, I have quite an open mind. I do not believe that a complete system of axioms required for arithmetic or analysis has ever been successfully laid down. I am sure this would have been done if it were easy: so I suppose it is difficult, and that the satisfactory and flourishing development of a subject need not wait upon the precise definition, if this is ever achieved, of the system of axioms from which the subject might (as an academic exercise) later be developed.

N. Keyfitz to Fisher: 18 *November* 1955

I was delighted to see your article [*CP* 261] in the number of the *JRSS* which has just come in. It makes clear some very important issues about which I have been puzzled; several of my colleagues here join me in thanking you.

If we understand you, the difference between the situation of an acceptance sampler and that of an experimenter is the difference between facing a problem of human devising and facing a problem presented by nature. The acceptance sampler who has to discriminate between two distributions may perhaps be compared with a student of probability discriminating between two urns in a textbook problem exemplifying Bayes' Theorem. To make the investigation of nature analogous to sampling a controlled production process apparently is only useful if the capacity of nature for offering surprises is of the same order as the production process. You presumably fear that to promote acceptance methods in science does more harm than merely distorting the meaning of some words like 'decision' and producing irrelevant mathematical embroidery; it helps hide from the investigator the tentative process by which alone he can uncover new facets of the real world. The ideological implications of a mechanical decision theory are terrifying if one thinks about them along the line to which your article points. . . .

Fisher to N. Keyfitz: 21 *November* 1955

I was very glad to get your letter of November 18th, and I may say that from other sources I learn that the article has led to a healthy reconsideration of what we are doing in some fields, and I hope it will lead to a recognition of when we are and when we are not attempting to add to natural knowledge.

I think it will be common ground in the technological fields generally that the technologist is constantly concerned with fact finding, or with seeking verification of provisional hypothetical suppositions of the scientific type. In quality control, for example, I imagine that the high level work is all of this kind, involving the whole range of concepts of experimental design, and that only the warning bells and red lights can usefully be mechanized in an automatic acceptance procedure.

What is particularly troublesome is that Neyman, in importing from Eastern Europe his misconceptions as to the nature of scientific research, should have chosen so ubiquitous a scientific tool as the test of significance as the subject on which to fasten ideas relevant only to the acceptance procedure. A typical test of significance is based on a probability statement derived from the hypothesis to be tested, and therefore existing only in the hypothetical world created by this hypothesis. Typically it leads to no probability statement in the real world, but to a change in the investigator's attitude towards the hypothesis under consideration, for which if we choose to use the word 'rejection' we must remember that the rejection is only provisional, and that our hypothetical calculations have shown that there would be a finite probability of our obtaining the observed level of significance even were the hypothesis true. In fact future observations may later demonstrate, with all the force of which scientific evidence is capable, that it is really true, and the the provisional rejection was due to an exceptional coincidence. Of course we may on occasions take irreversible action on the strength of such evidence, but such a possibility does not at all affect the logical status of the evidence upon which it is based. . . .

Fisher to T. Koopmans: 29 *November* 1935

. . . As I expect you noticed, in my paper, 'Two new properties' [*CP* 108], I dissent from the view expressed by Professor Pearson and Dr. Neyman that the results they have arrived at cannot be obtained by the theory of estimation.[1] Indeed, I think their remark was only made in self-defence from the obvious criticism that the only result of practical importance to which their methods had led was a restatement of an easily demonstrable property of sufficient statistics.

As I did not introduce the term 'confidence intervals' I cannot object to its use in any circumstances for which it was intended. By themselves they seem to be only another way of saying that, by a certain test of significance, some kinds of hypothetical possibilities are to be rejected, while others are not. This is not what struck me as a novelty when I first put forward the notion of fiducial probability, namely, that when properly defined, rigorously exact probability statements could be made about the values of unknown parameters, the whole set of such statements providing the fiducial frequency distribution of these parameters. If for this purpose non-sufficient estimates are employed, the statements of fiducial probability and the frequency distribution which comprehends them will be as definitely erroneous as if we had arbitrarily rejected or ignored a portion of the available data and based our conclusions upon the selected remainder.

With respect to the general form of the parent distributions, when sufficient estimation is possible you may be interested to see that I arrive at conclusions very like those which you give, in the section of the paper on 'Two new properties' preceding that to which you call attention, the section in which I develop the general sampling distribution of sufficient statistics for simple estimation. I suppose, though I have not myself verified it, that the simultaneous sampling distribution of a set of statistics sufficient for the estimation simultaneously of several parameters could be developed in the same way.

[1] Koopmans had written asking (i) whether there was a 'real difference between the approach of Professor Pearson and Dr. Neyman in their papers on hypothesis testing and that which you call "theory of estimation" ' and (ii) what were the grounds on which Fisher, in his contribution to the discussion of Neyman's 1934 paper on the representative method (*J.R. Statist. Soc.* **97**, 558–606), 'deprecated the use of "confidence intervals" based on non-sufficient statistics'.

Fisher to H.E. Kyburg: 12 *January* 1962

I have just read your letter of Oct. 17, which must have travelled a long way since it left your hand; moreover it is so kindly expressed that you must have been puzzled at not receiving a prompt reply.

I only realised the backwardness of many departments in U.S. purporting to teach mathematical statistics, when I put in a term at Michigan State (I fancy it was in the Fall of 1956 [1957], at least the first Sputnik went up while I was there). I was in fact very deeply shocked not only by the poor material being taught to the students, but by the spirit of ideological propaganda, in which other views were 'bawled down'. That seemed to me to have no place in a centre of learning. . . .

However, just in this last year, those in the States who thought there was some sense in what I had been saying, have been reinforced by two vigorous voices from across the northern frontier: Donald Fraser (Toronto), whom I have not yet met, and D.A. Sprott from a rather unknown University (Waterloo) whom I met last May in George Barnard's department at Imperial College. So the moulding of mass opinion has not been fully successful. Sprott thought that *no one* in U.S. would agree with him. I think he knows better now. . . .

Fisher to H.E. Kyburg: 22 *February* 1962

I believe in answering your most flattering letter I may have misunderstood you on one point. I supposed that your book was already published, and intended to procure a copy.

It now seems possible that this is not so, and that you were offering me the favour of reading typescript or proof prior to publication. I should at once accept this kindly offer; for though I am blind as any bat, I can still do some reading, and I am indeed immensely curious that an approach through Symbolic Logic should have so closely concurred with my own approach through experimentation.

I should hate you to think that I had ignored so friendly a gesture.

I expect you will have heard from Rao in Calcutta that he wants you to talk, if you can bear it, at a symposium of the International Statistical Institute at Ottawa in the Summer of 1963.

With luck I hope to be there.

[P.S.] I am at work on the rather difficult job of specifying when, and when not, the data supply the materials for fiducial probability inferences.

Fisher to H.E. Kyburg: 14 *May* 1962

After a long while I have now succeeded in obtaining your book on Rational Belief.[1] So far it seems to be as good as I had hoped, which would indeed be high praise.

As you know I am inclined to stratify Keynes' wide notion of degree of rational belief, according to the nature of the evidence, into Mathematical Probability in the sense of the 17th century mathematicians which does not include the whole of uncertainty as Voltaire and Montesquieu assumed, and a number of humbler states of uncertainty including the Likelihood Statements, almost always available in scientific work, and which have recently been rediscovered by Birnbaum, whom perhaps you know. I forget if I ever sent you a puzzling paper called the Underworld [*CP* 267].

Anyway let me know if ever you want any such thing of mine. I wonder if the Underworld could be translated into your logical symbolism. . . .

[1] *Probability and the logic of rational belief.* Wesleyan University Press (1961).

J. Neyman to Fisher: 9 *February* 1932

. . . I do not know whether you remember what I said, when being in Harpenden in January 1931, about our attempts to build the theory of 'the best tests'. Now it is done more or less. It follows from the theory that the 'Students' test, the '*t*-test' for two samples and some tests arising from the analysis of variance are 'the best tests', that is to say, that they guarantee the minimum frequency of errors both in rejecting a true hypothesis and in accepting a false one. Certainly they must be properly applied. You will see that these results are in splendid disagreement with the last articles of Professor Pearson in *Biometrika*.

Presently Dr. Pearson is putting all the results in order. They will form a paper of considerable size. We would like very much to have them published in the *Philosophical Transactions*, but we do not know whether anybody will be willing to examine a large paper and eventually present it for being printed. The paper contains much of mathematics and not all the statisticians will like it just because of this circumstance.

We think that the most proper critic are you, but we don't know whether you will be inclined to spend your time to read the paper. . . .

Fisher to J. Neyman: 12 *February* 1932

. . . I should be very much interested to see your paper on 'the best tests', as the whole question of tests of significance seems to me to be of immense philosophical importance, and the work you showed me was surely of great promise. It is quite probable that if the work is submitted to the Royal Society, I might be asked to act as referee, and in that case I certainly shall not refuse.

J. Neyman to Fisher: 16 *October* 1932

Reading the paper by A.T. McKay: 'Distribution of the coefficient of variation and the extended *t* distribution' in the *J.R.S.S.*, Vol. XCV, Part IV, p. 695, I have been alarmed by the information that you have been doing something with this distribution and that the results are published in the B.A.M.T.[1]

As a matter of fact the said distribution is being tabled in this laboratory and the work is already fairly advanced. Unfortunately our financial conditions do not allow us to purchase the B.A.M.T., at least in the nearest future, so I beg you to be kind enough to inform me what sort of tables concerning the distribution of t do they contain.

E.S.P. writes that you have recommended our paper for publication. Although it may be considered ridiculous to thank a judge, I have intense feeling of gratefulness, which I hope you will kindly accept. . . .

[1] British Association *Mathematical Tables*.

Fisher to J. Neyman: 24 *October* 1932

It is a great pleasure to hear from you again. About five years ago in order to encourage Airey to calculate the successive integrals of the Probability integral, I drew out some notes on the theory of the functions concerned, their connection with the Hermite functions, and their statistical use. My notes came with Airey's table into the hands of the British Association Committee, from whom, on account of their incompetence and extravagance I was later obliged to dissociate myself, and they published my notes [*CP* 91] in the Introduction without submitting a proof to me, or giving me an opportunity of obtaining offprints. E.S. Pearson is a member of the Committee and might have been able to tell you what the book of tables contains.

My note gives the theory only, giving the solution in terms of the *Hh* functions, which are pretty well tabulated by Airey. Tables of chosen percentiles for varying n and τ would of course be much more useful in application. I am surprised and interested to hear that you have been working on the same distribution; but before assuming this you ought to be sure that it is the same.

I was a good deal interested in your paper,[1] but I ought not to have passed your statement that a distribution is completely determined by its moments. Do you not know Borel's distribution with all its moments zero?

$$e^{-4\sqrt{x}} \sin {}^4\sqrt{x}\, dx.$$

You can always add a bit of Borel to any distribution without altering the moments, and if your distribution has tails less steep than $e^{-4\sqrt{x}}$, you need get no negative frequencies. . . .

[1] Eventually published as Neyman, J. and Pearson, E.S. (1933). On the problem of the most efficient tests of statistical hypotheses. *Phil. Trans.* **A 231**, 289–337.

J. Neyman to Fisher: 28(?) *October* 1932

Many thanks for your kind letter. I am glad to hear that the work of my students is not lost. On the other hand I am rather troubled by your criticism, which seems to be quite correct. The origin of our mistake is the same statement in P. Lévy's book, concerning the continuous distributions. I know of course that the set of moments does not necessarily determine the frequency distribution and that there are means to decide whether it does or not. /See f.i. the papers of H. Hamburger 'Ueber die Erweiterung des Stieltjesschen Momentenproblems', *Mathematische Annalen* Bd. 81, 82./

Unfortunately the criteria are not easy to apply, so — being very much occupied with the new notions of best critical region etc. — we believed too much in what said Lévy.

At present I see that a restriction is necessary.

Namely our theory holds good in cases when the set of moments of ϕ — if you remember the notation — leads to a unique frequency function, and this question may be solved by means of the criteria, given by H. Hamburger.

As a matter of fact, I think that in all practical examples we have considered, the theory holds.

Of course there is lot of questions to be worked out. There are too many restrictions.

I am often thinking that it would be very useful for me to work in contact with you. Unfortunately this requires considerable amount of money — without speaking of your consent — of course.

Fisher to J. Neyman: 1 *November* 1932

You may be sure of my consent, if ever it becomes possible for you to work with me. The difficulty you speak of arises I believe solely through working with the moments instead of with the characteristic function, in the Taylor expansion of which, if it has one, the moments appear. I do not think it affects your work on the best critical region.

J. Neyman to Fisher: 3 *December* 1932

A letter more! I am afraid you have already sufficient of correspondence with Poland. However the question seems to be important. We are considering with E.S. Pearson what sort of reference it will be necessary to give in the paper about your showing us the inaccuracy with the moments etc.[1] Can it be: Formerly we did not see that this restriction* is necessary and we are indebted to Dr. R.A. Fisher, who kindly called our attention to the fact that the set of moments does not necessarily determine the corresponding frequency function.

*Namely: 'in cases when the set of moments . . . etc.'

Will it do?

I hope to be in England just after the 20th of December. But the proofs may come any time. So perhaps you will be kind to answer — at least if you do not like the proposed text. . . .

[1] See Neyman, J. and Pearson, E.S. (1933) On the problem of the most efficient tests of statistical hypotheses. *Phil. Trans.* **A 231**, p. 315.

Fisher to J. Neyman: 7 December 1932

Do not think of thanking me especially for the little point I made about the moments. It really does not affect your paper and I only mentioned it casually because I was writing to you. You and Dr. Pearson are not indebted to me for anything more than, as it were, the correction of a verbal slip.

All the same, I like hearing from Poland. Best wishes for a Merry Christmas.

J. Neyman to Fisher: 13 *June* 1933

Dr. Pearson writes me that soon you will be Galton Professor at the University College, London. Very probably this means a general reorganization of the Department of Applied Statistics and possibly new people will be needed. I know that there are many statisticians in England and that many of them would be willing to work under you. But improbable things do happen sometimes and you may have a vacant position in your Laboratory. In that case please consider whether I can be of any use.

Fisher to J. Neyman: 17 *June* 1933

Many thanks for your letter of congratulation. You will be interested to hear that the Dept. of Statistics has now been separated officially from the Galton Laboratory. I think Egon Pearson is designated as Reader in Statistics. This arrangement will be much laughed at, but it will be rather a poor joke, I fancy, for both Pearson and myself. I think, however, we will make the best of it. I shall not lecture on Statistics, but probably on 'the Logic of Experimentation', so that my lectures will not be troubled by students who cannot see through a wire fence. I wish I had a fine place for you, but it will be long before my new department can be given any sort of unity and coherence, and you will be Head of a Faculty before I shall be able to get much done. If in England do not fail to see me at University College.

M.H. Quenouille to Fisher: 11 *March* 1957

Recently at a meeting in Imperial College, D.V. Lindley put forward an example which he alleged showed that Fiducial Distributions could not be used with Bayes' theorem. I was not present at this meeting, but I have managed to borrow the enclosed note circulated by Lindley. I understand that it is due to be published in *Biometrika.*[1]

The other enclosed note indicates how I think Lindley's arguments should be refuted. I should be grateful for any comments or suggestions that you may have on this. Would you please return the Lindley note when you have finished with it as I have this on loan?

There is a second matter which I should like to take this opportunity of raising with you. In 1952, I started a book entitled 'Foundations of Statistical Thought' which I expect is to be published this year. At the time Dr. Yates was kind enough to look over and comment upon the chapters concerning fiducial inference. My attitude on this agrees with your own apart from one or two relatively minor points.

The recent publication of your own book [*SMSI*] has made necessary several alterations in the text as originally planned, and I am still carrying this out. However, I should be very grateful if you have time to look over and comment upon the manuscript as finally prepared for press. I should explain that the book[2] is intended to be a concise account, for students, of the methods and, in particular, ideas that are in current use in statistics.

P.S. On page 135 of your book, the Bessel function should be $I_0(aR) = J_0(aiR)$.

[1] Published eventually as Lindley, D.V. (1958). Fiducial distributions and Bayes's theorem. *J.R. Statist. Soc.* **B 20**, 102–7.
[2] Published as *The fundamentals of statistical reasoning*. Griffin, London (1958).

Fisher to M.H. Quenouille: 12 *March* 1957

I was very much interested to have your letter and its enclosures. There have been so many variations in the notation used for Bessel functions that I am not yet sure whether I have, as you say, introduced yet another variant, or whether I have, as I thought, been following one of the reputed or semi-accepted notations; not, I think, that it matters much so long as the function concerned is independently defined.

As regards what you say about fiducial inference and Bayes' theorem, I imagine that Lindley's paper was put forward after conferring with George Barnard, who in turn had been discussing some aspects of these matters with me. Originally Lindley was inclined to attack the whole concept of fiducial probability on a wide front. I put it to George that the particular example in Chapter V [*SMSI*], in which I had used a fiducial inference in a Bayes-like manner, was objectionable in the first instance because I had ignored the requirement emphasized in Chapter III of sufficient estimation, and also for a more subtle reason, which may not appeal to everybody, namely that Bayes' argument was based quite explicitly on an independent prior act of sampling, and not only on the statement of probability derivable from that act. In fact, however, the parent population, and the act of sampling from it, cannot be inferred from the probability statement, and therefore contain a logically independent ingredient, which is lacking when only the probability statements are available.

Otherwise, of course, after deriving a probability statement from the sample, we could infer something about the parent population, which would be irrational since our information concerns only one population, and beyond that, nothing about any parent population from which it might have been derived by sampling, a point which I tried to make on page 120[1] in the paragraph overlapping page 121, which is perhaps more intelligible if the misprint of 'or' for 'of', at the end of the fourth line, is corrected.

Anyway, I am replacing example 6 (page 123) by a more competent one proving the same point in the second edition, which has already been asked for, and I presume that by the time Lindley came to give the paper you refer to at Imperial College, he thought it wiser to confine his attention to an example which, presumably, he already knew I wish to retract.

Do you know if 'Neyman's accredited representative' in this country is still opposing Behrens' solution of the comparison of the means of two normal samples on the ground that on 'repeated samples from the same population' his test gives a significant result less frequently in some special cases of the composite hypothesis tested than in others, or than the level of significance would imply for a simple hypothesis?

You might like to see the replacement section I have drafted. Please send it back and do not let it be made a means of attacking my work, until it is published.

[1] *SMSI*, p. 124–5.

M.H. Quenouille to Fisher: 28 March 1957

Thank you for your letter and comments. I am returning the draft of the replacement section which you kindly loaned me.

I am not sure about the reasons (in particular, the first) that you give for changing this section. It seems to me that if joint fiducial distributions are to be something other than the product of independent fiducial distributions, some step involving statistics that are not fully sufficient is required. My formula (1) seems to me to be useful (whether it is called Bayes' theorem or not) in that it serves to connect fiducial distributions based upon partial information with those based upon the whole. In particular, as might be expected where the difference between x and y lies in the number of independent observations n and m used in their calculation, the two integrals given in that formula are in the ratio $n : m$. Hence

$$f_d(\theta \mid z) = \frac{n+m}{m} \int_{z(x,y)=z} f_d(\theta \mid y) f_1(x \mid \theta)\, dx.$$

A particular instance where this might be used is in relating the fiducial distribution of σ^2, given $a = \bar{x} - \mu$, to that when a is not given. Here

$$f_d(\sigma^2 \mid S^2) = \frac{n}{n-1} \int_{nS^2=(n-1)s^2+na^2} f_d(\sigma^2 \mid s^2) f_1(a^2 \mid \sigma^2)\, da^2.$$

This seems to me to provide a justification for saying that fiducial statements of σ^2, given s^2, are similar in character to statements of σ^2, given S^2.

If you could manage it, I should welcome an opportunity to discuss this and other matters with you. The enclosed notes written in 1953, (with which I don't fully agree now) still show a number of points of difficulty, the major one being whether a distinction should be drawn between fiducial limits of estimation and fiducial limits of significance. It seems to me that a good case exists for doing so.

Fisher to M.H. Quenouille: 29 March 1957

I should be glad to discuss your recent note and your Private Report No. 2 on 'Problems involving restricted parameters' whenever you can make me a visit, as I hope you will sometime. From London it is not too inconvenient to come down perhaps and lunch with me, but for a longer stop I could, of course, put you up as my guest in Caius.

I believe all these problems are intrinsically simpler than has sometimes been thought, but that a few fundamental ideas have to be straightened out, if misapprehensions are not to intrude themselves.

My previous example of combining information at two levels was, I think, faulty in principle and in detail, and that any number of other simpler examples would illustrate better all that I have to say.[1] What seems to me now important, and what was indeed always obvious, is that a probability statement representing a certain type of partial knowledge, of the same kind as that which might have been derived from a previous sampling process, does not imply the historical reality of any such process, or of the bulk, or population, from which the sampling is supposed to have taken place. In fact, Bayes postulated more than probability *a priori*; he postulated a prior sampling process, and it was this that allowed him to combine the probabilities of two sampling operations by the product formula, for they were known to be independent. A probability statement, on the other hand, derived from observations is not independent of any subsequent observations, and the combination must be effected at the observational level, and not at the level of inferred probability statements.

Let me know when you are coming.

[1] See also Fisher's letter of 8 February 1958 to Barnard (p. 38).

Fisher to C.R. Rao: 6 February 1956

I have just seen your *Sankhyā* paper[1] 'On Fisher's lower bound', and it will certainly do nothing but good to bring the discussion of consistency into the open. You may have seen, for example, Anscombe's review[2] of the 12th edition of *Statistical Methods* in which, while complaining that there is nothing new in this old book, he shows that he cannot recognize anything new to himself by confusing the definition of consistency, made explicit for the first time in the 12th edition, with a mere stipulation of lack of bias, which since it is not invariant for functional transformation of parameters has never had the least interest for me.

I was, however, surprised by the statement at the foot of page 333 that I 'was considering only analytic functions of frequencies'. I do not believe I introduced the word 'analytic', for it seems to me that the properties of *analytic* functions, as developed in the theory of the complex variable, are really quite irrelevant. I did stress, not as an interpretation of my earlier definitions, but as a result of later reflection, that we are concerned in estimating only ultimately with functions linear in the frequencies, though I do not see why the estimate should not be a non-analytic function of such a function.

The introduction of the word 'estimator', I suppose by Maurice Kendall, seems well fitted in this connection to confuse the discussion by slurring over the distinction between a method of estimation on one side, and on the other the particular estimates, or class of estimates, to which this method leads in a particular case or class of cases.

However I am only concerned now to make it clear that the properties of analytic functions are to my mind not important in the stipulation that an estimate should be a consistent one. The suggestion to the contrary does suggest, as indeed you do on page 334, that I have changed my mind in this matter, a suggestion that will give great pleasure in Berkeley. I must rely on you to check their exuberance about this.

[1] Kallianpur, G. and Rao, C.R. (1955). On Fisher's lower bound to asymptotic variance of a consistent estimate. *Sankhyā* **15**, 331–42.
[2] *J.R. Statist. Soc.* **A 118**, 486–7 (1955).

C.R. Rao to Fisher: 10 *February* 1956

Thanks for your letter of 6 February 1956. Our main aim in writing the paper is to draw the attention of the Statisticians to your definition of consistency. We are sorry for the paras. at the foot of page 333 and the top of 334 of our paper. We are aware that you have not introduced the word analytic in your writings but I remember to have heard your mentioning about the analyticity of the statistic as a function of frequencies. Maybe it is based on a misunderstanding. We will publish a correction in the next issue of *Sankhyā*.

Of course, so far as the definition of consistency is concerned analyticity of the statistic is not needed. But it seemed necessary to assume continuity to prove probability convergence. We had to state this because there exist artificial examples of a Fisher consistent statistic which does not tend to the true value of the parameter in probability. We needed differentiability to establish lower limit to variance. Otherwise counter examples can be produced. We wanted to stress these two analytical properties.

Fisher to C.R. Rao: 16 *February* 1956

Thank you for your letter. I am still not quite clear what your ideas are, e.g. why you feel it necessary to establish a limiting property for large samples, when that property is well defined for all samples of finite size. Is this not a very long way round? I do not think any counter-examples are even conceivable if the matter is treated solely from the point of view of finite samples, nor would it matter if they were, for misbehaviour at the limit would be a very academic peril to a worker whose examples are, and always will be, finite. However, I may have misunderstood you. Perhaps you will let me see, before it appears, the correction you propose in the next issue of *Sankhyā*.

Fisher to C.R. Rao: 21 *February* 1956

I have read your letter of February 15th, with the note of correction to be published shortly in *Sankhyā*.[1] I believe its publication will minimize confusion, but that some writers will try to take advantage of the original passage.

[1] See *Sankhyā* **16**, 206 (1956).

C.R. Rao to Fisher: 6 *March* 1956

Thank you for your letters of February 16 and 21, 1956. I am glad to have the opportunity of corresponding with you and getting my ideas clear about this problem. I realise that limiting property is not important. But I am worried if somebody defines a function in the sample space such that on the curve of expectation it has the value θ, the value of the parameter, and arbitrary elsewhere. Strong consistency or what we have called Fisher consistency is satisfied but such a function defined above cannot be used as an estimate. The condition of continuity in the neighbourhood of the curve of expectation may be a reasonable restriction on the function. This implies the limiting property of convergence in probability which is used merely as an alternative condition and is somewhat weaker than strict continuity.

Fisher to C.R. Rao: 20 *March* 1956

I am glad to have your letter of March 6th, though perhaps I do not fully grasp what you are saying. If you consider a generalized space representing class frequencies observed, possible observations will fall on the lattice points throughout this space.

A continuous curve will represent multiple expectations for different values of the parameter, and therefore different points on this expectation curve are identified with the values of the particular parametric function used (of course more generally we may have a p-fold expectation space corresponding with simultaneous estimates of p parameters).

If a 'statistic' means a determined function of the observed values, this statistic may be the same, or very nearly the same, at different lattice points, and if it is expressed explicitly in terms of the observed frequencies, it will determine a space of appropriate dimensionality passing through and among the lattice points and cutting the curve of expectations at a point already identified with a particular value of θ. This same value of T is then assigned to the estimate common to the whole equi-statistical region, if the estimate is to be consistent. In fact the expectation curve is used to assign metrical values to any system of non-intersecting equi-statistical regions.

Consequently I am not sure what you mean when you say 'if somebody defines a function in the sample space such that on the curve of expectation it has the value θ and arbitrary elsewhere'.

Formally the problem is simplified by considering only functions linear in the observed frequencies represented by regions planar in the multi-dimensional coordinates, and intersecting only in the impossible regions in which some frequencies are negative. It is intuitively obvious that only one such system can be both '*consistent*', that is have its regions correctly labelled, and *efficient* in the sense of striking the expectation line orthogonally, when orthogonality is defined in a Poissonian sense in relation to the expected frequencies at the point of intersection.

The justification, which I think should be not a prior but a posterior consideration, for using demonstrations appropriate only to functions linear in the frequencies, is that all other functions lead to discontinuities in the statistics calculated, when minimal changes in continuous measurements lead two distinct values to coincide, for any accurate measurement must be regarded as a frequency of unity for the number of occurrences in an infinitesimal range of its continuum.

George Barnard has also been reading your paper with great interest.

C.R. Rao to Fisher: 29 *December* 1961

I could find some time this morning to give a first attempt.[1] The results are as follows.

u_1, u_2, u_3: independent normal variables, zero mean and unit variance.

v_1, v_2, v_3: indep. χ^2 variables, apart from constant multipliers, with $n-1$, $n-2$ and $n-3$ d.f.

Then there exist functions $f_i(x,y,z)$, $i = 1,2,3$, such that

$$u_1 = v_2 f_1(r) - v_1 f_1(\rho)$$
$$u_2 = v_3 f_2(r) - v_2 f_2(\rho)$$
$$u_3 = v_3 f_3(r) - v_1 f_3(\rho) + f_1(r) u_2^{\dagger}$$

† this term arises due to orthogonalisation of u_2, u_3.

$$f_1(r) = \frac{r_{12}}{\sqrt{1-r_{12}^2}} \text{ as in Fraser}^2$$

$$f_2(r) = \frac{r_{32.1}}{\sqrt{1-r_{32.1}^2}}$$

$$f_3(r) = \frac{r_{31.2}}{\sqrt{1-r_{31.2}^2}} \cdot \frac{1}{\sqrt{1-r_{12}^2}} = \frac{r_{31.2}}{\sqrt{1-R_{1.23}^2}}$$

$f(\rho)$ are corresponding functions of ρ.

This is capable of generalization but I don't know whether any simplification is achieved.

I have just started checking whether the fiducial distribution of ρ obtained from Fraser's equation is the same as yours obtained by inversion from the distribution of r. The answer is yes by 'general reasoning' but I must work out the details.

I hope to give some time to the other problem of fiducial distribution of a future vector observation $(x_1, \ldots, x_p)$ given a sample of n observations from a multivariate normal distribution with an unknown mean and covariance matrix. I mentioned in a paper about 8 years ago that this distribution is useful in classification problems but thought it would resemble Hotelling's distribution. Recent work of Cornish shows that there may be slight differences.

[1] See Fisher's correspondence at this time with Fraser (p. 106) and Sprott (p. 215).

[2] Fisher has written in the margin alongside these three equations, cot C, cot A and cot B/cot C, respectively.

Fisher to C.R. Rao: 23 *March* 1962

Did Fraser decide to visit you in India after all? I do not know whether you have been thinking of the trivariate problem. If you have made any progress you will not need this, but if not it may start a train of thought.

We want three pivotals each varying monotonically in one parametric function. I suggest

(a) a function of r_{23}, ρ_{23} such as Fraser's,

(b) a function of R^2 and $\pi_1{}^2$,

$$R^2 = \frac{r_{12}^2 + r_{13}^2 - 2r_{12}r_{13}r_{23}}{1-r_{23}^2}, \quad \pi_1^2 = \frac{\rho_{12}^2 + \rho_{13}^2 - 2\rho_{12}\rho_{13}\rho_{23}}{1-\rho_{23}^2}.$$

It is known that the distribution of R^2 depends only on π_1^2. For N odd the probability integral has been given explicitly in $(N-3)(N-1)/8$ terms. In my 1928 paper on the Multiple Correlation [*CP* 61] I also bring in ψ. I think we might define ψ as

$$\tan^{-1}\frac{(r_{13}-r_{12})\sqrt{1+r_{23}}}{(r_{13}+r_{12})\sqrt{1-r_{23}}} - \tan^{-1}\frac{(\rho_{13}-\rho_{12})\sqrt{1+\rho_{23}}}{(\rho_{13}+\rho_{12})\sqrt{1-\rho_{23}}}$$

with the simultaneous distribution

$$\frac{\{(N-2)/2\}!\,(1-\rho_{23}^2)^{(N-1)/2}(1-r_{23}^2)^{(N-4)/2}}{\pi^{5/2}(N-3)!\,\{(N-5)/2\}!}\,\mathrm{d}r_{23}\left(\frac{\partial}{\sin\theta_1\partial\theta_1}\right)^{N-2}\frac{\theta_1}{\sin\theta_1}$$

$$\times\ (1-\pi_1^2)^{(N-1)/2}(1-R^2)^{(N-5)/2}\mathrm{d}(R^2)\mathrm{d}\psi\int_{-\infty}^{\infty}\frac{\mathrm{d}z}{(\cosh z-\pi_1 R\cos\psi)^{N-1}}$$

[where] $\cos\theta_1 = -\,r_{23}\,\rho_{23}$.
The first line is dP, marginal, when $P(r_{23}, \rho_{23})$ is the probability of the variate r_{23} being less than any given value r_{23}.

This convinces me that a trivariate fiducial distribution *exists*. I doubt if I shall ever clean it up. You might find the introduction to the 1928 paper worth re-reading. It has been called obscure.

The human mind tends to think everything new to it, or from which it has anything to learn, obscure.

If it works the process would seem to generalize easily for stepping up from $t-1$ dimensions to t.

V.I. Romanovsky to Fisher: 17 *March* 1931

I thank you very much for the copies of your and your pupils' papers.

Your paper on inverse probability is very interesting but I cannot wholly consent with it. Let us take the simplest case: an event of the unknown probability p is observed $np'(0 < p' < 1)$ times in very large number n of trials. Let $f(\mathrm{p})\mathrm{d}p$ be *any* continued differential *a priori* distribution of p, then it can be shown (e.g. in my note in the *Comptes Rendus* of Paris on the *a posteriori* probabilities, for 1929, October) that

$$\int_{p'-\alpha M}^{p'+\alpha M} f(x)\,x^{np'}\,(1-x)^{n(1-p\,)}\mathrm{d}x \Big/ \int_0^1 f(x)\,x^{np}\,(1-x)^{n(1-p')}\mathrm{d}x$$

($M = \sqrt{2p'\,(1-p')/n}$, α — arbitrary positive number) is asymptotically the same with

$$\int_{p'-\alpha M}^{p'+\alpha M} x^{np'}(1-x)^{n(1-p')}\mathrm{d}x \Big/ \int_0^1 x^{np'}(1-x)^{n(1-p')}\mathrm{d}x$$

and the latter expression corresponds to the supposition that $f(x)$ = Const. I am sure that in many more complex cases we shall have the analogous results and then the theory of inverse probability is saved for these cases provided we have sufficiently large numbers of corresponding experiments. But it must be confessed that in the case of small samples this theory fails: then we cannot get rid of the functions like $f(p)$.

Moreover I think that your maximum likelihood is, if not logically, analytically very closely connected with the inverse probability. Indeed, the equivalence of the above written expressions and of all other analogons [*sic*] of them is based on the existence of certain maximums.

In conclusion I shall communicate you a theorem recently demonstrated by me.

Let $x_1^{(1)}, x_2^{(1)}, \ldots$ be a series of independent random variables such that $E\{x_i^{(1)}\} = 0$, $E\{x_i^{(1)2}\} = \sigma_x^2 = \text{const.}$, $E\{x_i^{(1)}x_{i+1}^{(1)}\} = 0$ $(i = 1, 2, \ldots)$, E being the sign of mathematical expectation. Consider

$$x_i^{(2)} = x_i^{(1)} + x_{i+1}^{(1)} + \ldots + x_{i+s-1}^{(1)}$$

$$x_i^{(3)} = x_1^{(2)} + x_{i+1}^{(2)} + \ldots + x_{i+s-1}^{(2)}$$

....

$$y_i \equiv x_i^{(n)} = x_i^{(n-1)} + x_{i+1}^{(n-1)} + \ldots x_{i+s-1}^{(n-1)}$$

(s — any integer $\geqq 2$) and

$$z_i = \Delta^m y_i = z_{i+m} - C_m^1 z_{i+m-1} + \ldots + (-1)^m z_i.$$

Then, if $m/n \to \alpha \neq 1$ as $n \to \infty$, the series $z_1, z_2, \ldots$ is obeying to a limiting sinus law with the period L defined by the equation $\cos(2\pi/L) = r_1$, r_1 being the correlation coefficient of z_i and z_{i+1}, that is to say, for n sufficiently great, with the probability arbitrarily near to 1, the series $z_1, z_2, \ldots$ will follow the above mentioned sinusoide as near and during as many consecutive periods as you wish.

This theorem is the generalisation of the special case ($s = 2$) demonstrated by Mr. E. Slutsky (*C. R.* of Paris, t.185, p. 169) and costed me much efforts.

It is evident that, with aid of this theorem, we can construct such series of dependent random variables that will follow an arbitrary prescribed limiting harmonic law. I find this most remarkable!

It is worth to mention that the stochastical scheme lying in the base of the series $z_1, z_2, \ldots$ is a simple discrete or continuous chain of Markoff. . . .

Fisher to V.I. Romanovsky: 31 *March* 1931

Thanks for your interesting letter of March 17th. I fully admit the truth in the limit $n \to \infty$ of the expression you give, but I do feel the difficulty that unless we set an upper limit to $d(\log f)/dp$ in the neighbourhood of the population value of p, we cannot say that our sample is large enough for the inverse inference to be justified. Sometimes certainly no such limit can be postulated. For example, the intraclass correlation for fraternities of k cannot be less than $-1/(k-1)$; hence unless there is a law of nature that k cannot exceed some determinate limit, the intraclass correlation in the population cannot be negative at all. But it may very well be zero. There is thus in this case a sharp discontinuity in f, and I feel doubtful whether in other cases, though we cannot prove the existence of the discontinuity, we have any right to assume its non-existence.

When the Mendelian ratios were discovered Mendel had many families giving approximately the ratio 3:1; the observations were equally compatible with a theory that the ratio was a transcendental number sufficiently close to 3. His inference that the true ratio was just the whole number must have been based upon the view that *a priori* the whole number ratios were more probable than the irrational ratios in their neighbourhood. Here again f is discontinuous.

Your new theorem strikes me as very remarkable, and I hope you will be publishing the whole investigation; at present, I do not fully understand it, but will be sure to return to the matter, as soon as I have less correspondence to deal with. . . .

L.J. Savage to Fisher: 10 *August* 1952

Thank you for your recent letter.

I hear that you are expected back in Cambridge on the 20th. I shall be here until the 25th, so there is some possibility that I shall have the pleasure of seeing you. My immediate purpose is to mention two technical matters I hope will interest you and about which I would like to hear your reaction.

The first concerns your 'problem of the Nile' (p. *27.257* in *Contributions* [*CP* 137]).

Suppose the land to be divided is represented by $0 \leqq x < \infty$ and that the yield of a plot E is given by $\int_E e^{-\lambda x}dx$, where λ is the reciprocal of the height of flood. I say there can be no equitable plot. Suppose indeed that E were equitable and that $f(x)$ is 0 or 1 according as x is or is not in E, then

$$\int_E e^{-\lambda x}dx = \int_0^\infty f(x)e^{-\lambda x}dx = \beta \int_0^\infty e^{-\lambda x}\, dx \text{ for some constant } \beta, 0 < \beta < 1.$$

$$\text{Therefore, } \int_0^\infty \{f(x) - \beta\}e^{-\lambda x}dx = 0$$

for all $\lambda > 0$, or (if you like) all λ in some interval.

It then follows, according to a very well known theorem, that $f(x) = \beta$ for all x, which is absurd.

Doesn't this example show that the 'problem of the Nile' does not always have a solution, or do I mistake the terms of the problem?

Second, I would like to turn to the method of calculating, or may I say defining, fiducial probability, given in the penultimate paragraph of your

[*CMS*] page 25.395 [*CP* 135].

Suppose x and y are sufficient for α and β and distributed according to the density $\phi(x,y;\, \alpha, \beta)$.

Let $F(x_0;\, \alpha,\beta) = P(x \leqq x_0|\alpha,\beta)$ and $G(x_0,y_0;\, \alpha,\beta) = P(y \leqq y_0|x_0,\alpha,\beta)$.

F and G are then uniformly and independently distributed from 0 to 1, so your process can be applied; and under favorable conditions it will lead to a probability density $\psi(\alpha,\beta;\, x,y)d\alpha d\beta$. If however the rôles of x and y are reversed, the process may lead to a *different* ψ. I shall give an example that, if I do not miscalculate, illustrates this danger.

Suppose one observes x and y subject to the density:

$$\phi(x,y;\, \alpha, \beta)dxdy = \frac{\alpha^2\beta^2}{\alpha+\beta}(x+y)e^{-(\alpha x+\beta y)}$$

for $x,y;\, \alpha,\beta$ all non-negative.

Clearly x,y are sufficient for α and β.

$$\int_0^y \phi(x,\eta;\, \alpha,\beta)d\eta = \frac{\alpha^2}{\alpha+\beta}\, e^{-\alpha x}\{(1+\beta x) - [1+\beta(x+y)]e^{-\beta y}\}$$

$$\int_0^\infty \phi(x,\eta;\, \alpha,\beta)d\eta = \frac{\alpha^2}{\alpha+\beta}\, e^{-\alpha x}(1+\beta x)$$

$$G(x_0,y_0;\, \alpha,\beta) = P(y \leqq y_0|x_0, \alpha, \beta) = \left\{1 - \left(1 + \frac{\beta y}{1+\beta x}\right)e^{-\beta y}\right\}$$

$$F(x_0;\, \alpha,\beta) = P(x \leqq x_0|\, \alpha,\beta) = \left\{1 - \left(1 + \frac{\alpha\beta x}{\alpha+\beta}\right)e^{-\alpha x}\right\}$$

$$dFdG = \psi(\alpha,\beta;\, x,y)d\alpha d\beta = \left|J\begin{pmatrix} F & G \\ \alpha & \beta\end{pmatrix}\right| d\alpha d\beta$$

$$= \frac{\partial}{\partial\alpha}\left\{\left(1+\frac{\alpha\beta x}{\alpha+\beta}\right)e^{-\alpha x}\right\}\cdot\frac{\partial}{\partial\beta}\left\{\left(1+\frac{\beta y}{1+\beta x}\right)e^{-\beta y}\right\} d\alpha d\beta$$

$$= \frac{xy\, e^{-(\alpha x+\beta y)}}{(\alpha+\beta)^2\,(1+\beta x)^2}\cdot\{(\alpha+\beta)^2 + \alpha\beta(\alpha+\beta)x - \beta^2\}\times \{(1+\beta x)^2 + \beta y(1+\beta x) - 1\}d\alpha d\beta.$$

In view of the two forms in which $\psi d\alpha d\beta$ is written, its integral is 1 and it is non-negative; therefore, formally, a probability density. But obviously if the rôles of x and y had been reversed in the calculation,

$$\psi^*(\alpha, \beta;\, x, y) = \psi(\beta, \alpha;\, y, x)$$

would have resulted, which is quite different.

Fisher to L.J. Savage: 23 *August* 1952

. . . About the Nile, you propose that fertility should depend apart from lambda, only on one coordinate x. In that case, the households receive plots stretching from $x = 0$ to infinity, of width y proportional to their needs. However, I have probably missed your point. I have not yet understood your second example.

L.J. Savage to Fisher: 23 *August* 1952

Though I expect to have the pleasure of conversing with you tomorrow, it may be useful for me to write down some remarks on the problem of the Nile suggested by your letter of this morning.

Not having taken your parable sufficiently literally, I thought it would be alright to imagine the terrain to be divided as one dimensional. Two-dimensional examples can be constructed, but whether they fall within your meaning, you alone can say.

1. Let $x(y,z)$ be a one-to-one measure-preserving transformation from the full (y,z) plane onto the half line $0 < x < \infty$, and let the value of a plot A in the (y,z) plane be given by

$$\int\int_A e^{-\lambda x(y,z)} dydz,$$

where λ is inversely proportional to the height of the flood. The argument I gave in my first letter shows that no A can be equitable (unless it be almost empty or almost the whole plane).

Perhaps you will dislike this as an example because $x(y,z)$ is horribly discontinuous.

2. Again let the terrain be the whole (y,z) plane but now take the value of a plot A as

$$\int\int_A e^{-\frac{1}{2}(y^2+z^2) + yh \cos h + zh \sin h} dydz$$

where h is the height of the flood, $0 \leqq h < \infty$.

Suppose, if possible, that A is equitable and worth an αth of the whole terrain, and let $f(y,z) = 1 - \alpha$ or $-\alpha$ according as (y,z) does or does not lie in A. This implies

$$* \qquad \int\int_A f(y,z) e^{-\frac{1}{2}(y^2+z^2) + yh \cos h + zh \sin h} dydz = 0$$

for all $h \geqq 0$. Consider any ϕ, $0 \leqq \phi < 2\pi$; let $h = \phi + 2\pi n$ for n a non-negative integer; and compute (*) for this special h, thus:

$$y = y' \cos \phi - z' \sin \phi$$
$$z = y' \sin \phi + z' \cos \phi$$

$$\int\int f(y' \cos \phi - z' \sin \phi, y' \sin \theta + z' \cos \phi) e^{-\frac{1}{2}(y'^2+z'^2)} e^{y'(\phi+2\pi n)} dy'dz'.$$

It follows by a uniqueness theorem for Laplace transforms that

$$\int f(y'\cos\phi - z'\sin\phi,\, y'\sin\phi + z'\cos\phi)e^{-\frac{1}{2}(y'^2+z'^2)}dz' = 0$$

for all ϕ and y', and thence, by Fourier analysis, that $f(y,z)$ is almost everywhere zero. This contradicts the hypothesis that A is equitable at level α.

Perhaps you intend that the value of every plot should be non-decreasing in h.

3. Such monotoneity is achieved if the terrain is restricted to $y^2 + z^2 < 1$ with the value of A given by

$$\iint_A e^{h(2+y\cos h+z\sin h)}dydz$$

4. If you don't like the terrain in (3) you can smoothly transfer it into almost any other, for example into the whole plane.

I hope one or another of these examples will shed some light on the problem.

L.J. Savage to Fisher: 24 *October* 1952

Certain points were left dangling after our correspondence and very pleasant conversation of last summer, and I am writing this in the hope that you will now have the leisure to help me put some of them out of the way.

The paragraph beginning at the bottom of page 4.20 of the tentative manuscript of my book, of which I believe you have a copy, reads thus:

> It has been countered, I think by R.A. Fisher, that if experience systematically leaves [?leads] people with opinions originally different to hold a common opinion, then this common opinon, and it only, is the proper subject of scientific probability theory.

Of course if it is an error to couple your name with this line of argument, I want to delete reference to you here. Even if the allusion is correct, it would be much better to have formal reference, and I wonder whether you can supply one.

You may remember my mailing you, when you were in America, what I thought was an example showing that the problem of the Nile cannot always be solved. You found a fault in my example, which clarified the problem for me, and I submitted a second somewhat more complicated one, which you had not yet had time to appraise when I left Cambridge. I wonder whether you have been able to give that second example some attention. If you find it correct and interesting, I think I'd like to publish it briefly, perhaps in the *Annals of Mathematical Statistics*.

Another technical matter you were not in a position to discuss with me when I was in Cambridge, but on which I hope you can make some remarks now, is what seems to me to be a serious ambiguity in the definition of fiducial probability in multi-parametric contexts. I wrote out one example illustrating this ambiguity, and orally called your attention to the thesis of a Mr.

Williams, which brings out the same point with different examples, and is now filed as Ph.D. 1675 in the University Library.[1]

[1] Williams, R.M. (1949). The use of fiducial distributions with special reference to the Behrens–Fisher problem. Ph.D. Thesis, University of Cambridge.

Fisher to L.J. Savage: 30 *October* 1952

Thanks for your letter. On your first point, I think my name has come in to your statement on page 4.20 in error. At first reading, I do not clearly comprehend the nature of the argument, and feel sure that I never originated or used it.

Please do not hesitate to publish your example showing that the problem of the Nile cannot always be solved. This has long been my opinion and for a long while I thought that Sam Wilks had a demonstration in a particular and interesting statistical case, for he has asserted that no test of significance in Behrens' problem fulfilling certain obvious conditions could exist for which the level of significance was independent of the variance ratio of the two populations sampled.[1] I was pleased to see this statement made, as it knocked the bottom out of the objection to Behrens' own solution, an objection that I have always felt to be based on the misapprehension that tests of significance always referred to unconditionally repeated sampling, and therefore without force. I should, therefore, much like to see a particular case clearly demonstrated in print.

I am still not clear about the supposed ambiguity in the definition of fiducial probability. Perhaps you could discuss it in relation to simultaneous estimation of the mean and variance of a normal population.

[1] Wilks, S.S. (1940). On the problem of two samples from normal populations with unequal variances. *Ann. Math. Stat.* **11**, 475–6 (abstract).

L.J. Savage to Fisher: 7 *January* 1953

A word on the problem of the Nile. A few months ago, I suggested that the function

$$\phi(x,y|\lambda) = \exp\{-\tfrac{1}{2}(x^2 + y^2) + \lambda x \cos \lambda + \lambda y \sin \lambda\}$$

was a counter example to the general solvability of that problem, because it is quite easy to show that there is no region in the plane such that

$$(*) \qquad \textstyle\iint_R \phi(x,y|\lambda)\, dx dy = \alpha \text{ for all } \lambda \text{ and for } 0 < \alpha < 1.$$

I hold myself responsible for the demonstration of (*), but I do wish you would tell me whether (*), if granted, does represent an instance of what you would call an unsolvable problem of the Nile.

It has just come to my attention that Feller gave what amounts to an n-dimensional example in *Statistical Research Memoirs*, *2*, (1935), 117–125. He gives in fact, for each n, a family of probability densities $p(x|\theta, n+1)$, (to use his own notation), such that

$$n!\int \ldots \int_R p(x_1|\theta, n+1) \ldots p(x_n|\theta, n+1)$$

cannot be a constant $\alpha(0 < \alpha < 1)$ for all θ, if R is contained in the 'valley' $0 \leqq x_1 \leqq x_2 \ldots \leqq x_n$.

May I turn also to my puzzle about fiducial probability, putting it in terms you may find more convenient. Let x and y be distributed according to the density

$$\phi(x,y|\alpha,\beta) = \frac{\alpha^2\beta^2}{\alpha+\beta}(x+y)\exp\{-(\alpha x+\beta y)\}; x, y, \alpha, \beta \geqq 0.$$

It can be shown that, for all α and β, the variables

$$S = \left(1 + \frac{\alpha\beta x}{\alpha+\beta}\right)\exp(-\alpha x)$$

and

$$T = \left(1 + \frac{\beta y}{1+\beta x}\right)\exp(-\beta y)$$

are independently and uniformly distributed between 0 and 1. This is most easily seen, I think, from the fact that

$$T = \int_0^\infty dw \int_0^x dv\, \phi(v,w|\alpha,\beta)$$

$$S = \int_0^y dw\, \phi(x,w|\alpha,\beta) \Big/ \int_0^\infty dw\, \phi(x,w|\alpha,\beta).$$

Your recommendation near the foot of p. 395 of Paper 25 of *Contributions* [*CP* 135] leads to a definition of the fiducial distribution of α and β. But the puzzlement is that

$$S^* = \left(1 + \frac{\alpha\beta y}{\alpha+\beta}\right)\exp(-\beta y)$$

$$T^* = \left(1 + \frac{\alpha x}{1+\alpha y}\right)\exp(-\alpha x)$$

leads to a different one. Is there some way out of this apparent ambiguity?

Incidentally, John Tukey has exhibited the same sort of thing in connection with the Behrens–Fisher problem. Has he shown his example to you, and if so what do you make of *it*?[1]

[1] See Tukey, J.W. (1957). Some examples with fiducial inference. *Ann. Math. Stat.* **28**, 687–95.

Fisher to L.J. Savage: 13 *January* 1953

Thanks for your letter with the various puzzles you have put into it. As regards your own difficulty of the ambiguity of the fiducial distribution, do you not think you should use one-valued function[s] so that an element of frequency on one representation shall correspond uniquely to one on another? I do not think that there is anything else of importance in the example you send.

L.J. Savage to Fisher: 23 *January* 1953

I am not sure I understand the following sentence from your recent letter: 'As regards your own difficulty of the ambiguity of the fiducial distribution, do you not think you should use one-valued function[s] so that an element of frequency on one representation shall correspond uniquely to one on another?' If the sentence means what I think it does, it must arise from an error in algebra on your part or mine. The functions S and T defined on p. 2 of my letter of January 7, are certainly single-valued functions of all their arguments. Also, according to my calculation, for each value of T there is exactly one β; and for that β, and any S, there is exactly one α; so the inverse mapping from S and T to α and β is single-valued — similarly for x and y.

Quite aside from fiducial probability, I do wish you would let me know whether the remarks I made on the problem of the Nile do in fact solve that problem, in the sense in which you understand it.

I'm afraid my questions bore you, though they concern matters in which you are the initiator, and indeed the only authority, and about which you have in the not remote past displayed great interest. I must also confess that I am haunted by the suspicion that you may share to some extent the unpleasant opinion voiced by Owen on my visit to your laboratory, namely, that any critical point raised by a mathematician is for the purpose of destroying what others have built. I do hope you, and he, will believe that that is not the case in connection with my questions. I think it more than possible that fiducial probability, like so many of your ideas, is a very good idea indeed. But I can scarcely form an opinion, as long as I believe that the only general definition formulated is quite ambiguous. Of course, in connection with the problem of the Nile, my questions are not critical in any sense at all. You have proposed a problem as to whether such-and-such can always be done — in the suspicion that it cannot. I have adduced instances in which I think that the thing required cannot indeed be done. I am simply asking you to confirm whether these are indeed instances, or whether there are aspects of the problem that I do not yet appreciate.

Fisher to L.J. Savage: 28 *January* 1953

I am sorry I did not make myself clear. The function you choose for S

expressed in terms of α is such that one and the same value of S corresponds with two different values of α. I am sure that you noticed this for yourself without my having to mention it, and indeed it seems to be the point of the whole trick.

I am sure that George Owen was not criticising mathematicians in general, he being himself a very highly qualified analyst. He may have implied that mathematical ingenuity could be employed either helpfully or unhelpfully.

L.J. Savage to Fisher: 4 *February* 1953

It does seem to me that the function I choose for S expressed in terms of α is such that the same value of S never corresponds with two different values of α (within the range of meaningful possibilities for α, i.e. $\alpha > 0$).

Indeed,

$$S = \left(1 + \frac{\alpha\beta x}{\alpha + \beta}\right) \exp(-\alpha x)$$

$$\frac{\partial S}{\partial \alpha} = \left\{ -x\left(1 + \frac{\alpha\beta x}{\alpha + \beta}\right) + \frac{\beta^2 x}{(\alpha + \beta)^2} \right\} \exp(-\alpha x)$$

$$= \frac{-x}{(\alpha + \beta)^2}\{(\alpha + \beta)^2 - \beta^2 + \alpha\beta x(\alpha + \beta)\}[\exp(-\alpha x)] < 0,$$

since by hypothesis α, β, $x > 0$. Thus S steadily decreases with increasing α, which seems to prove my point.

I hope you will find this calculation correct, and will reconsider my example.

Fisher to L.J. Savage: 11 *February* 1953

Do you mean to assure me that the Jacobian relating S and T to α and β is of constant sign over its whole range?

L.J. Savage to Fisher: 19 *February* 1953

I do assure you that, in my calculation, the Jacobian

$$\frac{\partial S}{\partial \alpha}\frac{\partial T}{\partial \beta} - \frac{\partial S}{\partial \beta}\frac{\partial T}{\partial \alpha}$$

is strictly positive for all positive (i.e. all meaningful) α, β, x, y. To facilitate any check you might wish to carry out, I find

$$\frac{\partial T}{\partial \alpha} = 0; \ \frac{\partial S}{\partial \beta} \text{ irrelevant},$$

$$\frac{\partial T}{\partial \beta} = \frac{-y \exp(-\beta y)}{(1+\beta x)^2}\{(1+\beta x)^2 - 1 + \beta y(1+\beta x)\} < 0,$$

$$\frac{\partial S}{\partial \alpha} = \frac{-x \exp(-\alpha x)}{(\alpha+\beta)^2}\{(\alpha+\beta)^2 - \beta^2 + \alpha\beta x(\alpha+\beta)\} < 0^1.$$

[1] No reply from Fisher to this letter has been found.

Fisher to L.J. Savage: 11 *June* 1957

Many thanks for your letter of June 5th mentioning the arrangements you have made with the State College at Ames to enable me to make a joint visit to your respective campuses.

This I shall very much hope to do, and fulfil the engagement as of the dates you mention. It will be nice to see old friends again at both campuses, and I hope you will give a moment's thought to the subject you would like me to discuss; perhaps at Chicago something like 'The validity of tests of significance for composite hypotheses', adverting to the manner in which the doctrines of Neyman and Pearson have led to a misunderstanding of the problem solved by Behrens.

Of course, there have been so many erroneous tests of significance for composite hypotheses developed from the same theories that it is hard to trace them all out, or 'cover the ground' in the way that would be necessary to prevent students and practical workers from falling into some of the many pitfalls that have been dug. I rather suspect the proliferation of non-parametric tests on this issue.

L.J. Savage to Fisher: 26 *June* 1957

It is good to hear, as definitely as would be possible at this early date, that you will be visiting us at Chicago and Ames in the fall.

The talk on tests of significance that you suggest would be splendid, as would any of the important themes in your new book. For example, a clear statement of what the fiducial argument is and is not would be extremely helpful to all of us, students and faculty alike. Again, a thorough discussion of ancillary statistics would be interesting, but I suspect that we will probably get that in any event, because it is likely to be relevant to whatever you do discuss. In short, talk about what you like, and you will have a good and appreciative audience.

L.J. Savage to Fisher: 3 *February* 1958

When we first met in Cambridge, I mentioned a certain one-parametric family of distributions for which the problem of the Nile is not solvable. I think that I may never have given you a formal statement or demonstration of the example, and I enclose one now in case it is of interest to you. For my part, I would be glad to hear what the present state of the problem is. Would my example be interesting enough to publish, or are such examples now a dime a dozen? . . .

[Enclosure]

An example of an insolvable problem of the Nile.

On the square $|x| \leqq 1$, $|y| \leqq 1$, there is clearly a family of probability densities of the form

$$p(x,y|\lambda) = \alpha(\lambda) \exp\{x\lambda \cos \lambda + y\lambda \sin \lambda\} \text{ for } \lambda > 0.$$

The existence of a similar-region implies the existence of a function f essentially different from 0 and carried on the unit square such that

$$\int\int dx\, dy\, f(x,y) \exp\{(x \cos \theta + y \sin \theta)(\theta + 2\pi n)\} = 0$$

for all θ and all integers n.

The 'distribution' of $x \cos \theta + y \sin \theta$ under the signed measure induced by f is thus 0 for each θ. It follows, to reproduce a well-known argument, that

$$\int\int dx\, dy\, f(x,y) \exp\{i\rho(x \cos \theta + y \sin \theta)\} = 0$$

for all real θ and ρ, so $f(x,y)$ is 0, contrary to assumption.

It is of course trivial so to modify the example that the unit square is expanded to the whole plane.

Fisher to L.J. Savage: 7 February 1958

Thanks for your note and the example. I have, as you know, long thought that such examples could be, or had been found. The interest of publication would, I am sure, lie in the technique of the demonstration of the non-existence of a solution, which should therefore bring out conditions of non-existence recognizable, perhaps, beyond the single example.

Since it was put forward in 1936 [*CP* 137] it has, I believe, always been these conditions of solubility which constitute the problem. Naturally I have never been impressed by the characteristically phoney 'solution' offered by J. Neyman,[1] which Pitman[2] recommends to the American Statistical Association.

[1] Neyman, J. (1934). On the two different aspects of the representative method. *J. R. Statist. Soc.* **47**, 558–675.

[2] Pitman, E.J.G. (1957). Statistics and science. *J. Am. Statist. Ass.* **52**, 322–30.

Fisher to F.D. Sheffield: 11 *February* 1955

I am sorry my reference in *The Design of Experiments* was wrong.[1] I should have referred to part of a series of articles on 'experimental error' in the same journal, namely 1929, vol. 6, pp. 5–11. I asked Maskell to what paper of his I should refer for this point, and he must have given me the wrong reference. I remember he was early confident, being less inhibited than some of us mathematicians, of the propriety of making probability statements about the unknowns behind experimental data. It has now been proved completely that this point of view is right, although I believe there are places in the States, particularly Berkeley, much concerned to avoid this conclusion!

[1] Sheffield had questioned the accuracy of the reference to E.J. Maskell's work in introducing fiducial limits given in Section 62 of *DOE*. Maskell, a plant physiologist who had worked with Fisher at Rothamsted, published a series of papers in *Trop. Agric.* **5** and **6** (1928–9) under the title 'Experimental error: a survey of recent advances in statistical method'. In one of these (*Trop. Agric.* **6**, 7) after discussing the use of 'Student's' *t* in testing the significance of differences between variety means, he wrote, 'In general, however, the exact probability for each individual difference is immaterial: we wish usually to know what differences we may accept and what should be rejected as not established. For this purpose it is convenient to calculate, from the standard error and the table of *t*, values which may be called '*significant differences*', which would be exceeded by chance only once in twenty trials ($P = 0.05$) or, if we wish for greater certainty, only once in one hundred trials ($P = 0.01$).'

Fisher to W.A. Shewhart: 2 *February* 1940

I have just received your letter of January 16th, with the news that you are sending me a complimentary copy of your monograph on *Statistical method from the viewpoint of quality control*. This I shall be very glad indeed to have, as I had already heard a good deal about it, and, having followed your earlier work on the subject with great interest, it will be good to see how you have developed it further.

There is a technical point of some importance with respect to quality specifications, which has had rather a devious history in the literature, and may have reached you indirectly. Some time ago I studied one of the natural extensions of 'Students's' distribution, namely, the distribution of t as inferred from a finite sample for a numerical value, the true deviate of which is τ in relation to the population sampled, i.e.

$$\bar{x} + st = \mu + \sigma\tau$$

where, for the sake of algebraic simplicity, I have taken the old fashioned estimate

$$s^2 = S(x - \bar{x})^2/n.$$

Then I showed that the distribution of t for given τ depended on τ and the size of the sample only. I did not publish the solution until 1931, since it involved functions not at that time tabulated, but it appears in the introduction to

Vol. I of the *Mathematical Tables* published by the British Association [*CP* 91], which contains a fine table by Airey of the integrals of the probability integral, there called the *Hh* functions, which are the functions required. The point is that one might wish to give a guarantee, under penalty, that not more than a certain proportion, specified by τ, of a consignment would fail in some well-defined way; then one could use a value t applied to a random sample experimentally tested in advance to find what limit could be safely specified at a given risk, this being determined by the position of t in its known distribution for the value of τ required. Alternatively, of course, one could find the risk corresponding to any t found to be the value corresponding with a specification required by the buyers.

Technically, it is obvious that this distribution, if adequately tabled, would have very important applications in your field. I pointed out its importance to Neyman,[1] some time before he took his post with Pearson, and was surprised to see later a paper in the *Comptes Rendus* of the French Academy of Sciences by a Polish pupil of Neyman's in which my solution was given without indication of its source, accompanied by a totally inadequate table. I questioned Neyman on the circumstances, and he assured me that his assistant's paper had originally referred to my work as the source of the solution, but that the reference had been excluded when the paper was shortened by the editors![2]

Whatever plausibility this tale may have had at the time, it is certain that Neyman has since referred to this solution, for example in his contribution last year to the Geneva Congress on applications of the theory of probability,[3] as one of the important fruits of his own theory of testing hypotheses, and not as a result available years before that theory was heard of.

During last Winter Mr. Stevens in this Department has been going ahead with the preparation of more adequate tables, which I think may interest you considerably when they can be completed. You may be glad to know of this, as also that misleading statements are liable to be made as to the origin of a method, the relevance of which to industrial contracts I had in mind from the first, as perhaps my early correspondence with you may have shown. . . .

[1] See Fisher's letter of 24 October 1932 to Neyman (p. 190).
[2] Compare *CP* 181, p. 143.
[3] Also published in *J. R. Statist. Soc.* **105**, 292–327 (1942).

H.F. Smith to Fisher: 6 *August* 1954

I have long been under the impression that when I was at the Galton Laboratory (and not yet able to appreciate the points involved), I saw a short paper of two or three pages by you claiming that for purposes of a regression analysis it would always be valid to treat the independent variable as fixed even although it might have been observed as a random sample of some variate population. The point is frequently cropping up, although not usually of much importance, and I am unable to quote your reference if it exists or to track it down in any bibliography. I have always meant to ask you when we meet but invariably forget. I have asked many other statisticians and none know the reference, although you are frequently quoted as having given the conclusion as a verbal opinion. . . .

Fisher to H.F. Smith: 27 August 1954

I was glad to get your letter, and particularly so as you take up an exceedingly important point. Probably the earliest place in which I refer to the matter is in *J.R.S.S.* LXXXV, page 599 [*CP* 20] where I am discussing the rather wider question of the goodness of fit of regression lines. Its application to the sampling distribution of estimates of the coefficients comes on page 609.

I am inclined at the present time to stress the importance of the principle involved because it cuts at the root of the fallacious approach to tests of significance introduced by Neyman and Pearson, and usually expressed in some such phrase as, 'the frequency found in repeated samples from the same population'. This, as the case of the regression coefficient shows rather clearly, would be quite the wrong question to ask. I believe no one has adopted a mistaken test in this case, but on at least two occasions entirely competent mathematical statisticians, namely E.B. Wilson at the Harvard Medical School, and G.A. Barnard at Imperial College London, have derived erroneous levels of significance for the analogous case of the 2 × 2 table where, as you know, I recommend the calculation of levels of significance from a reference set of samples all having the same margins, and not such a set of samples as might have been derived by a repetition of the actual process by which the particular sample under consideration was obtained.

It is noteworthy that both Wilson and Barnard, when the method was challenged, and after a detailed consideration of the logical position, have frankly agreed that they were mistaken, and have taken the trouble so to express themselves in print.

Reliance on 'repeated sampling from the same population' was certainly responsible for Bartlett's criticism of Behrens' test of significance, and to the succession of more or less indefensible attempts that he has made to obtain an alternative solution.

The relevant phrases to which I refer you are on page 599, '. . . we do not in practice ignore the size of the array', and on the previous page, 598, '. . . we have not attempted to eliminate known quantities, given by the sample, but only the unknown quantities which have to be estimated somewhat inexactly'.

I have only realised recently, chiefly owing to discussion with Barnard, to what extent Neyman and Pearson's approach, including this question of 'repeated samples from the same population', is due to their thinking of a test of significance as though it were a kind of acceptance procedure in which the repeated samples have an objective reality, and are not, as they are with tests of significance, constructs of the statistician's imagination.

Fisher to D.A. Sprott: 13 *January* 1962

I have delayed answering your letter, partly by reason of receiving other exciting correspondence. Even from that darkest Continent, N. America!

Do you know the name of H. Kyburg of the Rockefeller Institute 21 N.Y.? His line seems to be abstract symbolic logic, but he has recently caught fire on the fiducial argument and indeed may be exaggerating its importance.

No doubt you already know Donald Fraser your neighbour from Toronto, who has recently been bitten all over by mad Neymannians in California, but seems to be undefeated. I fancy he has answered the question with which you end your letter, in the negative, by eliminating s_1, s_2 and σ_1, σ_2 and obtaining[1]

$$w = \frac{r}{\sqrt{1-r^2}}\chi_{n-2} - \frac{\rho}{\sqrt{1-\rho^2}}\chi_{n-1}$$

an equation between ρ and r involving three random variables, for w is normally distributed about zero with unit variance. For any given ρ this gives the distribution of r as I gave it in 1915 [*CP* 4], while for any given r it provides the fiducial distribution of ρ, in a variety of new forms, but agreeing with that I gave in 1930 [*CP* 84], and used in my book [*SMSI*] (1956, 59).

I have been telling him that in the multivariate case, starting with the known distribution of ρ_{12} (marginal) he should get the simultaneous distribution of ρ_{13}, ρ_{23} conditional on ρ_{12}, and thence step up to t variables.

Mauldon seems not to have a glimmer, and Quenouille, who is often quite acute, seems to confuse Likelihood with Fiducial Probability in spite of the fact that the Likelihood has no measurable Reference Set, such as probability must have.

Try to cheer Donald up (and Kyburg too). It would be a pity if people of his quality were driven to suicide by the concerted yell of these Yanks.

[1] This result was in fact derived by Sprott. See p. 853 of Fraser, D.A.S. (1963). On the definition of fiducial probability. *Bull. Int. Inst. Statist.* **40**, 842–56.

Fisher to D.A. Sprott: 29 *January* 1962

I have been working at your letter of Jan 24, and it cheered me up, whatever this may do to you. However my eyes are so bad that when my pen fails to write I do not notice the difference!

Fraser sent me what I took to be an invitation to spend a year at Toronto, which you might think a step in the right direction. I told him I should love to see the De Lurys again, but he has not followed the matter up, so perhaps there is a financial crisis in the Dept of Maths.

I confess Kyburg's letter gave me quite a new idea of symbolic logicians. His very general idea of probability with non-differentiable symbols sounds like the opposite pole to my own approach, which is to particularize the different types of uncertainty which a rational mind can experience. Yet his letter seemed sincere, and not a bit demented.

If I decipher you aright you say that my (1956) simultaneous distribution of σ_1, σ_2, ρ does not give the correct distribution of $\rho\, \sigma_2/\sigma_1$, the true regression. What do you compare it with? The distribution for a sample with given $S(x-\bar{x})^2$, or what? But I probably put your mind on to the wrong road, by trying to say that only when the measures of scale s, σ have been eliminated, do you get down to that of ρ. Perhaps I am wrong, but sticking to the 5-parameter distribution in my book, which I think is impregnable, I should like to see the procedure of getting it out of a 5 fold continuous group of transformations.

I have the t-dimensional distribution of r_{ij} $\{\frac{1}{2}t(t-1)$ values$\}$ in the handy form[1]

$$\frac{\pi^{-t(t-1)/4}\, 2^{-t(N-3)/2}}{[(N-3)/2]! \ldots [(N-t-1)/2]!} \left|\rho_{ij}^{*}\right|^{(N-1)/2} \left|r_{ij}\right|^{(N-t-2)/2} dr_{ij}\, F_{N-2}\,(\gamma_{ij})$$

in which $\gamma_{ij} = -r_{ij}\,\rho_{ij}^{*}$, $\rho_{ij}^{*} = \rho_{ij}/\sqrt{\rho_{ii}\,\rho_{jj}}$ and

$$F = \int_0^\infty \ldots \int_0^\infty (u_1 \ldots u_t)^{N-2} \exp -\tfrac{1}{2}(u_1^2 + + -2\gamma_{ij}u_iu_j - -)\, du_1 \ldots du_t.$$

I have it all written out, but wanted to know whether it would invert to a distribution of ρ_{ij}. In principle I accept that it is not axiomatic that such a distribution exists.

Simultaneous exhaustive estimation need not be a sufficient condition. The example I gave in *Sankhyā* last year shows that strictly it is not a necessary condition. By the way I should like your reaction to that example, arising out of Behrens. The case rather surprised me by suggesting that the fiducial argument can work without exhaustive *estimation senso strictu*.

For bivariate ρ, I get

$$\frac{1}{\pi\,(N-3)\,!}\,(1-r^2)^{(N-2)/2}\ \ (1-\rho^2)^{\,(N-3)/2} \times$$

$$\int_0^\infty\!\!\int_0^\infty (uv)^{N-3}(N-2+\gamma uv)\exp -\tfrac{1}{2}(u^2-2\gamma uv+v^2)\,du dv,\ \gamma = r\rho$$

which has a suggestive little factor under the integral.

On the whole the only proofs satisfying me in this field involved probabilities of inequalities monotonic over a double range of variables. A continuous group is all right if it leads to such inequality statements. If not, I do not see the reasoning.

[1] Compare *CP* 288, p. 4.

Fisher to D.A. Sprott: 13 *February* 1962

I have just seen your letter, but not yet reflected on it, as I thought it better to lend it to Rao for the day or so before I fly to Adelaide.

You must apply for a personal assistant typist at your next University but the bigger ones are not always better. Think of where the Cambridge Stats Dept has gone since Daniels left it!

You probably have considered that the data needed for the regression do not imply that x is Normally distributed, but those for ρ and thence for σ_1, σ_2 certainly do.

In the case of one variate it does not seem possible to ball up the integration of the bivariate μ, σ distribution, though undoubtedly Bartlett about 25 years ago must have thought so, for he reacted vigorously though confusedly to the idea of such a distribution, denying even that you could get $\mu + \sigma$, although the more general form $\mu + \lambda\sigma$ had been published earlier. Claims have been published putting frequently a merely faulty analysis; though errors of logic are commoner.

I like inequalities because one is not tempted to use them in transformations which are not monotonic, while a certain amount may be hidden in a Jacobian.

e.g. If $P(\rho, r)$ is the probability of getting a correlation less than r from a population with correlation ρ, r being an exhaustive estimate, and P monotonic in both r and ρ unconditionally, then there exists a function $r_P(\rho)$ such that

$$Pr\{r<r_P(\rho)\} = P.$$

Equally we can define $\rho_{1-P}(r)$ so that

$$\rho_{1-P}\{r_P(\rho)\} \equiv \rho \text{ identically;}$$

then using this function as an operator,

$$Pr\{\rho_{1-P}(r)<\rho\} = P$$

or

$$Pr\{\rho<\rho_{1-P}(r)\} = 1-P$$

all since $\rho_{1-P}(r)$ is unconditionally monotonic in r and ρ. The reference set for these statements of probability consists of pairs (ρ, r) for all samples of given size from all Normal populations.

This was the first, and I think may fairly be called the typical fiducial argument. I fancy only a small minority fail to see it for a single parameter, and I have long been sure that there are conditions and stipulations required for more than one.

Fisher to D.A. Sprott: 22 *February* 1962

I believe I ought to clear up what I meant by the use of inequalities.

Pivotal functions must be functions of a parameter (only one to each such quantity) and of a set of statistics exhaustive for that parameter.

They serve as a basis for probability statements provided

(i) their sampling distribution is independent of all parameters,

(ii) their value is Monotonic with the parameter in question.

If there is more than one parameter each such function provides a marginal distribution for the parameter used,

e.g. $t = (\mu - \bar{x})/s$, or s/σ for a Normal sample.

For simultaneous distribution you may have

(iii) a set of pivotals each involving one parameter, and jointly a set of statistics exhaustive for all,

(iv) the simultaneous distribution of these pivotals is independent of all parameters,

(v) each is monotonic *uniformly* for all variations of the others,

e.g.

$$\frac{r}{\sqrt{1-r^2}}\,\chi_{N-2} - \frac{\rho}{\sqrt{1-\rho^2}}\,\chi_{N-1}$$

has a standardized Normal distribution and is monotonic in ρ. It therefore supplies the marginal distribution of ρ

but

$$\frac{r_{12}}{\sqrt{1-r_{12}^2}}\,\chi_{N-2} - \frac{\rho_{12}}{\sqrt{1-\rho_{12}^2}}\,\chi_{N-1}$$

and

$$\frac{r_{13}}{\sqrt{1-r_{13}^2}}\,\chi_{N-2} - \frac{\rho_{13}}{\sqrt{1-\rho_{13}^2}}\,\chi_{N-1}$$

are not I suppose independently distributed, or to take the crucial condition their simultaneous distribution is not independent of $\rho_{12}, \rho_{13}, \rho_{23}$.

The last chapter in *Statistical Inference* is an attempt to illustrate the use of marginals to climb higher.

I think some of your pseudo-fiducials and all of Tukey and Savage's examples are the result of ignoring some of these requirements.

Fisher to D.A. Sprott: 8 *March* 1962

I have just seen your letter of Feb 28.

Of course I do not think the range of application of probability statements to parameters has even nearly been explored. The first examples I put forward were met with such mud-headed incomprehension, combined with rather rancorous hostility, that the statistical world was diverted from the fruitful but difficult job of thinking, to the more enjoyable game of 'taking sides'. The fiducial argument has suffered more from its defenders or rather equivocators than from opponents.

I think the conditions I put down, if I put them all down, are demonstrably *sufficient*; probably some are not *necessary*. e.g. If $t_1, \ldots t_r$ are pivotals with a simultaneous distribution independent of $\theta_1, \ldots \theta_r$, parameters each involved in one pivotal, then for any limits α_i, $i = 1, \ldots r$,

$$Pr\{t_1 < \alpha_1, \ldots, t_r < \alpha_r\}$$

is calculable as a function of $\alpha_1, \ldots \alpha_r$ and of the statistics in $t_1, \ldots t_r$ but not of θ.

If each pivotal is *uniformly* monotonic in its own parameter, this probability statement is *identical* with a simultaneous inequality statement about $\theta_1, \ldots \theta_r$ whenever this set of parameters has a simultaneous distribution depending only on the statistics of the sample. If, and only if, these are exhaustive, it is impossible to find any function of the observations capable of defining a sub-set to which different probability statements apply.

I fancy this argument can be applied to any particular case which fulfils these conditions.

One would like to know (a) on what conditions pivotals with these properties can be found, (b) are there cases in which the simultaneous distribution of t parameters being accessible, many marginals of not more than s parameters are accessible. If you have 100 Mackinaw Indians and take 200 measurements on each, it is obvious that taking head breadth does not diminish your information about head length, which is a t distribution. (It is I suppose obvious to you and me but not to friend Roy who perhaps out of loyalty to Hotelling gives results which imply the opposite). A simultaneous distribution can be put up for the racial means of any 99 characteristics taken simultaneous, but the data give nothing further about the simultaneous distribution of any 100.

I shall be sorry if it proves impossible to get a simultaneous distribution of $\rho_{12}, \rho_{23}, \rho_{31}$ for a number of one-dimensional margins and Rao tells me some two dimensional margins are obtainable.

I mentioned climbing up from the margins to explain the point of my last example in *Statistical Inference*. Parameters can exist in strata, with the corresponding sets of exhaustive statistics.

I wonder if there can be more than 3 strata.

[P.S.] I think Box argues that assumed Prob. *a priori* though erroneous makes very little difference. Edgeworth argued in this way in 1908. My point is then what is this assumption doing in our reasoning? Why not see what can be said without it?

J. Tukey to Fisher: 27 August 1954

It now begins to look as if I may be in England briefly this fall. If so, I look forward very much to seeing you again. I hope to find you in Cambridge.

Perhaps you will remember a little document on the 'Purposes of Fiducial Inference', a copy of which I sent to you some time back. Both you and Owen were kind enough to send back some comments, to which I have not made organized reply. I am afraid, if what I hear by devious routes is correct, that you and Owen misinterpreted my intent in giving a number of examples. All but the last type of example were intended to illustrate the need for the restrictions currently imposed on pivotal functions (e.g. in Owen's paper[1] in *Sankhyā*. The *last* type of example was intended to illustrate, and still illustrates to my satisfaction, how these restrictions *do not* imply uniqueness, *even* in the Behrens–Fisher case.

I look forward hopefully to discussing this matter with both Owen and yourself.

[1] Owen, A.R.G. (1948). Ancillary statistics and fiducial distributions. *Sankhyā* **9**, 1–18.

Fisher to J. Tukey: 1 September 1954

. . .It would not be surprising if there were a theoretical ambiguity in Behrens' problem, since there is no sufficient set of estimates appropriate to the three parameters, i.e. the two variances, and the distance between the means. The amount of information dropped, however, must be quite minute, and any ambiguity of a serious magnitude, e.g. such as the difference between my solution and those of Bartlett and Welch, could not be explained in this way.

J. Tukey to Fisher: 22 April 1955

You may remember our previous correspondence about the nonuniqueness of fiducial probability. I suspect that you will class me as 'still unregenerate', but I still adhere to my belief that examples 6 and 7 are relevant. I have finally revised the material with a view toward publication, separating the more factual material on nonuniqueness from the more 'opinionated' material on the place and support of fiducial inference.

I enclose a copy of a draft of the material on nonuniqueness. I should not like to misrepresent your views on what restrictions should today be attached to pivotal quantities, so I send this in the hope you will care to look it through for misrepresentations. If you and/or Owen would care to look at examples 6 and 7 any further, that too would be fine. . . .

Fisher to J. Tukey: 27 April 1955

I am interested to get your long screed about probability statements concerning parameters derived from observational data. I think I can see what you are trying to do, but I do not think you are getting very warm, or are trying the right way.

If you must write about someone else's work it is, I feel sure, worth taking even more than a little trouble to avoid misrepresenting him. One safeguard is to use actual quotations from his writings; better still a series of comparative quotations, for it may be that, like yourself, he is still capable of revising his forms of words.

On page 7, you ascribe to me, without reference, a belief that I have never held. I have rather often emphasized my belief that statements of fiducial probability are in fact statements of probability, and can be verified by observational frequencies in exactly the same way as the statements of probability of the eighteenth century writers, Montmort, de Moivre, or Bayes. Clearly you have no support in my own writings for what you say I 'admit and, in fact, advertise'. Could you let me know from whom this mischievous statement was obtained?

[Start of page 2 of letter.]

It occurs to me that you may be referring to my objection to identifying the level of significance with certain frequency criteria. This is because such criteria have been arbitrarily or unsuitably chosen, and in some cases because no such criteria can be found for all cases of a composite hypothesis. This of course throws no doubt on the verification of any probability statement by observed frequency, if the appropriate frequency is observable, as you would see if you could ever get your bull-headed mind to stop and think. A level of significance is a probability derived from a *hypothesis*, not one asserted in the real world. A paper of mine in the *J.R.S.S.* [*CP* 261] may help you here. I have only seen the proof so far.

I expect you know really, whatever you may pretend, what the process of reasoning was for which I coined the term 'The fiducial argument'. You know too that I have, since 1930, laid stress on it as a form of reasoning of a particular kind. Probably you could reconstruct its steps. The probability integral of the exact frequency distribution, in finite samples, of an exhaustive statistic is used to form a continuum of probability-statements, of the form

$$Pr\{T < T_P(\theta)\} = P,$$

and using the monotonic property of the functions T_P for all P, this is transformed to the equivalent

$$Pr\{\theta_{1-P}(T) < \theta\} = P,$$

a complete set of probability statements about θ, in terms known for a given sample value T.

[Start of page 3 of letter.]

If you are sensitive to inductive processes of reasoning you will not think that the argument stops there. It requires also the observation that though the particular pair of values (T, θ) relevant to the experimenter has been shown to belong to a reference set in which the frequencies of all such inequality statements as the above are exactly known, yet the possibility of recognizing a sub-set having a different probability has to be excluded, and that this can be done if T is exhaustive, and in the absence of knowledge *a priori*, or from sources other than the sample under discussion.

In speaking of fiducial probability in connection with your numerous examples, I am sure you would have done better to examine in each case whether the material does or does not exist for the development of an argument of this type. This would at least spare you from arriving at contradictions, for the argument can be used quite rigorously. In examples 6 and 7, I should not expect to be able to complete such an argument, but it may be you will be more successful, if you actually try to see whether a fiducial statement can be inferred or not, instead of asserting that it can.

You may like to verify in a couple of lines that it can be inferred in the only case in which I have given a bivariate frequency distribution, namely that for the two parameters of the Normal distribution, so sternly rejected by Bartlett!

[P.S.] p.15. If you have a frequency distribution, of course you can derive a set of Tests of Significance. I suppose this is what you mean by a family of confidence sets. It is the converse argument which fails.
p.13. I have never spoken of the 'desirability' of uniqueness. Uniqueness is guaranteed by the fiducial argument when conducted with sufficient care to be rigorous. It is not guaranteed by short-cuts which by-pass the argument. Lack of uniqueness is a good check on them.

J. Tukey to Fisher: 10 *May* 1955

Thank you very much for your letter of 27 April. I will endeavor to do what I can to get my 'bull-headed mind' to stop and think. I would suggest, however, that what I was engaged in when I sent this document to you was 'taking even more than a little trouble to avoid misrepresenting you'. I shall give the material in your letter a very careful thinking, and I am sure that this will enforce various changes in the document. I hope you will not mind if I send the next version on too so that you may kick it around again if you think it needs it. . . .

J. Tukey to Fisher: 12 *July* 1955

I have been slow in getting back to your letter of 27th April . . .

I appreciated your letter very much, and have been trying to get things through my 'bull-headed mind'. There are some points which appear to me to require careful consideration and on which I cannot feel sure of your position from your letter. So I am venturing to write again.

(1) The 'mischievous statement' about 'admit and in fact, advertise' was not obtained from anyone else and was solely my own responsibility.

(2) You say that you have emphasized your belief (and presumably continue to believe) that 'statements of fiducial probability are in fact statements of probability, and can be verified by observational frequencies'. Under what circumstances can I verify the fiducial probabilities resulting from the numerical results of the Behrens-Fisher procedure as observational frequencies? (And are these circumstances generally reasonable in connection with actual experiments?)

(3) If I understand the top paragraph of your second page correctly, I agree. Therefore let us postpone any discussion here.

(4) The lower paragraph of your page 2 certainly must be accepted by anyone.

The emphasis on statements of inequality and the 'monotonic property of the function T_P' causes me to ask the following questions.

Define ψ as a $1-1$ function of θ by

$$\psi = \begin{cases} \theta, \text{ when } \theta = \{\theta\}, \text{ that is when } \theta \text{ is an integer,} \\ \{\theta\} + (1 - (\theta - \{\theta\})), \text{ otherwise,} \end{cases}$$

and then let, for example, x be a sample of 1 from a normal distribution with average value θ and variance 1. Can we then make fiducial probability statements about ψ? I cannot predict your answer.

(5) I come now to the top paragraph of your page 3, where you say: 'yet the possibility of recognizing a sub-set having a different probability has to be excluded . . .'. The words are here, but I am not sure that I have caught all the thought.

Do you mean that the formal inferential procedure is unsatisfactory for your purposes, unless it has been so restricted as to produce a unique result? And that if it is unsatisfactory, it should not be called fiducial probability? If your answer is 'yes', then I would assume that each proposed application of fiducial probability would be incomplete until a uniqueness theorem had been proved. Would this be your position?

It is of course true that monotonicity provides additional uniqueness in one-dimensional cases, and in some two-dimensional cases. I am not sure whether you rely on monotonicity in cases involving more than one parameter. Do you? And if you do, what definition of monotonicity do you use?

(6) Later on in the same paragraph, you say 'in the absence of knowledge *a priori*, or from sources other than the sample under discussion'. Before reading these clauses, I would have felt that you believed that it was always possible, and often wise, to summarize the inferences from any single body of

experience. Therefore I would have assumed that you would have been willing to begin such a statement 'in the actual, or assumed, absence. . .' and thus encompass both senses. But your detailed words now leave me in doubt. What is your position here? If there exists a further, entirely separate, body of experience, may we properly make statements of fiducial probability (i) about either body of experience separately, *and* (ii) about both bodies of experience jointly?

(7) Until the top paragraph of your page 3 is clearer to me, I don't think that I can usefully discuss the next paragraph.

(8) You then suggest that I verify something for the case of the two parameters of the normal distribution. What may I verify? That some specific formal procedure gives a unique answer? Which formal procedure? I am truly at a considerable loss here!

(9) You say in that paragraph 'in the only case in which I have given a bivariate frequency distribution'. Am I in any sense to interpret these words as disclaiming the application of fiducial probability to any other example with two or more parameters? What about the bivariate normal with known second moments? What about the between variance component in a simple analysis of variance? What about the Behrens - Fisher problem? You have me at another loss!

(10) In your discussion of the one-parameter case you laid great stress, properly in my view, on the monotonicity of distribution functions. As we are both well aware, a distribution of one statistic is monotone in a handy way, while distribution functions of two or more statistics maintain a positive increase in terms of higher differences which are not nearly so handy to deal with. What use would you plan to make of monotonicity in situations involving more than one statistic? . . .

Fisher to J. Tukey: 18 *July* 1955

I believe you are trying to understand what I write, though this would be easier if you were more familiar with what I have written in the past. If I could be sure of that, I could assume that you were using words in senses conformable with my own usage, or, in case you thought my usage incorrect, that you would draw my attention to a better way of speaking.

I see from your letter that the word 'probability' does not mean to you even approximately what it means to me, and I am not unfamiliar with the mathematical and scientific literature involving this word. In the first place you seem to express surprise at my belief that 'statements of probability can be verified by observational frequencies'. It would have been more explicit had I said 'statements of probability in the Natural Sciences' or 'statements of probability referring to the real world', for it is possible as an abstract mathematical exercise to make statements which refer to nothing on earth. Your view apparently is that no statements asserting a mathematical probability are capable of such experimental verification, which seems to imply, in the Natural Sciences, that all such statements are meaningless.

In the second place you do not seem to grasp the central characteristic of the concept of mathematical probability, namely that it enables a statement of uncertainty to be made with rigorous exactitude. This requires a specification not only of what is known, or can be validly asserted on the data, but also of what is unknown, in order that the probability statement should be distinguishable from a statement of certainty. To specify what is unknown requires, naturally, an exhaustive scrutiny of the data; this, of course, is not necessary when 'certain' inferences are deduced, since for them only a portion of the axiomatic material available may be required, and other axioms may be ignored.

I do not see how anyone even slightly acquainted with what I have written on statistical topics could not have seized my meaning of these points.

Having accepted page 2, if I understand you rightly, you ought to read the first sentence of page 3, when perhaps the rest of the paragraph will become clear to you. There is not a word out of place, and scarcely one that is not needed for an exactly rigorous statement. If you cannot understand the first paragraph of page 3, then I am afraid you cannot know what the phrase 'fiducial probability' means, and as you want to write about this, you do owe the effort necessary to grasp its meaning.

I do not know what your trouble is about monotonicity. Is there more than one definition for me to choose from, for you ask me what definition I prefer? The function T of θ and ϕ may, of course, be monotonic for θ uniformly for all ϕ. Is this the kind of definition you are asking for? The point of monotonicity was, as you would have seen from my letter, merely to establish the validity of the transformation of an inequality statement. Do you claim it is impossible validly to transform simultaneous statements of inequality? You have exhibited, of course, that it is rather easy to make bogus, or invalid, transformations of this kind. If the probability that $x>a$, and simultaneously that $y>b$, is given by a numerical value P over any well-defined field of possibilities, or reference set, then I believe that the same probability may be applied to the simultaneous inequality $a<x$ and $b<y$. I suggested that you should verify, what is not difficult, that the simultaneous distribution of the parameters of a normal population demonstrably had the simultaneous distribution I gave for them long ago, and that the method outlined on pages 2 and 3 supplied a rigorous form for such a demonstration. The case you mention of the five parameters defining a bivariate sample also offers no particular difficulty. If you could discipline your mind to carry out the argument step-by-step and with exactitude, even in a single case, you would, of course, put yourself in a position to recognize the many cases in which a parallel argument cannot be developed.

I was wrong in saying that the bivariate frequency distribution of the normal parameters was the only case I had published; I have also published more extensive problems. I do not suppose that you are such an idiot as 'to interpret these words as disclaiming the application of fiducial probability to any other example with two or more parameters'. The process of addition in arithmetic is widely applicable, even though neither you nor I have published every possible example of its use. In Behrens' problem I confirmed that Behrens had given a valid and precise test of significance. There is no Sufficient simultaneous estimate of the three parameters, so that a three-fold fiducial distribution is not obtainable.

But surely you can see these things for yourself!

J. Tukey to Fisher: 25 *July* 1955

1. Your immediate, though partial, reply to my letter about fiducial [probability] is a great help to me, but I still am in certain difficulties, which I shall endeavor to explain below.

2. I certainly believe I am trying hard to understand what you write. I agree that a maximum of familiarity with what you have written in the past would be a good thing, but even the entropy of the universe only tends toward a maximum. In this connection, however, may I remind you of your letter of 27 April, in which you suggested that, in writing about someone else's work, it is well to use 'a series of comparative quotations, for it may be that, like yourself, he is still capable of revising his forms of words'. I hope that I am engaged in following this advice.

3. I will plead guilty to the charge that probability means many things to me, but will assert that one of them is what you had in mind. Moreover, I did *not* intend to express surprise that 'statements of probability can be verified by observational frequencies', for surely such a 'probability' is a probability in the most useful and esteemed sense. It was my intention to accept this as the meaning of the word 'probability' for this part of the discussion, and then to inquire 'Under what circumstances can I verify the (fiducial) probabilities resulting from the numerical results of the Behrens–Fisher procedure as observational frequencies?' (I have now added parentheses around fiducial in view of your comments of 18 July.) I should like to repeat this question in what I hope is now a far more clearly defined context. Under what circumstances?

4. You go on to say that I do not seem to grasp 'the central characteristic of the concept of mathematical probability, namely that it enables a statement of uncertainty to be made with rigorous exactitude'. This is, I suspect, a place where even carefully chosen words may have two meanings, and where I have been likely to have missed the intended point in the past. There are two ways in which, as it seems to me, the words 'rigorous exactitude' may be construed:

(1) as meaning that the asserted probabilities will be exactly confirmed by observational frequencies, if and when we observe under the specified circumstances,

(2) as meaning not only (1) but also that among the statements with property (1) the statement being made is, to a suitable extent, rigorously unique.

(I will maintain that either construction is logical and grammatical). I inquire whether I am now correct in believing that (2) is what you really mean.

5. We come now to the first sentence of page 3 (your letter of 27 April). Does the preceding paragraph suggest that this and its paragraph are now clear(er) to me? (There remain some more formal questions, which I propose to take up below).

6. I am still unclear as to your position on where the line may be drawn and still have the result fiducial probability. May I quote you from two places for comparison:

'. . . In following his example it is not necessary to deny the existence of knowledge based on previous experience, which might modify his result. It is sufficient that we shall deliberately choose to examine the evidence of the sample on its own merits only. This has not only the advantages of giving simplicity and definition to the problem, it has the profoundly important effect that modern tests of significance, treating each body of data as unique, can thereby derive from them *independent* evidence which may be compared, knowing it to be independent, with evidence from other sources. In applying this principle, there is, of course, nothing to prevent us from combining the evidence of several different samples. We can do so and at the same time treat the whole body of available material as a unique body of data. Without methods of treating unique samples, we should have no real guidance in these more complex cases.' (*Annals of Eugenics*, Volume 9, page 175) [*CP* 162].

'. . . and that this can be done if T is exhaustive, and in the absence of knowledge *a priori*, or from sources other than the sample under discussion.' (letter of 27 April, page 3).

Am I to interpret the phrase in the letter as meaning that we *must* always include *a priori* information and other samples, or should I follow the longer quotation and hold that we are entitled to a choice as to whether we include it or not?

7. As I interpret your statements in the letter of 18 July: 'I do not know what your trouble is about monotonicity'. 'The point of monotonicity was, as you would have seen from my letter, merely to establish the validity of the transformation of an inequality statement'. You do *not* feel that a monotonic connection between the parameters and the sufficient statistics is essential. Is this correct?

8. You state 'If the probability that $x>a$, and simultaneously that $y>b$, is given by a numerical value P over any well-defined field of possibilities, or reference set, then I believe that the same probability may be applied to the simultaneous inequality $a<x$ and $b<y$.' Let me ask your opinion about a specific application of this sentence. Suppose that

(*) u is normally distributed with mean μ and unit variance,
(**) v is normally distributed with mean υ and unit variance,
(***) u and v are independently distributed,
(****) α is a fixed real number,
(*****)

$$x=\sqrt{(u-\mu)^2+(v-\upsilon)^2}\quad\cos\left\{\tan^{-1}\frac{v-\upsilon}{u-\mu}+\frac{\alpha\mu}{1+(u-\mu)^2+(v-\upsilon)^2}\right\}$$

(******)

$$y=\sqrt{(u-\mu)^2+(v-\upsilon)^2}\quad\sin\left\{\tan^{-1}\frac{v-\upsilon}{u-\mu}+\frac{\alpha\mu}{1+(u-\mu)^2+(v-\upsilon)^2}\right\}$$

Then we have

$$P=\frac{1}{2\pi}\int_a^\infty e^{-\frac{1}{2}x^2}\,dx\int_b^\infty e^{-\frac{1}{2}y^2}\,dy$$

and, if $|\alpha|\leqq 1$, I have shown that the simultaneous inequality $x\leqq a, y\leqq b$ can be inverted into a statement concerning μ and υ, the inversion being by the solution of (*****) and (******) for μ and υ in terms of x and y, for u and v fixed, with this solution being unique, continuous and differentiable. *Does your statement apply here*? If not, why not? If it does apply, is any one value of α to be distinguished from another? If no value is distinguished, then, since the result depends on α and is thus not unique, must not all results fail to give the fiducial distribution of μ and υ given u and v (even the one with $\alpha=0$)? (In this case, would not the means of two independent unit normals fail to have a fiducial distribution?) If some value of α is distinguished, by what general principle is it distinguished?

9. You suggest that I 'should verify, what is not difficult, that the simultaneous distribution of the parameters of a normal population demonstrably had the simultaneous distribution I gave for them long ago, and that the method outlined on pages 2 and 3 supplied a rigorous form for such a demonstration'. At this point, I am still baffled as to just what is to be shown. Consider the following statements:

(A) There exists a pair of inequalities whose simultaneous probability is independent of the parameters.

(B) There exists a unique inversion of the functions appearing in these inequalities.

(C) These inverse functions are continuous.

(D) These inverse functions are differentiable.

(E) There exists no essentially different pair of inequalities with property (A).

(F) There exists no essentially different pair of inequalities combining properties (A) and (B).

(G) There exists no essentially different pair of inequalities combining properties (A), (B), (C).

(H) There exists no essentially different pair of inequalities combining properties (A), (B), (C) and (D).

Just what combination (or one of which combinations) would I have to establish in order to be sure that I had a fiducial probability distribution for the two parameters of a general univariate normal?

10. In your last paragraph, you say that there is no three-fold fiducial distribution in Behrens' problem because 'There is no sufficient simultaneous estimate of the three parameters . . .'. I believe I may possibly understand the facts of the situation, but let me recapitulate them so that we may see if we agree:

(A) $\bar{x}_1 - \bar{x}_2$ is a sufficient estimate of $\mu_1 - \mu_2$.

(B) s_1^2 is *not* a sufficient estimate of σ_1^2, *nor* is s_2^2 a sufficient estimate of σ_2^2, *nor* are s_1^2 and s_2^2 jointly sufficient estimates of σ_1^2 and σ_2^2, *because* $(\bar{x}_1 - \bar{x}_2)^2$ provides information about $\sigma_1^2 + \sigma_2^2$ in the special case where $\mu_1 = \mu_2$ by assumption.

(C) However, $\bar{x}_1 - \bar{x}_2$, s_1^2 and s_2^2 *are* jointly sufficient *statistics* for $\mu_1 - \mu_2$, σ_1^2 and σ_2^2.

Apparently, the subtle distinction between 'sufficient statistics' and 'sufficient estimates' is very important here. Is this your view??

11. The following simplified version of the Behrens' problem should prove illuminating:

(A) $\{x_{1i}\}$ is a sample from a normal population of (unknown) mean μ_1 and (known) variance 1.

(B) $\{x_{2j}\}$ is a sample from a normal population of (unknown) mean μ_2 and (unknown) variance σ^2.

(C) We desire either
   (i) to make a valid test of significance concerning $\mu_1 = \mu_2$,
   (ii) to make a valid test of significance concerning $\mu_1 = \mu_2 + \theta$ for some preassigned θ,
   (iii) to establish the fiducial probability distribution of $\mu_1 = \mu_2$ (if this is legitimate).

As I understand the situation under (i), evidence concerning σ^2 is provided both by s_2^2 and by $(\bar{x}_1 - \bar{x}_2)^2$ and, if we are to make exhaustive use of what the observations bring us, we should combine these in some way. Is this your view?

Fisher to J. Tukey: 29 July 1955

Your long letter shows that you have been thinking, but still you do not easily take notice of what I have written. Instead you keep reiterating any irrelevant puzzle which in the past you have been unable to get clear about, as though you were unable to think clearly about our interchange of ideas until everything had been restated to you to your own satisfaction. If you would concentrate on one question at a time I am sure you would make better progress.

I am glad that you do not now challenge my statement that probabilities (as the word is used in the natural sciences) can be verified by observed frequencies. That is one point gained. The next step might be to consider the second point in which you do not (or apparently did not) understand the word in the same way as myself. However, you break in with a confused remark about Behrens' test, knowing, I believe, exactly what my answer is, but anxious to avoid the serious consideration of what you are overlooking in the semantics of the word probability.

On the (irrelevant) Behrens question, you know how to verify 'Student's' t distribution experimentally. Do this with two t distributions simultaneously, with n_1 and n_2 degrees of freedom; for any assigned area marked off on a (t_1, t_2) diagram compare the number of points observed with the number calculated by integration.

You seem to dispute my statement, in my last letter, that there does not exist a simultaneously sufficient set of estimates for the common mean and true variance of a pair of independent samples. So you are still confusing this case with those in which sufficient estimation is possible. This is a fair test case: Do you only want to argue, or to make some progress in understanding?

I hope now I have set aside your obstacles, doubtless erected under a subconscious urge, to your consideration of the 'central characteristic'. 'Rigorous exactitude' is not an unfamiliar idea. The phrase would be quite intelligible to you if you wanted it to be. Neither of your 'two ways' is near the mark. For (i) please note that I am not harking back and reiterating what was said about observational frequencies; I hoped that that point had been done with. I was making a *second* point with a new paragraph and page. About (2) I suppose all rigorously exact statements are unique. Correct me if you disagree about this, but it was not what I was saying, for, as I tried to suggest, I was not using unnecessary words.

The words following 'rigorous exactitude' now deserve your attention. The passage is less than four lines long. Probably you are unwilling to agree with it, and will resolutely shut you eyes to its meaning. The paradox you attempt to build from it is phoney, as you perfectly well know. There is no contradiction between the two quotations, though written 20 years apart or so. At the risk of Dame-School simplicity, may I suggest that it would be reasonable to say 'On the data as it existed before 1950 the probability that x exceeded a was over 90%', or 'evidence that has accrued in recent years has led to very different views of these probabilities'. If you can understand statements in this form, you will be able to distinguish, as men of science do, between statements referring to the real world, and statements referring only to those hypothetical worlds created by hypotheses. When speaking about the real world we must, if we are making a statement involving uncertainty with the care such a statement requires, always include all the data available for the inference. In a hypothetical world also we must use all that, within that hypothesis, is available.

You will have a better chance of getting this if I stop here without discussing your very clever paradoxes.

Fisher to J. Tukey: 9 August 1955

Thanks for your note[1]. At the risk of being tedious, I am sure I ought to suggest again that you try to carry out a strict fiducial argument based on inequalities, instead of relying merely on the pivotal transformation. I feel sure you might find in this way where your difficulty lies.

[1] Tukey had written thanking Fisher for a pleasant time during their recent discussions in Cambridge.

J. Tukey to Fisher: 18 *October* 1955

Your note of 9th August has just reached me . . . In it you suggest that I 'try to carry out a strict fiducial argument based on inequalities, instead of relying merely on the pivotal transformation'. I am afraid that I have my old trouble with this remark — wondering which of the several meanings that seem plausible to me you meant — and fearing that you meant still another.

(1) Do you mean that following through a chain of inequalities is likely to bring to the surface subtle misbehaviours which are easier to overlook in the pivotal analysis?

(2) Do you suggest that cases where inversion is in closed form are safer? Or better in principle?

(3) Would you feel that a correct pivotal argument which showed that the mapping from the pivotal plane to the parameter plane was everywhere smooth and one-to-one was not enough?

(4) Or have I missed the point again?

P.S. The day with you was a great enlightener, and the effect has steadily increased.

Fisher to J. Tukey: 22 *October* 1955

No, I do not mean any of the three things you say, in so far as I can understand your numbers 1, 2 and 3. I only meant what I said, namely that you should 'try to carry out a strict fiducial argument based on inequalities, instead of relying merely on the pivotal transformation'. In fact I am sure that the pivotal transformation as used by you, and without safe-guards, which if you were a little forward-looking you might be able to think up, is thoroughly misleading.

The direct use of a strict fiducial argument, which I recommend, I have illustrated rather elaborately and specifically for your benefit on page two of my letter of April 27th, so it is really disappointing that you should not know what I meant by my suggestion.

J. Tukey to Fisher: 10 *January* 1956

There appears to be a certain chance that I could be in England for a while this summer. I found the day I spent with you last summer very stimulating and helpful. Consequently, a major factor in my decision about the coming summer, should a trip to England become possible, would be the possibility of more extended contact.

Do you plan to be in Cambridge this summer? Would you desire to spend a substantial amount of time in discussion, educating me further? (Please do not hesitate to say 'no' if you feel so inclined. I should not like to impose on you!)

Fisher to J. Tukey: 16 *January* 1956

I expect to be in Cambridge during the summer, at least until about mid August and perhaps later, though there will be a short interval for a congress on human genetics in Copenhagen.

Of course I should like to resume talks with you, and these could be fruitful if you can get over the sort of caginess or inhibition which seems to be preventing you from hearing or seeing anything that could possibly remove any misapprehension.

At least I expect I have got you to understand that I do not use, or recommend, the process of mechanical transformation by the Jacobians, in which you and Savage seem to have been exercising yourselves following the earlier work of Segal and Diananda. I know that an unguarded phrase of mine has been read to this effect, but this misinterpretation has not only led to you barking eagerly up the wrong tree, but to your missing the point I was trying to make, namely the immense analytic abbreviation made possible by recognition of the fiducial argument. One of the best examples of that is its application to *Behrens' problem*.

J. Tukey to Fisher: 11 *February* 1958

. . . I am writing to you to ask your advice and sanction on a matter of terminology. I have been asked to give the Wald lectures at the Institute of Mathematical Statistics meeting next fall and have accepted. The purpose of this set of lectures is to give a connected account of some branch of statistics which is not easily available as such in the literature. It is my present intention to say what I can about those matters connected with fiducial inference which I think I understand. (And perhaps go on and make suggestions rather than statements about other matters.) I am not at all sure, as always, that what I would say would be exactly what you would like to see said. This perhaps offers a good reason for using a name slightly modified from 'fiducial' such as, for example, 'fiduciary'. On the other hand, the material which is to be discussed is built on a foundation of your own ideas. From this point of view, it would be desirable to retain the word 'fiducial'.

Do you have any wishes, desires or advice in this matter? I should be glad to try to be guided by any preference you might have as to which choice of terminology I should adopt. . . .

Fisher to J. Tukey: 17 *February* 1958

I have your letter of February 11th and I suppose its purpose is to know whether I like the word 'fiduciary' as a variant of 'fiducial'. The answer is no, not a bit. Of course I knew of the use of the word 'fiduciary' in financial circles and was therefore anxious to avoid its use in another connection. This use fiducial is as an adjective qualifying the word 'argument' in order to contrast it with a Bayesian argument, though in particular examples they may be brought indistinguishably close together, as perhaps you have seen in Section 6 of Chapter V (*Statistical Methods and Scientific Inference*).

It is not an adjective qualifying the word 'probability' save in the sense of a probability derived by the fiducial argument. I use the word 'probability', unlike Neyman, Wald, and most of your friends, in the sense in which it was used by the old masters, Fermat, Pascal, Leibnitz, Bernouilli, Montmort, de Moivre, and Bayes. These all thought that you could make probability statements about parameters characteristic of the real world. They did not think that such probabilities were necessarily zero or unity. So far your publications on this subject have not clarified it.

Fisher to S.S. Wilks: 15 *August* 1939

I was hoping you would visit us again before leaving London. However, I expect you found that you had a great deal to do in your last days here.

I am sending a short note,[1] submitted for the *Annals of Mathematical Statistics*, written chiefly to correct the impression given by Bartlett's recent paper[2] that I have confirmed the validity of his own approach.

As I think it was just this so-called 'exact inference of fiducial type' which led him into all his other criticisms, it may be worth making clear that I have written nothing which could be taken as confirming its validity. The relation of tests of significance to the various possible hypothetical succession of repeated samples may, I hope, be made a little clearer by the latter part of my note.

[1] Subsequently published as 'A note on fiducial inference'. *Ann. Math. Stat.* **10**, 383–8 (1939) (*CP* 164). Wilks was editor of the journal.
[2] Bartlett, M.S. (1939). Complete fiducial distributions. *Ann. Math. Stat.* **10**, 129–38.

S.S. Wilks to Fisher: 27 October 1939

In reading your 'Note on fiducial inference' several points have occurred to me which I should like to pass on to you. Perhaps you can straighten me out on them. I am glad that you have indicated on the first page that when you introduced the concept of fiducial probability, you had in mind its use in connection with the theory of estimation, and hence in making significance tests. It seems to me that this point cannot be emphasized too strongly and that it might well be brought in at several points, for the phrase 'inference of fiducial type' has apparently reached a stage of wider interpretation (perhaps unfortunately) than you originally had in mind. To take a rather trivial and extreme case, if one has a sample of n objects from a normal population and considers the quantity

$$t = (x_1 - a)/s \tag{1}$$

where x_1 is one of the items taken at random from the sample, then t has the 'Student' distribution with $n - 1$ degrees of freedom. One can make 'inferences' of 'fiducial type' about a in this wider sense from t given by (1). The inference about a is of course far from being satisfactory from an estimation point of view as compared with the inference which can be made by using the mean $\bar{x}$. In practice, one would obviously fix up a function (a t function in this case) that would enable him to make inferences about a as close as possible. Now in this wider sense, I think everyone will agree that Bartlett's T and T' (the choice determined at random) furnishes a method of making an 'inference of fiducial type' but, as you have pointed out, the inferences by this method from the estimation point of view are unsatisfactory on account of the peculiarity of the functional form of T'. In practice, one would not want to use T' for the purpose of making inferences about the difference between the population means. Now my own feeling, based on what I know of the controversy between yourself and Bartlett together with what a number of others have said about it, is that readers will misinterpret the first sentence in your summary by regarding your statement as being equivalent to the statement that Bartlett's test does not provide an 'inference of fiducial type'. This will also in my opinion leave open the way for another rebuttal from Bartlett. In other words, is not the following statement true: Bartlett's test provides a basis for making an 'inference of fiducial type' about the difference $(a_1 - a_2)$ between the population means in the wider sense, but the inference is not satisfactory from an estimation point of view, as you have shown?

On page 5 . . ., the first three lines of the last paragraph are not clear to me. I am probably misinterpreting your statement. Apparently $\bar{x}$ will exceed $st_n/\sqrt{n'}$ with the chosen frequency regardless of whether σ^2 is known or not. Of course, if the value of σ^2 is actually used, we can say that $\bar{x}$ will exceed $\sigma t_\infty/\sqrt{n'}$ with the assigned frequency. I believe you will confirm me in saying that this latter statement is not inconsistent with the statement that $\bar{x}$ will exceed $st_n/\sqrt{n'}$ with the chosen frequency. . . .

Fisher to S.S. Wilks: 6 *November* 1939

Much of my note on fiducial inference is devoted to the question 'Did Fisher assert that Bartlett's distribution provided an exact inference of fiducial type?' Much of your own letter, on the other hand, is concerned with the different question, whether Bartlett's T and T' furnish a method of making an inference of a fiducial type. The second question may be the more important, and would be, if there were agreement as to the use of the word 'fiducial'; but it was the first, which is at least capable of being settled definitely, that I was concerned to settle in my note.

The case which you take on your first page is not nearly so extreme as the case of T and T'. Your case is of the familiar type, where the fault of the method is equivalent simply to ignoring part of the data. I have always protested against speaking of such inferences explicitly or implicitly as inferences from the data, though I can have no objection to speaking of them as fiducial inferences from a selected portion of the data. In Bartlett's case, on the other hand, I do not think that the distribution leads to a test of significance at all.

In your allusion to my statement on p.5, that it is highly improbable that $\bar{x}$ will exceed the limit assigned with the frequency chosen, I think you have missed the point of the distinction I was drawing. In any fixed population the limit will be $\sigma t_\infty/\sqrt{n'}$ and this will only be equal to $st_n/\sqrt{n'}$ for a particular value of s. It is highly improbable that this particular value will be the value of s observed. As I say later, on p. 6, the distribution of t will be so verified. This implies that $\bar{x}$ exceeds $st_n/\sqrt{n'}$ with the correct frequency when each observed $\bar{x}$ is associated with its own observed s. . . .

I want you to remember that the occasion of my note is that Bartlett has ascribed to me a view which I do not hold. It is always rash in controversy to express the views of an opponent in any other way than by specific quotation. Misrepresentation, I always think, is particularly to be avoided, since, in political and forensic controversy, no artifice is commoner than that of mirepresenting an opponent's views in order to create prejudice. To do so in scientific controversy seems to me a very serious offence against good taste, since we are each doing our best to express the truth clearly. . . .

S.S. Wilks to Fisher: 21 *November* 1939

I am enclosing proof and *ms* of your note. You will note that I took the liberty of making a few minor changes which, however, I think are consistent with the understanding contained in our recent exchange of letters. I still feel that the first sentence in the second paragraph may be a bit too strong and not sufficiently impersonal but to leave out the word misleading may leave the statement too empty.

On p. 2 of your letter (first paragraph),[1] I think I understand the distinction you are drawing, but there is some question in my own mind as to whether the statement might be improved in wording. I simply mean this: If for each sample we compute $\bar{x}$ and s, we can make the statement that $\bar{x} > st_n/\sqrt{n'}$ and be correct with the frequency assigned, say 0.025, understanding of course that $\bar{x}$ and s are inserted for each sample. If s for a particular sample is used over and over, then obviously we cannot quite say that $\bar{x} > st_n/\sqrt{n'}$ and be correct with the chosen frequency (the only thing changing from sample to sample being $\bar{x}$). On the other hand if σ is known we can say that $\bar{x} > \sigma t_\infty/\sqrt{n'}$ and be correct with the chosen frequency, where the same value of σ is used repeatedly. The whole question hinges on whether one has in mind using the value of s for a particular sample repeatedly for various samples. The statement in my other letter on this point presupposed that in considering $\bar{x} > st_n/\sqrt{n'}$, $\bar{x}$ and s changed from sample to sample according to the sampling law of s and $\bar{x}$. Presumably, you do not mean to change s from sample to sample.

. . .

[1] i.e. third paragraph of Fisher's letter.

Fisher to S.S. Wilks: 8 *December* 1939

I have sent back the corrected proof by air-mail, the main alterations being in the punctuation, especially where formulae are concerned, and one important error, which I hope you will see is not messed up, namely where S^2 is used for s^2.

Of course, in the case of 'Student's' problem, the practice of obtaining a fiducial limit from a single sample, e.g.

$$st_n/\sqrt{n'}$$

and then complaining that a population of samples drawn from a fixed population will not be divided by this limit in the ratio specified by the level of significance chosen is a ridiculous one, in that it shows a complete misunderstanding of 'Student's' test. It is, however, actually this argument which has been used, or implied, as though it were an effective criticism of Behrens' test, the critics being, I suppose, genuinely misled by the fact that, in 'Student's' test, we can use actual tables of a quantity t whose distribution is known, whereas, in Sukhatme's case, the simultaneous distribution of t_1 and t_2 cannot be used directly to show the integral over any relevant area, so that it is necessary, for purely practical purposes, to have a secondary table of d. I imagine that Bartlett's puzzlement arose from him thinking that d was the analogue of t, not only in being used for tabular reference, but in its logical position in the fiducial argument. Actually, of course, a number of different functions, such as d' defined by

$$d' = d\sqrt{s_1^2 + s_2^2} \;/\; f(s_1, s_2)$$

might be used for tabular purposes in place of d, giving an exactly equivalent test of significance where f is any homogeneous function of the first degree.

Fisher to E.B. Wilson: 20 *May* 1935

I have now looked through W.E. Deming's paper[1] . . . What first Treloar and Wilder,[2] and later, it seems, Deming failed to see is that, in setting our level of significance at any value such as 1%, we are choosing voluntarily to make the mistake of rejecting the hypothesis when it is true in this proportion of cases: and as, on the hypothesis discussed, we are always equally wrong in such rejection, it is a matter of complete indifference, provided the proportion is kept right, in which particular samples this mistake is made.

Obviously, of course, when the population variance is known, the test which utilises this knowledge is preferable to one using only an estimate, and the reason for this preference is not that we should prefer to be deceived by one sort of sample rather than by another.

It is curious that the simple history of the problem should not have kept the author straight. From the beginning of the Theory of Errors, it must have been clear that, knowing the population variance, one could use the normal curve to test the significance of a mean. The test can easily be extended to a weighted mean, or to the regression coefficient, as soon as this is recognised as merely a weighted mean.

In the majority of practical problems, it was easily perceived that the population variance was not provided in the data, but, if the sample were large, could be estimated with some confidence, so that the problem was discussed by Gauss, what estimate it was best to take. Gauss satisfied himself that the mean square error, with allowance for degrees of freedom, was most proper, though, under the influence of Peters, astronomers were largely persuaded to use the mean error instead. As soon, however, as any estimate is used rather than a known value, the question of its errors of estimation arises; and throughout the nineteenth century it was usual to say that we needed some large number, e.g. 50 observations, before relying on the test. 'Student's' work showed that, using the mean square error, the frequency of rejection of the hypothesis, when true, can be calculated exactly, and for small samples was larger than that derived by substituting the estimate in the formula for the normal curve. 'Student's' tables, in fact, show how to make the classical test exact. . . .

[1] See also Fisher's letter of 19 September 1935 to Deming (p.82).
[2] See Fisher's letter of 13 May 1935 to Immer (p.266).

Fisher to E.B. Wilson: 8 *March* 1955

. . . A good deal of modern work, in fact, prompts in me the question that it is often very difficult for a mathematician to recognize the difference between what we do know, and what we do not know, a distinction which is not difficult to the worker in the natural sciences, but being irrelevant to a good deal of mathematical work, mathematicians are often surprised to find that it matters.

I find that the word 'fixed' is often used by statistical writers to claim for a quantity supposedly unknown, the properties of one actually within our cognizance.

Fisher to E.B. Wilson: 26 September 1955

. . . I do not think the existence of innate or spontaneous emotional reactions to logical situations should prevent there being rational analysis, for we can calculate and communicate the extent of the evidence or data having such effect, and the clear recognition of this extent is at least a rational as well as a communicable effect of the evidence.

I think the typical example is a simple test of significance in which we normally experience a reluctance or resistance to a belief involving a very long chance. The emotional state accompanying this resistance may be quite sub-rational, but we can invite any other rational being to re-calculate the length of the odds, and verify the nature of the evidence available to create such a rational reluctance. In most cases, indeed, there is no means of going further, e.g. of deriving a probability statement concerning the probability of any hypothesis subject to which such long odds have been calculated. I am inclined, therefore, to speak of the hypothetical improbability as affording a rational basis for disbelief in a degree measured by the smallness of the improbability calculated.

Fisher to E.B. Wilson: 8 August 1956

You will be glad to hear that my new book is on the point of emerging from the binders under the title of *Statistical Methods and Scientific Inference*, and I hope you will be amused by my various hair-splittings in the attempt to get clarity on a subject which has been so gravely confused in the literature.

A great deal of the thought is new, even to me, and I have only just realized that Keynes's excellent phrase, as I have always thought it, 'the degree of rational belief', is not really well applied to the concept of mathematical probability, in its classical sense implying the possibility of verification by observations of frequency, but can be seen to be more appropriate to the more primitive type of inference represented by those tests of significance from which no statement of probability can be derived.

I have discussed and, I think, made this clear, in essence though not in terminology, in the opening section of Chapter III, but I am sure that I have often commended this qualitative definition of Keynes's as appropriate to probability *senso strictu* and I now realize that it is not really the concept of mathematical probability to which it is appropriate.

I hope I may look forward to your comments some day when you have glanced at this little book. . . .

Fisher to E.B. Wilson: 28 *March* 1957

It was nice to see your letter of March 16th.

There were a sprinkling of new things in my book on scientific inference on which I should certainly like your opinion.[1] Section 4 of Chapter II, on 'The meaning of probability', is, I think, of some permanent importance as stressing an aspect of the meaning of probability which seems to have been overlooked by the leading twentieth century writers, such as Kolmogoroff and Feller. The statement of the fiducial argument in Section 3 of Chapter III is certainly more complete than anything I have done before, and I believe clears up the slight confusions left by Kolmogoroff's and Jeffreys' discussions of the subject. As a mathematical *beispiel*, the asymptotic application of the fiducial argument to discontinous data has, I believe, not been before attempted, and leads to some interesting comparisons. I hope to get the misprints in this analysis corrected for the second edition, which has already been asked for. However, from you I would particularly want to get the reader's reaction to the forewords that I have inserted before some of the chapters, such as that of Chapter V, pages 106–110, on axiomatics and Gödel's theorem, which may well amuse you and perhaps help others. The example on page 169 gives the correct fiducial inference for a case which has been mauled over a good deal, I think by one of the Princeton men, for there has been quite a cult, led I fancy by Savage of Chicago and Tukey from Princeton, for exhibiting garbled versions of the fiducial argument without any reference to the strict application of that argument itself.

On your point about the 2×2 table,[2] when making an exact calculation I always use the single tail, and if I want to compare significance with cases where both tails are used, I simply double the value obtained, without regard to the question of how lumpy the other tail may be. Usually, indeed, I think that the single tail is appropriate, though of course not always. For example if, being new to the subject, I test 500 students with phenyl-thiocarbamide and find there are more non-tasters among the men than among the women, I should only be sure there was a sex difference if on doubling the probability the result was still small enough to satisfy me, but having heard that *X*, *Y*, and *Z* had found fewer non-tasters among women than among men, I should be satisfied that my population was telling the same story as theirs if the probability for the single tail was satisfactorily small; but in the first case I would not feel at all concerned with the discontinuities at the tail other than that which had been observed.

No, I do not discuss this in the new book, where in Chapter IV I am principally concerned to get across the important point that it is frequencies within the least recognizable subset of hypothetical possibilities that are relevant to scientific inference. The 2×2 table gives a fairly good preliminary example of this, though in such a discussion I suppose I ought to emphasize that if, without my meaning to perform an experiment, three rabbits die unexpectedly, and on enquiry I find that a mistake has probably been made leading to their food being poisoned, I believe any rational man would think it probable that the cause of these unexpected deaths had been revealed, even if the particular poison suspected had not previously been shown to be lethal to rabbits. As a recommender of scientific procedure, I should say that at that point it would be reasonable to perform a definitive and well-controlled experiment capable of verifying this inference, that in such an experiment there would be an adequate number of untreated animals, and verification would only be claimed if there was a significant *contrast* between the reactions of the treated and the untreated, and that contrast in the expected direction. In fact I would distinguish for this purpose between being sure, and being in a position to claim an experimental demonstration.

[1] Wilson had written to Fisher that he had read *SMSI* with great pleasure but had not found much new material in it.
[2] Wilson had asked about the use of one or two tails when calculating exact probabilities with 2 × 2 tables.

Fisher to F. Yates: 27 *July* 1955

I am sending herewith the last chapter,[1] including something about estimation and the use of ancillary values. It will be mostly old stuff to you, apart perhaps from section 12, which I put in because some Americans, Savage at Chicago and John Tukey, have been rather working themselves up about the uniqueness of simultaneous fiducial distributions, and have been trying to use Jacobians in which some of the elements, or minors, are not uniform in sign. I fancy it is one more example of the difficulty felt by many mathematicians at the idea that the kind of conclusion which can be drawn depends on the analytic form of the data. Anyway, it occurred to me that in my early work the logical notions used in this book had been developed chiefly in connection with thc thcory of cstimation, amd there were one or two things, such as the definition of 'consistency', in which I had not been very consistent myself, and in which, inevitably, people like Kendall had chosen the less satisfactory definition, without a hint that the idea had been expressed in any other way!

[1] A draft of chapter 6 of *Statistical methods and scientific inference.*

Fisher to F. Yates: 13 *October* 1955

Quite a long while ago I sent George Barnard a note on Pearson and Hartley's table 11, and he has taken some time to mull over it, and perhaps has not yet sent you the copy I sent him. Anyway, enclosed is a more complete copy, with various corrections and additions in the form in which I should, at present, be inclined to print it. Together with this I am sending a letter from George Barnard, written yesterday, and a copy of my reply.[1] This interchange may be useful as a means of getting the point of the problem clearly realized. I think George shows some confusion on his second page, for in the case of Behrens' values as against Welch's, so far as I know both are aiming to test the significance of the same hypothesis, and both are using to that end identically the same statistic. They differ only in respect of the sampling distribution of that statistic which is regarded as relevant for testing the hypothesis.

However, confusion is to be expected, for in the course of time every sort of lunacy has been written about this problem.

Could you send me back the copy of my note that I am herewith enclosing, including the diagram? I have other copies of George's letter and my reply.

[1] See p.29.

F. Yates to Fisher: 1 *November* 1955

I am now returning your paper.[1] By and large I am in agreement with it, but the fact that people such as Barnard find difficulties suggests that some further discussion may be worth while.

The crux of the matter is, I suppose, what is meant by 'Once in ten (or $1/P$) trials' (your p.8). Welch can claim for his test that this property holds in repeated sampling from the populations with the same σ_1/σ_2, but only by ignoring the actually observed ratio s_1/s_2 (or should I rather say, using it to enter the table with, but ignoring the other information it contains).

Logically I think every right minded person would agree that all relevant information should as far as possible be taken into account when making a test of significance. The snag is in the phrase 'as far as possible'. We are, I believe, fully agreed that vague and indefinite information (such as that of the *a priori* distribution type) is best excluded from formal tests and taken into account in the subsequent overall assessment (or should I say, considered judgement) which rational beings normally make after doing their figuring. But once it is admitted that certain information can be excluded the question arises where to draw the line? In the analysis of the results of an experiment we agree to ignore the information provided by the actual pattern of treatments in that particular experiment and merely to utilize the information that the design was selected at random from a set of designs having certain properties. This is justified by just the argument which you condemn in your reply to George Barnard, and the purist, using the same argument, might well say the fact that he gets too small a chance of significance when he was lucky enough to select a Knut Vik 5×5 Latin square[2] is not compensated for by the fact that he gets too high a chance in the diagonal square. And that, if that is the best the statistician can do, he proposes to use Knut Vik squares in future, which have been shown statistically (as common sense would indicate) to be the more accurate. (I know there are other arguments against this course, but I don't think they are relevant to the present issues.)

I bring up this point because it does show that the statistician considers himself free to ignore inconvenient information, and also emphasises, and rightly I think, that some[3] value is attached to getting $1/P$ significance in repeated sampling from the same population. The fact that in some common cases, such as the ordinary t test, this occurs while at the same time taking into account all relevant information, has rather blinded people to the fact that it does not *necessarily* occur. But it might be worth while giving examples in this paper — I have particularly in mind the linear regression case, and the 2×2 contingency table. The second paragraph of your letter of 13th is also very relevant.

On the paper as it is at present I have little detailed comment, except that I am not very happy about your comments on the quotations. James, I think, might well be dropped. As I see it the reason for his footnote was that he had himself cast doubts in the text on the mathematical accuracy of Welch's solution. I don't think it is *a propos de rien* — indeed, I might well have added just such a footnote myself in similar circumstances. The other people have taken the stand they do because they regard $1/P$ significance in repeated sampling *the* desirable feature. I doubt if Bartlett will now regard Welch's solution as in the least intolerable. Perhaps he ought to, but that is a different matter.

The paragraph which you added contains the phrase 'the error of these calculations'. The error is surely that of refusal to take into account relevant information, and not of calculation. . . .

[1] Subsequently published as *CP* 264.
[2] Fisher has written in the margin, 'Is there a test valid for Knut Vik?'
[3] Fisher has underlined 'some' and put '!' in the margin.

Fisher to F. Yates: 2 November 1955

I thought you must have got into some very deep entanglement from the time you have taken in replying, and I see that the analogy which has misled you is that of randomization.

Over the period in which I was putting forward this recommendation to experimenters, I naturally gave a great deal of thought to the effects of this procedure and the purposes that it could usefully fulfil; and, so far as statistical methods are concerned, what I was constantly pointing out was that its object was to guarantee the validity of the test of significance; that is to say that with all the great advantages of the Knut Vik square, if the results of using it were reduced by an analysis of variance, or one of the cruder techniques that preceded it, the probability statements obtained in the z test would be erroneous, whereas if proper randomization were applied, as I think you and Eden once demonstrated experimentally, the z test was made to be reliable.

Of course if a method were available to give a reliable test of significance for the use of the Knut Vik square, there would be no advantage in wider randomization in this respect. In the analogous case of eliminating blocks in a randomized block arrangement, or rows and columns in a Latin square, we do, and I think you will agree, properly and inevitably consider an experiment laid out in randomized blocks as one of a population arranged subject to this restriction, and not as one of the larger population, to which it also belongs, in which there is indiscriminate randomization regardless of blocks. If you can be clear in your own head as to why an experimenter who knows not only that the plots did in fact fulfil the conditions of a Latin square, but also that they were laid out by choosing one such arrangement out of all possible, by such a random process as you, yourself, have explained, would it not be simply erroneous, the error being due perhaps to prejudice or ignorance, if he insisted on drawing conclusions as though the plots had been distributed at random over the whole area?

In my view it would be simply erroneous in exactly the same way as a rain maker who claimed significant success by comparing the frequency of rain following his experiments with that of the annual frequency observable in his neighbourhood, although it is within his knowledge that the frequency of rain is greater than the annual frequency in that part of the year during which his experiments were carried out. Of course we do not know a probability unless we know it, and it is only when it is within our knowledge, that it is erroneous to substitute for it a less appropriate probability. It is when we lack this knowledge, that randomization provides the safeguard.

I do not quite know what you mean about Barnard, but perhaps you have had some correspondence with him. What he said in his letter of October 17th was 'Thank you for your letter of the 14th, and that of the 13th, which make all clear, on Welch's test'. He refers to having quoted Yates' 1939 paper indicating why one was concerned with fixed s_1/s_2, but I have not looked this up so I do not know if, and if so, why you thought at that time that s_1/s_2 needed to be fixed. From my point of view this 'fixing' is not a voluntary act to be done or abstained from, but a fact to be recognized, as Behrens clearly perceived, in obtaining the appropriate test of significance.

The tests put forward by Behrens and by Welch respectively are (a) attempts to solve the same problem, and (b) attempts using exactly the same statistic, the appropriate notation for which was, I fancy, first fixed in Sukhatme's paper. The two tests differ only in the frequency distribution ascribed to this statistic. Such a difference can surely be resolved without reference to the intangible elements of judgement, but if you do employ horse-sense on this problem, do you not feel any difficulty in Welch's value being actually less than the value of t for the number of degrees of freedom supplied by the two samples jointly? What hit me in the face when I first realized this was the idiocy of supposing that a pair of samples showing no significant difference between the means when tested by 'Student's' test, on the assumption that the sums of squares could properly be pooled, should provide significant evidence that the means are unequal to a man who professes complete ignorance, apart from the evidence supplied by the samples, as to the relative precision of the two empirical means. ‘Thank God I am not absolutely sure’, he says, ‘that these two varieties have the same variability, else I could not claim that one gives significantly a larger yield than the other’. . . .

2
Statistical Theory and Method

C.I. Bliss to Fisher: 13 *September* 1938

. . . I enclose an extract from my correspondence with Snedecor, upon which I would appreciate your opinion.

Letter from Snedecor dated Sept. 6, 1938:

'I found that in applying the inverse sine transformation to Moore's data (the numerical example in my Russian paper containing the table of angles),[1] the relative size of the means was reversed in two of the stages. That is, stage A had a larger mean percentage than stage B, but in the transformed values mean A was smaller than mean B. In that case what are you testing, the significance of the difference between percentages or between inverse sines? This seems to me to raise a question as to the validity of the inference made from the transformed values.'

My reply:

'Your question concerning the inverse sine transformation raises an interesting point. I take it that we are testing the significance of the difference between inverse sines on the assumption that these are the better index to the biological differences in which we are really interested. From the analysis of this experiment we know that the fumigations differed markedly, so that the successive estimates for any one stage form a heterogeneous series rather than samples from a single population. In averaging a series of this type, one would probably weight each contribution to the total by the information it contains. This the transformation does automatically, so that in testing the significance of a difference between inverse sines we are in effect testing the difference between corresponding weighted mean percentages. It is to be expected that some differences may be reversed by the transformation, especially when they are well within the sampling error as in the case you cite. In other cases they may change in magnitude from non-significant to significant or vice versa as a result of the transformation. If this did not occur, one would question the utility of using the transformation in the first place. At least this is the way it appears to me although I may have missed the point of your criticism.'

Is my interpretation correct?

[1] The passage in parentheses is presumably an insertion by Bliss.

Fisher to C.I. Bliss: 6 *October* 1938

I was interested in your correspondence with Snedecor, and enclose my own reaction to it . . .

'Shorter catechism' on transformations

A. Are you testing the significance of the *mean percentage*, or of the mean of the corresponding *angle*?
B. Strictly speaking I am testing the hypothesis that mortality is unaffected by a certain specified change of treatment.
A. But that implies that both the mean percentage and the mean angle are unaffected.
B. Yes.
A. Then how can it matter which you use?
B. I believe the mean angles may be the more sensitive.
A. What do you mean by more sensitive?
B. That the experiment tested with the mean angles will more probably show a significant deviation, if in fact the hypothesis is untrue.
A. Do you mean that every possible sort of deviation from the hypothesis will be more obviously apparent using the mean angle?
B. No, that would not be true of every conceivable sort of deviation. From my experience of variation in the experimental material, and of experimental conditions, I judge that the change in treatment under examination will, if it has any effect, cause more nearly a constant increase in the angle than a constant increase in the percentage.
A. Then your test of significance is chosen by judgement based on experience. Have not Neyman and Pearson developed a general mathematical theory for deciding what tests of significance to apply?
B. I believe that is the aim of much they have written. The absence of reference to experience seems to me a serious flaw in their work. Their method only leads to definite results when mathematical postulates are introduced, which could only be justifiably believed as a result of extensive experience. They do not, unfortunately, discuss the nature of the evidence for these postulates. If they did so, they would find that in practice it could amount to no more than the experience that one test of significance gave more frequently significant results than another. In general we may come to the conclusion that this is so, partly by direct trials on analogous material, partly from our general concepts as to how the observable effects arise. The introduction of hidden postulates only disguises the tentative nature of the process by which real knowledge is built up.

C.I. Bliss to Fisher: 2 October 1941

. . . In regard to table XX [in *Statistical tables*], one possible application has occurred to me which probably is based upon a misconception but would be extraordinarily useful if it were legitimate. It often happens that in varietal trials involving many varieties one wants to use the pooled variance for error in testing the significance between varieties arranged *a posteriori* in order of yield. In this case, of course, the just significant difference as usually computed has serious limitations. It has occurred to me, however, that it might be possible on the basis of the intervals between successive scores in Table XX to form an adjusted series of mean yields to which the just significant difference might be applied. On this scale, varieties near the ends of the distribution would be moved toward the center and those in the middle of the distribution would be spaced more widely as determined by the intervals in Table XX. Hotelling gave me no encouragement when I suggested this possibility to him last July, in part because no paper is yet available describing just how Table XX was computed and the conditions under which it may be used. I have tried it myself with scores in a collaborative study on techniques for analyzing soils where 31 soils were ranked by 20 different observers. The transformation seemed to work nicely and its logic seemed perfectly obvious and convincing, but whether it could be extended to these other uses where tests of significance are particularly unsatisfactory at the present time is a problem for which you may have the answer. . . .

Fisher to C.I. Bliss: 24 *November* 1941

. . . In respect to the type of varietal trial you have in mind, I think the varieties giving intermediate yields, although in fact they will be crowded together, have, properly speaking, the same precision as those with the highest and lowest yields. The function of Table XX is to make the best use of observations in which, as in a race, order may be recorded more easily than any numerical measure can be obtained. I had thought that the formula on p. 14 and the explanation starting on p. 13 gave the theory of the thing sufficiently. . . .

C.I. Bliss to Fisher: 19 *July* 1956

Whenever I get really stuck on a problem and no one hereabouts seems to have the answer, I remember the ease with which you have found solutions for me in the past. Here is a problem that I believe has potentialities but is not treated effectively in any place that I have looked, although I suspect that it has been solved more than once in the past.

In a recent paper in *Science* (June 1, 1956), Peter D. King lists the number of births in each hour of the day in five hospitals. At my request, he has sent me the complete record for each of the five hospitals. The three largest, comprising some 86 percent of the data, differed less drastically in the number of cases. I have computed the analysis of variance on the attached sheet, the unit being the number of births per hour per hospital.

In fitting the data, I have followed in part Aitken's *Statistical Mathematics*, where he describes Fourier analysis under the heading 'Periodic Regression'. The first two terms seem to fit quite well as you can see from the data and diagram. Since the mean square for the interaction of series by scatter ($s^2 =$ 360.10) is even less than the mean ($\bar{y}$ = 398.99), I have assumed that the variation is essentially Poisson. Although the average scatter about the fitted curve shows a high significance, the distribution of the plotted points does not suggest any advantage in adding additional terms to the Fourier analysis. The observed small number at 7.00 a.m. coincides with the hour at which most hospitals change nursing shifts, and the new and old shifts meet to discuss current nursing problems.

The problem on which I need your help is to determine from the equation of the fitted curve two biologically meaningful statistics and their standard errors or fiducial limits. (1) At which hour is the number of births a maximum or a minimum? (2) What is the amplitude of the cycle in number of births (or percentage)? These questions may need restatement but they represent the statistics which should be most interesting, I believe, to the biologist.

Perhaps you can refer me to a paper where all this is set out explicitly in terms that I can follow. If by any chance the problem should intrigue you sufficiently, I would be delighted to have you take over the data and such calculations as I have made and write a paper for *Biometrics* describing the analysis. . . .

Hour of birth in man. Data from Sparrow Hospital (A), W.C.A. Hospital (B), and Jamestown General Hospital (C) over 3 to 9 year periods. P.D. King, *Science*, *123*: 985–986, 1956.

Hour starting	Births per hour, y A	B	C	Total T_h	cos θ a_1	sin θ b_1	cos 2θ a_2	sin 2θ b_2	Mean $\bar{y}$	Predicted Y
12 Mt	370	447	421	1238	1	0	1	0	412.7	401.31
1 am	349	447	415	1211	0.966	0.259	0.866	0.5	403.7	416.81
2	357	475	434	1266	0.866	0.5	0.5	0.866	422.0	431.08
3	411	508	447	1366	0.707	0.707	0	1	455.3	443.16
4	422	469	440	1331	0.5	0.866	−0.5	0.866	443.7	452.24
5	457	498	474	1429	0.259	0.966	−0.866	0.5	476.3	457.70
6	415	505	427	1347	0	1	−1	0	449.0	459.14
7	398	436	448	1282	−0.259	0.966	−0.866	−0.5	427.3	456.50
8	425	490	450	1365	−0.5	0.866	−0.5	−0.866	455.0	449.92
9	435	478	470	1383	−0.707	0.707	0	−1	461.0	439.88
10	406	501	415	1322	−0.866	0.5	0.5	−0.866	440.7	427.06
11	382	429	420	1231	−0.966	0.259	0.866	−0.5	410.3	412.33
12 M	381	422	388	1191	−1	0	1	0	397.0	396.67
1 pm	357	427	337	1121	−0.966	−0.259	0.866	0.5	373.7	381.17
2	339	413	356	1108	−0.866	−0.5	0.5	0.866	369.3	366.90
3	308	375	300	983	−0.707	−0.707	0	1	327.7	354.82
4	355	401	353	1109	−0.5	−0.866	−0.5	0.866	369.7	345.74
5	295	355	344	994	−0.259	−0.966	−0.866	0.5	331.3	340.28
6	296	416	331	1043	0	−1	−1	0	347.7	338.84
7	292	340	302	934	0.259	−0.966	−0.866	−0.5	311.3	341.48
8	331	434	339	1104	0.5	−0.866	−0.5	−0.866	368.0	348.06
9	339	441	365	1145	0.707	−0.707	0	−1	381.7	358.10
10	309	383	380	1072	0.866	−0.5	0.5	−0.866	357.3	370.92
11	331	456	365	1152	0.966	−0.259	0.866	−0.5	384.0	385.65
Total	8760	10546	9421	28727	0	0	0	0	398.99 =	$\bar{y}$

$\Sigma(a_1T_h) =$ 83.502

$\Sigma(b_1T_h) =$ 2165.474

$\Sigma(a_2T_h) =$ 34.316

$\Sigma(b_2T_h) =$ −143.434

$\Sigma a_1^2 = \Sigma b_1^2 =$ 12.000168

$\Sigma a_2^2 = \Sigma b_2^2 =$ 11.999648

i	$\Sigma(a_1y_i)$	$\Sigma(b_1y_i)$
A	−130.455	726.644
B	107.951	635.635
C	106.006	803.195
T_h	83.502	2165.474

Predicted $Y = 398.99 + 2.3195\, a_1 + 60.1538\, b_1$

Source	*DF*	Sum squares	Mean square	$\chi^2 = SS/\bar{y}$	*P*
Between hospitals	2	67 949.20	33 974.60	170.30	<0.001
Hours: 1st term	2	130 449.57	65 224.78	326.95	<0.001
" 2nd "	2	604.21	302.10	1.51	0.5
" Scatter	19	17 837.88	938.84	44.71	<0.001
Hospitals x hours, 1st term	4	4 304.77	1 076.19	10.79	<0.05
" " 2nd term	4	453.44	113.36	1.14	0.89
" " Scatter	38	13 683.92	$s^2 = 360.10$	34.30	0.64

Fisher to C.I. Bliss: 26 *July* 1956

The problem you sent me looks to me very like the topic I discussed in *The Design of Experiments* under the heading of Section 64 ('Wider Tests based on the Analysis of Variance'). That is to say one should not, I fancy, be satisfied with a neat statement of fiducial limits for any one particular parameter, without regard to the other.

If one were to apply the method of that section, it would go somewhat like this:

The sum of squares for 2 degrees of freedom, which you call the first term, is, I fancy, $\{(83.502)^2 + (2165.474)^2\}/18$. Had you been considering not the hypothesis that these two components should be zero, but that your measures of them should have true values α and β, this sum of squares would be replaced by $\{(\alpha-83.502)^2 + (\beta-2165.474)^2\}/18$. To decide how large this expression could be without over-turning the hypothesis, consider your error for 19 degrees of freedom, which has a mean square 938.84, a natural logarithm 6.83393, and one-half the natural logarithm 3.41696, to which can be added the tabular *z* for 5% significance with 2 and 19 degrees of freedom, namely 0.6295, bringing the total to 4.0465, which is the natural logarithm of 57.2 approximately. The variance for 2 degrees of freedom would then be $(57.2)^2 = 3271.84$ and the sum of squares for the 2 degrees of freedom would be 6543.7, so multiplying by 18 one would have $(\alpha-83.502)^2 + (\beta-2165.474)^2 < 117786$, if the hypothesis is not to be contradicted at the 5% level.

Graphically this puts the hypothetical first harmonic pair of terms within a circle, or rather a series of concentric circles, for different levels of significance. I think I should be inclined to express the apparent precision of the observations in some such graphical form, rather than to particularize as to the fiducial limits within which the phase angle and the amplitude separately lie. Of course, for the phase angle separately I should use the method of the ratio of two means, as in Section 62.1, while I suppose the amplitude treated in isolation from the phase angle would be dealt with exactly by Mahalanobis's D^2 (of which, perhaps, tables have been worked by R.C. Bose), but as you cannot help having a multiplicity of hypotheses in view, it seems to me more rational to treat them all simultaneously. . . .

C.I. Bliss to Fisher: 3 *February* 1958

Thanks for the offprint of 'The underworld of probability' [*CP* 267] . . .

Now that I have worked on periodic regression, I am discovering additional series where it would seem to apply. Also I am a little frightened by the many papers which discuss serial correlation. It seems to be the main hazard in applying the analysis of variance to periodic data. I hope I am not courting disaster in omitting this aspect which still seems to me a blind alley. . . .

Fisher to C.I. Bliss: 17 *February* 1958

Thanks for your note . . .. I do not believe you need worry about serial correlation. It leads, I think, to stochastic processes and general frustration, but perhaps there are better uses for it which I do not know about. . . .

Fisher to C.I. Bliss: 21 *August* 1958

Thank you for your letter. . . .

The trouble with variance components[1] seems only to have arisen through Tukey thinking that the matter involved simultaneous decision functions which in many cases would be exceedingly complex, as Duncan particularly has shown. If, without having heard of John Tukey, your own researches had given you measures of soporific power of seven different drugs showing significant differences, though of course every chosen pair would be significant at a different level, do you think that you would want to carry out a lot of further calculations on the seven values before you? Of course, if the second on the list was very much cheaper than the first, you might well want supplementary information, perhaps using these two only, from a further experiment.

[1] In his letter Bliss said he was 'struggling with a chapter on variance components' for a book that he was writing.

G.E.P. Box to Fisher: 19 *May* 1955

I have at some time heard or read that you said that whereas for a properly designed experiment we can perform a valid analysis, for an experiment which is not properly designed we can often only carry out a post mortem to decide what the experiment died of.

The above are not the exact words, and I would be very grateful if you would be kind enough to tell me if you did say something like this and if possible give me the correct version because, with your permission, I would very much like to quote it.

Fisher to G.E.P. Box: 21 *May* 1955

Yes, that was the kind of thing, and very much the *ipsissima verba* of what I used to say a goodish while ago at Rothamsted, say 1930–33 or so.[1] It served to bring home the fact that colleagues might take an interest in an experiment not only for its findings, but for its technique and even for its pathology.

[1] See *CP* 159, p.17.

W.G. Cochran to Fisher: 6 *September* 1938

I am interested to see your transformation for dealing with ordinal or rank data,[1] and should like to trouble you with a few questions about it. I should like to have seen an example of its practical use contained in the introduction. A good deal of labour has been and still is being devoted to the construction of tests of significance for ranked data, which your transformation, when its usefulness has been realised, will obviate. Can you refer me to any such practical example?

For amusement, I applied the transformation to the example *X X O X O O X O O O* which you discussed in your lectures. Comparing the difference between the mean score of *X* and *O* with the standard deviation (8 degrees of freedom) computed from differences within *X* and within *O*, I find *t* equal to 2.024, so that the probability of a better result is 0.039. According to your combinatorial solution, six cases were better, one hundred and ninety worse and fourteen neither better nor worse. If these are classed as better or worse according to the transformation, one finds that three of them are equal, one better and ten worse. This gives a significance level (for as good or better) of 10/210 or 0.0476, which shows a reasonable agreement with the *t* test.

Did you find any simple way to compute the values in the table? This at first sight seems to me rather a tedious business. Further, assuming that the data were originally normally distributed, what is the loss of information produced by replacing each value by its mean position?

[1] See Table XX in Fisher and Yates's *Statistical tables*, first published in 1938.

Fisher to W.G. Cochran: 7 *September* 1938

Thanks for your letter. I had made a few trials to satisfy myself that the *t* test would work well, and I am glad to see that you have done the same. The exact test of significance is, of course, that based on permutations, and from this point of view the scores are only conventional.

I think I arrived at them first by considering the problem: given the ordinal series for two variates drawn from uncorrelated normal distributions, what weighting would give an efficient estimate of a suspected correlation. The correlation of course is just the sum of products of tabulated values divided by the sum of squares given in the second table. This also, except for coarse grouping in small samples, naturally agrees well with the correlation from normal deviates.

The computation of the table is considerably simplified by expanding the series of values for each value of n in odd orthogonal polynomials. The series terminates, e.g. for $n = 2$ or 3 the linear term is exact; for $n = 4$ or 5 the cubic term is exact, and so on; but the diminution of the numerical values is also very helpful, e.g. I suppose that the 13th degree is very nearly right, apart from the last value, for numbers up to 20 or 25. The coefficients of these polynomials are a good deal simpler than the original integrals, being free from the variable r.

Years ago — I write from memory — I found that the loss of information due to replacing variates by ordinal values is not very large, if it is remedied by replacing ordinal values by mean variates. My impression is that the sum of squares of the true mean deviates, which differs a little from what we have tabulated, is $(n - 1)$ times the efficiency. If this is so the percentage loss of information falls off proportionately to $(\log n)/n$ and is moderately small when n exceeds 10.

If you are interested you might look in some time and I will try to hunt up what I have in the way of algebra, as I have a strong impression that a good many pretty, though intricate, results could be obtained by proper algebraical treatment.

E.A. Cornish to Fisher: 31 *August* 1938

I should be very grateful if you would be so good as to give me some advice on the following problem.

I have three mutually correlated normal variables which I may designate x, y and z, and I wish to perform a separate analysis of variance on the observations of each variate derived from a rather comprehensive randomized block experiment. The various values of x are independent of each other, similarly with y and z.

Now we know that apart from variation due to blocks, treatments, etc., a certain amount of the variation in x can be explained in terms of y and z. We can, therefore, obtain from the analysis of variance and covariance of x, y and z the correction to adjust the residual term in the analysis of variance of x. May we now proceed in that same way using y as the dependent variate and x and z as the independent variates, and then with z as dependent and x and y as independent? Some doubts as to the validity of the procedure are in my mind owing to the assumptions about absence of error in the independent variates.

Fisher to E.A. Cornish: 12 *September* 1938

. . . With respect to your 3-variate problem, what you can logically do is to choose one of them, say x, and perform an analysis of variance without regard to y and z, then to do the same for y, making this time allowance for variation in x, and finally, to analyse z making allowance for x and y. These three analyses will give independent information, e.g. if the order of the variates is so chosen, you might find that the treatments had a significant effect on x, but that after this is allowed for they had no significant effect on y or z, jointly and severally, or you might find that there was also an independent effect on y, but no further independent effect on z. Since one wants the simplest interpretation, it is desirable to discover if it is true that one or more variates are unaffected by treatment when allowance is made for the variates on which the treatments had some real effect.

To allow for the other two variates in all three cases gives non-independent results, which might, indeed, all be non-significant even though each variate singly was largely affected.

H.E. Daniels to Fisher: 24 *September* 1937

. . . I have been interested recently in the problem of how the z test is affected by a slight correlation between the members of the sample. The question arose out of an attempt to detect periodic variations in twist from a series of consecutive twist tests on a length of yarn, trial periods p being taken in the usual way, the totals T_r of $r, r + p, r + 2p, \ldots$ found, and the variation between the T_r tested against the variance within the groups. The consecutive twist tests can be shown to be slightly positively correlated, and this must affect the significance of the results.

It had always seemed to me in an intuitive way that the effect of the correlation would be in a sense to reduce the 'degrees of freedom' by a fractional amount, but a tentative investigation, outlined below, seemed to show that neglecting squares of small quantities, the effect of the correlation is to retain the z distribution with the same number of degrees of freedom, but to introduce a bias into z. . . .

Fisher to H.E. Daniels: 27 *September* 1937

Thanks for your letter. A slightly different, but analogous, specification may be relevant to your problem, e.g. if, instead of considering a correlation between successive values, i.e. a specification in which $x_{p+1} - \rho x_p$ constitute a series of independent normal deviates, you take the not very different specification, that the normal and independent deviates are deviations of each observation from the expected mean ζ from a run of p consecutive values to which it belongs, these means themselves showing some variability, the effect of correlation reduces itself to precisely a bias in z, i.e. then $z - \zeta$ satisfies the z distribution where ζ depends on the variability of the successive expected means.

The main point of this is that you can sometimes clear your head and simplify the algebra by choosing a specification involving the features you want and differing only in non-essentials from the one you are concerned with.

Fisher to C. Edelstam: 9 December 1948

Thank you for your letter of December 5th on the limits of significance for binomial and Poisson distributions. I rather think, however, that the tabulation you suggest would appear to be a step backward rather than forward.[1]

In the introduction to the book of tables you refer to, Example 1 shows how to get single tail values of 1% and 5% for the Poisson series, and Example 2 for the binomial, using the *z* distribution as tabulated.

In most problems of this kind very decent accuracy is obtained by interpolating for log *P*, if, and this is usually very exceptional, the experimenter wants to be particular as to what level of significance he chooses for making his test.

This is all part of a rather bigger question lying in the past between cases in which utility is best served by a compact table, which requires some expert knowledge to use, such as Stevens has aimed at in Table VIII.1, and large comprehensive tables, which are justifiable when in particular lines of work a very large number of separate entries have to be made, in which case it is convenient if the tables can be made almost foolproof. A great deal of scientific money has in the past been sunk in such large comprehensive tables for various purposes, and the corpus of such tabular publications which a comprehensive mathematical library can now acquire is indeed a very large one.

The situation has been in recent years greatly changed first by the introduction of convenient computing machines of desk size, which, for example, often make it easier to make a special tabulation for one's own use *ab initio*, rather than to seek by mathematical transformations to reduce the problem to expressions in terms of known functions which have already to some extent been tabulated, and then to use existing tabulations of the latter for one's own special purpose. A still larger change in the same direction appears to be impending in the construction in many parts of the world of high power electronic digital computers, capable of making rapidly *ad hoc* tabulations appropriate to particular problems.

I feel, therefore, that at the present time one should be very sure that a real need will be met before sinking labour and money in the production of largely expanded tables of functions which many may feel are already sufficiently available.

[1] Edelstam had suggested the preparation of an expanded version of Table VIII.1 in *Statistical tables*, incorporating more values for *a, N,* and *P*.

Fisher to D.J. Finney: 27 September 1954[1]

. . . I suppose it must be to assuage some uneasiness of conscience under this head that one finds in statistics, unlike let us say chemistry, an urge to warn students against genuinely accurate methods using various rather rhetorical grounds, instead of pointing out that some care is usually needed in choosing the amount of technical labour it is appropriate to give to any given job. I have been impressed lately by A.E. Mourant in the admirable book that he has written on the distribution of blood groups, putting in his chapter on computational methods quite a lot of counter-propaganda against the use of the method of maximum likelihood. It does not seem to occur to him that it is a purely factual question whether, at any stage of enquiry, doing more sums or collecting more blood is the more economical way of adding to our information. Rather it seems to be settled on emotional grounds by passionate appreciation of the supposedly insuperable difficulties which the more exact methods involve. Is the difference simply that we do not usually expect to do our chemical analysis for ourselves but that many people who have not the equipment, the time, or the technical training appropriate for the task, think that they ought to do their own statistical analysis, or does the trouble run more deeply than this?

[1] For the first part of this letter, see p.95.

D.J. Finney to Fisher: 4 *October* 1954

. . . Of course your comments on the economic aspects of information wasting are important, and I have often emphasized this point. I have not yet seen Mourant's book, but am sorry to learn of his timidity in respect of maximum likelihood. I would have said that the effort of collecting human genetic records was so great that, even if M.L. estimation took twice as long as some cruder method, it was worth while for as little as an extra 5%–10% of information. It is strange how the biologist who will cheerfully spend weeks on field work is scared by the thought of a few hours of computing.

I imagine that the simplicity with which electronic machines can cope with elaborate iterative procedures will shift the economic balance still further in the direction of extracting all the information from data. . . .

Your last sentence provokes the question: should we attempt to train biologists and others to do any statistical analyses for themselves, or should we seek to reserve their time and abilities for their own field of work and pass all statistical work through bigger and better Statistical Departments? I do not think the matter is quite dealt with on the analogy of chemical analysis, because I suspect the investigator can learn truths about his material from his own statistical analyses that the statistician might not observe or might fail to transmit to him, and this is perhaps less true of chemical analyses. Unless and until the statistical analyses of a much greater range of problems become purely mechanical tasks, isn't the answer that the best procedure depends upon the particular problem and upon the particular people involved?

Fisher to D.J. Finney: 7 October 1954

. . . You raise a very big question in speaking of the influence of high power computing machines. What needs, I think, to be safe-guarded is the practice of the scientist or, at least, a statistical consultant giving personal consideration to what is to be learnt from his data as to the hypothesis (or model) in terms of which they are to be interpreted. Short of that it seems to me that the geneticist, or ecologist, or whoever manages to gather the data, has virtually put himself into the position of asking the machine's masters what he has to observe for them. I suppose the answer is that no one can stop the human race from selling themselves into slavery, still one would like somehow to try.

I had a Ph.D. thesis to examine recently in which the young man, impressed by the large numbers of observations needed to distinguish ratios different to a practically important extent, mobilised the aid of a giant computer, but still passed through to it for programming the approximate methods which he had been accustomed to use when labour saving was really important to him. I think the point of this story is that he thought it was the job of the machine only to do the work for him, instead of realizing that it was a competent method of finding out whether his approximative processes were sufficiently accurate for the work required.

So I do think we should try to train biologists to do their own work with at least enough competence to know when further consultation is needed. I rather think the chemists have succeeded in this respect.

D.J. Finney to Fisher: 3 April 1956

I should be very interested sometime to have your opinion on what constitutes an analysis of variance. I think that authors of papers and reports frequently include details of these analyses unnecessarily, but obviously when questions of statistical technique are under discussion inclusion is essential. I find an increasing tendency to present, under the title of 'analysis of variance', a tabulation of degrees of freedom and mean squares *only* for the various components, the column for sum of squares being omitted. Since it is the total sum of squares of deviations that is analyzed, omission of this and its components seems to me very unfortunate. I regard the preparation of a full table of the analysis of variance as an excellent discipline to impose on computations, and the author of an expository paper who suggests that parts of it can be omitted may be doing a dis-service to those who follow him.

This is particularly relevant to papers submitted for a journal such as *Biometrics*. Do you think that an author should be asked to show the full table, or are you content to see merely the mean squares? I have heard it said that you dislike the practice of adding a further column showing ratios of treatment mean squares to error. I would never include these myself as I think they encourage excessively rigid interpretation of significance tests but I am not sure how strongly one ought to discourage them. A paper in front of me for refereeing this moment has a column headed *P* showing the probability levels for the ratios of mean squares to error mean square. What are your views on this?

Scarcely any statistical technique is more widely useful than the analysis of variance, but the novice today is confused by many styles of presentation. Some put the total at the top and some put it at the bottom. Some put degrees of freedom before the sum of squares and some reverse these. I am conservative enough to prefer the order that you have always used, but I have difficulty in deciding how far editorial pressure should be exercised in persuading other people to follow a standard pattern.

Fisher to D.J. Finney: 16 *April* 1956

. . . About the analysis of variance, I agree with you that such analysis must include a column for sums of squares, or its equivalent, and of course to the early workers, like myself, one of its main attractions lay in the automatic verification that everything had been accounted for by the check, or checks, provided by the totals in this column. Now that the procedure is widely used, I believe academic exposition should insist, far more than I have done in more or less broaching the subject in *Statistical Methods*, on tearing the sums of squares to pieces, in as many ways as possible, to see that every aspect of the data is behaving approximately as it should; you do not need to identify the microbe before using an antiseptic! I think editors could very usefully make this point, subject of course to the author's right, either to present an analysis of variance *senso strictu*, or a summary of an analysis, stated as such, or an elaboration with additional columns, such as you mention, with variance ratios, *P* values, or what you will. . . .

Fisher to D.J. Finney: 7 *November* 1956

. . . I notice in a review, in *J.R.S.S., A*, the last number, of a recent book by Federer, the reviewer reproves the author for quoting, apparently without disputing, the allowance I calculated for the loss of information about a mean owing to using an empirical estimate of error instead of the true, but unknown, variance.[1] As I think you were in these discussions, perhaps you could enlighten me as to what the reviewer can refer to as 'alternative methods'. Of course, if additional data are supplied, e.g. the value of *t* expected, the problem is reduced to 'Student's' test itself. My calculation is integrated over the whole *t* distribution, and seems to me appropriate in the realistic case in which the value of *t*, or probably the whole crop of values of *t*, for which comparisons are as yet unknown, and the planner of the experiment is concerned to judge how much he should sacrifice in order to have say 20 degrees of freedom, in place of 10, for the estimation of error. I believe confusion may have been caused by trying to bring in the sampling distribution of s for given σ whereas in reality the experimental planner knows that only s will be available, and that σ, if thought of at all, will have a rigorous fiducial distribution appropriate to that s.

[1] See also Fisher's letters to J.A. Nelder (p.280).

D.J. Finney to Fisher: 10 *November* 1956

I too thought that the reviewer of Federer's book had expressed himself rather badly over the point you mention, though I feel in considerable sympathy with him over the irritating manner in which the book presents its methods.

I imagine that the reviewer has in mind the kind of thing that Cochran and Cox discuss in their Chapter 2, and particularly on pages 26–29. They give a brief but sensible general summary and state clearly that the answer depends upon the use to which the results are put; I imagine that you would agree with this sentiment, though it might perhaps be better expressed by saying that the results depend upon the question that is asked. I don't know whether anyone has suggested alternative answers to your question, but clearly it is possible to frame alternative questions bearing on the same subject, as you point out in your letter. . . .

Fisher to D.J. Finney: 12 *November* 1956

Thank you for your reference to Cochran and Cox pages 26–29. I believe if you will re-read the final paragraphs of this section (2.31) you will find that, after considering what Neyman and Welch had written, these authors conclude, 'For general purposes it is suggested that this table' (the one embodying my form for the correction) 'be used to take account of the differences in degrees of freedom for error in two designs that are being compared.' They then give an example and an explanation of the cases in which such an expression might be useful.

The table on page 27, which I suppose is Neyman's or Welch's, clearly treats the question which conceivably could arise if the level of significance were known in advance, although in my opinion this problem had been completely solved by 'Student' himself.

Anyone reading Nelder's review could see, and doubtless you can see, that the reviewer was criticizing Federer for recommending what Nelder regards as an unsound method, for in addition to suggesting, as he doubtless believed, that alternative solutions had been given the same problem, he complains that my demonstration of it is peculiar, and that I have not used the method on other occasions; whereas the formula I used for assessing the amount of information about a parameter distributed in a known distribution was in my 1921 paper 'On the Mathematical Foundations' [*CP* 18], and had been illustrated in the same book in almost step by step parallelism in the previous section.

The question I was trying to answer was what allowance should be made in anticipating the precision of an experimental design for the number of degrees of freedom on which estimates of error are to be based. Apart from the question of alternative methods of solution of this problem, which, so far as I know have not been put forward, you say that 'clearly it is possible to frame alternative questions bearing on the same subject' and imply that this is in some way a justification for scolding Federer for advocating, however stupidly, the only method so far suggested, for what is in fact a question which does arise in practice.

I am not concerned with the merits of Federer's book, but with the use made of a review for an oblique attack on a method which the reviewer has not perhaps considered very deeply, but has been led to think, in the way that prevails in centres of statistical teaching, is unsound, not for any tangible reason but because someone else disagrees with it.

D.J. Finney to Fisher: 20 *November* 1956

. . . When I returned here, I reread the letter from you that arrived in my absence. I am in complete agreement with you about the misleading nature of the remarks in the review of Federer's book. The further comments that I made in my previous letter were not intended in any way to modify this opinion which relates to the factual content of the book and of the review. Since I had myself reviewed the book, however, I was disposed to criticize it on quite different grounds of style of presentation, not in any unique instance but in the manner in which the author presents references to a great number of statistical writers.

M. Fréchet to Fisher: 8 *February* 1936

. . . Je sollicite donc de votre obligeance l'envoi (à mon adresse: 12, Square Desnouettes, Paris, 15e) d'une Note indiquant — en quelques lignes, et sous forme de règles concises — en quelles circonstances et sous quelles réserves le coefficient r peut être utilisé *pour repérer* la rigueur d'une dépendance fonctionnelle. (Pour éviter tout risque de malentendu, je vous serais très obligé de bien vouloir laisser *en dehors* de votre note les applications bien connues de ce coefficient r à la représentation mathématique des lois statistiques, à l'ajustement, à diverses questions de physique, etc. . . .). A cette courte Note pourrait être adjointe, s'il était jugé nécessaire, des explications plus développées mais traitées séparément. . . .

Fisher to M. Fréchet: 19 *February* 1936

I must begin my reply to your circular letter of February 8th by a purely verbal difficulty, namely as to the exact meaning which should be attached to the word 'repérer', underlined in the middle of your second page. Should it be taken to mean simply to discover? In my ignorance I had recourse to my copy of Petit Larousse where I find 'Marquer des repères', while a repère is 'Marque faite à différentes pièces d'assemblage pour les reconnaître et les ajuster plus facilement'. This literal definition cannot be what is wanted, but I have missed the metaphor which should make it intelligible.

In the sense of simple discovery I have seen the correlation coefficient used with success, e.g. a soil physicist studying the plastic qualities of agricultural soils might find in the samples examined no association with the total percentage of calcium carbonate in the soil, but be delighted to find that a significant association appears when particles of calcium carbonate, too large to pass through holes 1 mm in diameter, are excluded from the analysis. The search for such a significant association among the many variables, simple and compound, which may be examined, is carried out not inconveniently in terms of the correlation coefficient, though certainly also not inconveniently in other ways. It is used, I think, in preference to other measures chiefly because the method of calculation is widely known and easily remembered, and because relatively precise tests of significance are easy to apply, owing to the availability of published tables.

The test of significance, i.e. the test of the hypothesis that there is no real association in the material from which the existing observations are regarded as a random sample, is necessarily equivalent to the test whether the coefficient b in the equation

$$y = a + bx$$

fitted to the data by least squares, as a means of predicting the value of y from that of x, differs significantly from zero. This is often the more appropriate and fruitful way of viewing the matter, but since the tests of significance are equivalent, and since, in the business of *discovery*, we are only concerned with significance and not necessarily with estimation, or even with knowing what we want to estimate, the correlation coefficient may be legitimately used in this way, even when its numerical value is without scientific meaning beyond that given it by its mathematical definition as a calculable function of observable quantities.

To say that this procedure is legitimate is, of course, not to say that it is the best that can be recommended. In particular cases other tests may be much more sensitive, that is to say they may be capable of demonstrating a significant association on the basis of fewer or less careful observations. This is an important advantage in statistical theory, but it is of little interest to the experimenter who, having found a significant correlation, is led, perhaps, to abandon a false theory by which he has hitherto guided himself, or to undertake a new line of research with different aims and different instruments.

M. Fréchet to Fisher: 22 February 1936

I thank you for your interesting answer to my circular letter.

I. I want to explain that the word 'repérer' was intended as a substitute for 'to measure'. I wanted to avoid the idea that a 'measure' could be found for the degree of strictness of dependency — as there is no meaning for the sum of two degrees of such strictness. I should think that the verb 'to graduate' is the proper equivalent of 'repérer'; for instance we can 'repérer' (graduate) but not 'mesurer' (measure) the physical feeling of heat.

II. Your answer, being based on the approximate translation of 'repérer la rigueur d'une dépendance' by 'discover the existence of a dependence' relates to a more restricted problem than the one I meant. Still, this restricted problem is important.

However, I do not see well your exact position and, if I am not too troublesome, I should be glad if you could precise [*sic*] the following point. For, your first page (first example) should evoke the impression that your soil physicist operated this way. As it is intended as an example of the uses of the coeff. of correlation, I understand that in his first experience he found this coefficient small and that it is this way that he concluded that there was no association between . . . . Then in the second experience, he found the coeff. near 1 and it is that way that he found a significant association in the second experience. These conclusions were obtained, as you say, not inconveniently in terms of the correl. coeff. as they might have been got other ways.

In your second page and end of third page, I gather the impression that a test of significance should be limited to the case when r is near 1, and that no conclusion from the value of r should be derived when r is near 0. The first page should correspond to what is very often done, the second one to a more safeguarding method.

But I have perhaps been mistaken in both cases, and a few words of comment would be much appreciated.

P.S. In my recent travelling in Russia, Prof. Kolmogoroff — who is one of the half dozen best 'probabilists' in the world — observed to me that you have recently come to a differ. equation[1] without probably knowing that it has been met before in a physical problem by Fokker and Planck. It may be interesting to you to hear it is called Fokker–Planck's equation by the physicists and that it may be written:

$$\frac{\partial p}{\partial t} = - \frac{\partial Ap}{\partial x} + \tfrac{1}{2} \frac{\partial^2(B^2p)}{\partial x^2}$$

[1] See *CP 86*.

Fisher to M. Fréchet: 5 March 1936

You ask me to explain more clearly the kind of case I had in mind as an example of the detection or discovery of an association by means of the correlation coefficient. This is easy, for in both cases to which I referred I had the same criterion in mind, namely whether the correlation coefficient does or does not exceed the least value which should be judged significant in regard to the number of observations available. Thus, if the soil physicist had 22 samples of different soils available, the properties of which he had examined and measured, he would probably choose the value 0.4227 as the least observed correlation which should be regarded as significant, the reason for this choice being simply that from uncorrelated and normally distributed populations this is the value that would be exceeded by chance just once in twenty trials. In order to judge a correlation to be insignificant, therefore, it is not necessary that the observed value should be very small or near to zero. He would ignore it with confidence coming from so small a sample if it were between plus and minus 0.35. On the other hand, a value as high as 0.55 would only occur by chance once in several hundred trials in samples of this size in a population from which correlation was absent, so that the value of the correlation need not be near to unity in order to satisfy the experimenter that the value he has determined does demonstrate the existence of a real connection. It is the reality of this association which is vital to the experimenter, for on his confidence in its existence he is willing to spend, and perhaps to waste, the work he is capable of doing for perhaps several years.

What is important in this is that he should be provided with appropriate and sufficiently exact tests of significance. It is a matter of no consequence that in making these tests he should use the particular function of the observations (statistic) which we denote by r. In fact, we arrive at an exactly equivalent test if, ignoring the correlation coefficient, we calculate the regression of either one variate on the other by the familiar formulae of least squares and compare this with its standard error,using the exact procedure introduced by 'Student' for allowing for the limited number of degrees of freedom upon which the estimate of error has been based. It is, in fact, only the test of significance in its entirety which is of value to the experimenter, and not the particular statistic in which he happens to find it convenient to carry out the test.

M. Fréchet to Fisher: 13 *March* 1936

I thank you very much for your answer. It throws light upon some points which your letter *published in my enquiry*, left obscure for me. In that former letter you treated simultaneously the two very different points: is there a dependence; what best can be done to *describe* the simultaneous variability of 2 quantities. Now in your last two letters the first point is treated separately, and then I can see that:

your method is based on the fundamental assumption that we have to deal with quantities satisfying to the so called normal law of errors (with 2 variables).

On this assumption your method — very different from the usual practice — looks very satisfactory.

And I think I will be able to recall it in my report.

Now, I am afraid to retain much of your time, but there remains the cases — very frequent ones—: 1st when the observer does not know whether the variation is normal or even suspects it to be slightly different; 2nd when the observer knows that the variation is far from being normal.

Is your position such that: I. we *must not* use the coeff. of correlation r in such cases *as far* as concerns the *existence* of a dependence. II. or that you possess in such cases a method to use it for *this same purpose*, a method necessarily different from the one you mentioned?

Fisher to M. Fréchet: 18 *March* 1936

The test of significance mentioned in my letter is somewhat less restricted in its application than you suggest in your reply of March 13. I mentioned that the formal conventional test of significance of the correlation coefficient was, in fact, equivalent to the test whether the linear regression of one variate y on the other, x, differed significantly from zero. The test is, therefore, exact not only for the case of a normal bivariate distribution, but also for one in which one variate is distributed in any manner whatever, while for given values of x the second variate y is distributed normally about a mean which is a linear function of x. The marginal distributions of both x and y may, therefore, both be far from normal, and for the test of significance it is immaterial which of the variables is regarded as the independent, and which as the dependent variate.

M. Fréchet to Fisher: 22 *March* 1936

I understand that the test you mentioned in your second letter is applicable in a wider range of cases than the cases in which there is a normal bivariate distribution, but that it still requires that one of the 2 lines of means be linear (and even it requires a little more).

Now my question is:

as far as concerns the cases in which none of the 2 lines of the means is a straight line, do you think that the use of r (for judging whether there is a significant dependency) is incorrect?

or do you know, for these cases, of a method — necessarily different from the one you mentioned — enabling to use correctly the coefficient of correlation for the same precise purpose?

Your previous letters were very useful to me; I hope that you can answer the above questions and so help me to define your position in the previous enquiry.

Fisher to M. Fréchet: 24 *March* 1936

I will try to answer your further questions of your letter of March [22].

The investigator who has obtained a single correlation and who merely infers that the variables are not independent is quite immune from criticism on the ground that the regression may be non-linear; for this [proposition] if true, would only [decrease] the sensitiveness of his test, and not increase the frequency with which, in the absence of association, high values of r would be observed. The investigator who finds no significant correlation should always be cautious not to say that no correlation exists, for it may be, as in the case of the physicists of whom I first spoke, that he has failed to detect an important relationship, through using a physical quantity less suitable than he might have used, e.g. total calcium carbonate instead of the calcium carbonate in small particles only. He may have failed equally for the reason you have in mind, namely, through seeking for correlation with a linear function of his observed value, instead of using a function of more complicated kind which would have revealed what it is in his interest to detect. In fact, I should say that the choice of one form rather than another for the regression equation to be examined is in the same sense a matter of the individual judgement, or intuition, of the investigator as is the choice of the physical attributes to which he is to devote his attention.

It may be worth noting that an investigator who feels more sure of what physical quantities to use than he does of what functions of them would be the most appropriate, can satisfy himself that the data before him deserve no more elaborate tests than he has applied by making a test of linearity of regression (Blakeman's criterion is, of course, quite incorrect). For, if there is no indication of departure from linearity in a test specially designed for detecting such departure, he has good ground for confidence that any curvature which exists will not have appreciably disturbed his judgement based on a linear regression function.

Fisher to L. Goossens: 28 *April* 1950

Thank you for your kind letter which I have just read. . . .

I think one can best get an idea of the argument[1] in paragraph 21.1 [*SMRW*] by putting the problem to oneself quite independently of the χ^2 test. Suppose we are told that a hypothesis is suspect, though difficult to test directly, and that efforts to do so by summarising groups of data which should be to some extent sensitive to the truth of the hypothesis have led on different occasions, using entirely independent data, to reliable values

$$p_1, p_2, p_3, \ldots$$

for the probability of obtaining a more extreme deviation calculated on the supposition that the hypothesis is, in fact, true.

On this hypothesis, therefore, these values p must be distributed between zero and unity in a rectangular distribution.

If the hypothesis is false we expect an excessive number of small values and a deficiency of large values compared with this standard rectangle. Arbitrarily, therefore, we might choose to multiply the values of p together to determine the sampling distribution of the product and to base a test of significance on whether this product appears to be significantly smaller than one would expect it to be by chance.

One may then bring in the idea of χ^2 for two degrees of freedom as supplying an easy way of solving the problem of the sampling distribution of the product of such a group of values p. Of course, such an approach is only valid if the distribution of p is truly rectangular and continuous, as it is effectively in many problems.

I hope this approach will remove your difficulties.

[1] i.e. on the combination of probabilities for tests of significance.

Fisher to H.W. Heckstall-Smith: 25 *July* 1957

Thanks for your letter of 24th.[1] The remark in *Statistical Methods* was, of course, paradoxical and intended to make people think, as it obviously has in your case. Of course the phrase 'In these cases the hypothesis considered (etc.)' means the hypothesis 'that the data under discussion were derived from a particular hypothetical system of causation'. In fact, logically the data are a part of the hypothesis. In most cases the hypothesis is more complex than this because its full specification involves assertions about independence, which are essential to the mathematical specification of the expectations, but often pass unnoticed by the mathematician discussing scientific questions. In the case of your coins one might imagine that the performer counted the result of the first throw, say a head, and then the next time tails turned up, and then the next heads, and so on, omitting all repetitions. As you say, the data should be rejected on the grounds that they were not obtained by independent random trials, and, as I have put it, the hypothesis that the data were obtained by independent random trials with a true chance of fifty–fifty (or any other for that matter) is to be rejected.

[1] Heckstall-Smith had written querying the statement in Section 20 (p.81) of *SMRW* that in tests of goodness of fit where P is over 0.999, 'the hypothesis considered is as definitely disproved as if P had been 0.001'. He suggested instead that 'either the evidence must be rejected or else the hypothesis must be considered as disproved'. He asked in particular about a hypothesis that a coin is unbiased and supposed that on tossing the coin, he found 500 heads and 500 tails; should he then reject the hypothesis that the coin is unbiased?

Fisher to F.R. Immer: 13 *May* 1935

I have not happened to see Treloar and Wilder's paper,[1] nor have I heard it mentioned in England. I presume it must be a re-hash of the objections based on departure from normality of the parent population which were much fostered in the Statistical Department of this College,[2] some years ago, but which seem now to have subsided, perhaps as a result of my dealing sometimes with these questions in lectures.

Logically, the complete answer to this objection, in respect of any particular sample of data, is to show that the sample will yield a test of significance very close to that given by t or z, without introducing any normal distribution at all, but using merely the known arithmetical distribution produced by randomisation.

In the case of the z test, you may perhaps remember Eden and Yates[3] taking a set of 32 values for 4 varieties in 8 blocks, and finding the z distribution for 1000 randomly chosen reassignments within the blocks. Their object was to examine the validity of the z distribution in material which was rather more skew, and more unequally variable than most observational material, and in this their test was completely successful. It has, however, a rather deeper interest logically since, on the hypothesis to be tested, that the 4 varieties are without effect, the experimenter could know that the 1000 values of z obtained by sampling were, in fact, all equally probable, whatever the nature of the parent population. And if the value of z actually observed falls among the top 50 of the sample, he knows, as an arithmetical fact, and without any statistical theory, that values so high as this are sufficiently rare to be regarded as significant. Anyone, therefore, who wishes to criticise a conclusion as to significance drawn by the t or z tests, on the ground that the population sampled is not normal, or not equally variable, should be invited to see for himself, by direct, though tedious arithmetic, whether the conclusion indicated by the statistical test is valid or not, when no assumption whatever as to normality is made.

I have explained and illustrated the process in detail in a book now in the press, called *The Design of Experiments*, which I think may interest you also in other ways.

[P.S] I have now seen Treloar's paper, which does not seem to be so intelligent as I had guessed. Of course t is a ratio, that is what it is for. I do not think it has ever been overlooked. He is merely comparing a test of significance t with an *estimate* $\bar{x}$ which appears in the numerator.

[1] Treloar, A.E. and Wilder, M.A. (1934). The adequacy of 'Student's' criterion of deviations in small sample means. *Ann. Math. Stat.* **5**, 324–41. Immer had written seeking Fisher's opinion of this paper. (See also Fisher's letter of 20 May 1935 to E.B. Wilson, p. 237.)

[2] University College London.

[3] Eden, T. and Yates, F. (1933). On the validity of Fisher's z-test when applied to an actual example of non-normal data. *J. Agric. Sci.* **23**, 6–16.

Fisher to J. O. Irwin: 28 *May* 1941

... About χ^2, the distribution is certainly unaffected by making a sub-selection at random of the contents of each cell (plate).[1] I recall giving a good deal of attention at the time to this subnormal variance, which I think we never cleared up experimentally. The kind of factor which would produce such an effect is competition. The distribution of passengers in compartments of a train is notably subnormal, because late-comers try to avoid the more crowded carriages. A bacterial species producing a toxin injurious, at least at germination, to the same species would tend to occupy points on the plate on a lattice, the scale of which is independent of the size of the plate. Consequently, sufficiently large plates and sufficiently dense suspensions would give subnormal χ^2.

[1] Irwin had enquired about the occurrence of subnormal χ^2 in bacterial plate counts and the suggestion by Fisher, Thornton, and Mackenzie (1922) that this was 'indicative of some defect in the composition of the medium' (*CP* 22, p.358). Irwin said that whilst χ^2 would be reduced if a set of frequencies were cut down proportionately, he could not see that χ^2 would be affected if the 'mortality' were itself subject to sampling error.

H. Jeffreys to Fisher: 19 *May* 1938

... I take it that we should all agree that however we make a selection, there will occasionally be cases where the selection will be correlated with something that we are trying to find, and either the estimate or its standard error will be biased. General rules for assigning at random presumably express a belief that they will introduce this trouble as seldom as possible. Tippett's numbers cannot be random in the sense that I should understand in my theory, because once assigned they are known and we have information about them that is relevant to future trials. They can only be random once. But there may be valid grounds for expecting that they are uncorrelated with anything that we may reasonably expect to want to test. Thus a test of their usefulness really requires consideration of the kind of additional functions that we may want to test in practice and of their correlation with the Tippett numbers. Now there are, I gather, about 40 000 of the numbers. If a selection of 100 in turn are made, and analysed for a linear variation, this will exceed the 1 per cent limit about 4 times. Bad luck if one picks one of these sets of 100, but it could be avoided if these ranges are specially indicated in the list. Again, it will happen fairly often that the means of two sets of 100 differ by more than the 5 per cent limit. In either case a real linear variation in the thing we are measuring will lead to a bias in the mean of the whole. So I think that there may be some point in testing what Kendall calls local randomness. But the danger is distinctly subtle. As far as the means are concerned, it requires both a correlation between the numbers and something genuine in the population sampled, and a non-orthogonality in the normal equations that will push the bias of selection into a further parameter. As to the standard errors, the correlation may lead to the treatment of something genuine as error, and there will be a second-order bias in the error; but this could presumably be eliminated by estimating the genuine effect and allowing for it. My general feeling is that the things Kendall is looking for are likely to enter in practice in such an indirect way that departures from randomness in the numbers can be tolerated at much higher levels of significance than we should use in testing anything that interests us directly.

Are you saying anything in reply to 'Student's' last paper?[1] There was a point that had rather bothered me about Latin squares, which may be related to his. Suppose you take Cartesian coordinates about the centre of the square. Then as you eliminate rows and columns completely you are virtually supposing that the fertility can be expressed in the form

$$F = a_0 + a_1 x + \ldots + a_4 x^4 + b_1 y + \ldots + b_4 y^4$$

with no cross terms like xy, such a form being capable of being chosen in a 5×5 square to fit every row and column total exactly. Now just what are you asserting about the higher terms? A linear gradient is presumably a known danger. But if the x^2 and y^2 terms are genuine, then except for special orientations there will in general be a term in xy, which will be treated as error. In some conditions this might be serious. Thus it might turn out that xy and the treatment means together account for the whole variation in the square, and the error is 0; but if xy is left in the error it may be important and real differences in the treatments means may be hidden.

The question seems to me to be one of fact, and that rules cannot be laid down from theory. I should say that if the x^2 and y^2 terms are worth eliminating, then xy is worth eliminating and x^3 and y^3 possibly are, possibly not; if they are not worth it, then xy, x^3 and y^3 can all be treated as error. The question seems to be whether analysis of row and column totals does in fact often lead to significant coefficients for x^2 and y^2. If it doesn't, your method will sacrifice somc information about the error, though I suppose that you have enough anyhow for this not to matter much.

I haven't the remotest idea how important these considerations are in practical cases, but I should be interested to know.

[1] 'Student' (1937). Comparison between balanced and random arrangements of field plots. *Biometrika* **29**, 363–79.

Fisher to H. Jeffreys: 30 *May* 1938

. . . It [randomization] is, as it seems to me, a tribute to our ignorance of the nature of the errors to which our results will be liable. Thus, if I want to test the capacity of the human race for telepathically perceiving a playing card, I might choose the Queen of Diamonds, and get thousands of radio listeners to send in guesses. I should then find that considerably more than one in 52 guessed the card right; also that of those guessing wrong more than half got the colour right, and probably a number of such favourable indications would be obtained. On the other hand, if I choose the 8 of Spades, I should expect to get just the opposite result.

Experimentally this sort of thing arises because we are in the habit of making tacit hypotheses, e.g. 'Good guesses are at random except for a possible telepathic influence'. But in reality it appears that Red cards are always guessed more frequently than Black.

For years agricultural experimenters made the similar unconscious hypothesis, 'Errors on different plots are distributed independently in the normal curve'. Actually the normal curve is good enough, but the errors are very far from independent; consequently any systematic arrangement *may* contain factors in common with the actual pattern of natural fertility. This difficulty is so fundamental that one has to consider the problem in a extreme form. Let the Devil choose the yields of the plots to his liking; his only restriction is that he may not change his mind after I have chosen where the different treatments are to fall. If, now, I assign treatments to plots on any system which allows any two plots which *may* be treated alike an equal chance of being treated differently, in the different ways in which this is possible, then it can be shown that both the experiment is unbiassed by the Devil's machinations, and that my test of significance is valid.

Things are not really so bad as in this game. We know nothing in detail about the errors, but experience does indicate certain components which are very often important, and such components one does not leave to chance but completely eliminates. Thus, if one had equal areas in two fields, one might legitimately assign pairs of plots, one from each field, and toss up between treatments *A* and *B*; then it would be annoying if chance (or the sequence of random numbers) put 7 *A*'s running in the same field, which is the sort of thing liable to happen. If the same sequence of numbers were used for adjacent plots, its locally systematic character would not matter, a run of alternate treatments might be quite good. Though, again, if we suspected alternation in field fertility, such as is known to occur sometimes, or even a steady gradient, one might prefer to randomise entire sandwiches *A B B A*, or *B A A B* by a single act of randomisation.

I think you hit the nail on the head in saying that a sequence can only be random once. Hence one must insist on a fresh randomisation each time when a set of identical experiments is laid down. For the same reason it is desirable that different centres should, as is already found convenient, each use their own tables for those purposes for which tables save time.

In principle I agree with you about the Latin square. Rows and Columns usually take out a useful lot of error, but I should not claim that the elimination of the same number of degrees of freedom, 12 in a 7 × 7 square, could not take out more. If, however, you impose the limitation that the components eliminated shall be orthogonal to those used for treatments, the combinatorial problem becomes greatly involved. I should, however, have no objection in theory to the claim that better arrangements can be found. One fact which makes the Latin Square work well in practice is that on any given field agricultural operations, at least for centuries, have followed one of two directions, which are usually those of the rows and columns; consequently streaks of fertility, weed infestation, etc., do, in fact, occur predominantly in these two directions. Streaks in other directions introduce bias in systematic squares, but validly estimated errors in random squares.

H. Jeffreys to Fisher: 22 September 1938

. . . I like your note on 'Student'.[1] Curiously, when I rediscovered his rule in *Sci. Inf.*[2] he was only a name to me; and when you went for me in the *Proc.Roy.Soc.* [*CP* 102] I thought that you were attacking this rule! But there are a few points that I don't quite agree with. . . . If randomization was necessary I think you would consider assigning all 25 plots at random in a 5×5 square as the ideal procedure, getting two of a treatment in one row and none in another. 'Student' misses the point of the Latin square, I think, when he says that it is both balanced and random. I should say that it is balanced for ground effects known to be often large (possibly indeed more balanced than is really necessary) and random for others that might mount up if the design was repeated in different squares; the balance and the randomness refer to different features of the ground variation. The real question, it seems to me, is, what ground effects matter in practice? If there is no correlation at all between plots it doesn't matter what you do. If there are correlations and systematic variations it is legitimate to arrange the work so as to estimate them and allow for them, also allowing of course for the degrees of freedom lost on the way. If they are doubtful, deliberate randomization will provide an alternative treatment and easier arithmetic. I am not sure but I think I have seen discussions on the point that don't notice that if there is anything to be dealt with by a systematic design, its proper analysis will involve the separation of additional degrees of freedom, the variance associated with which is neither treatment nor error variance. The row and column variance in a Latin square is a particular case; so would be the fertility gradient along the row in a randomized block experiment, which 'Student' suggests would be worth taking into account. That is, I think that in the Latin square design you admit his main contention; but I have not gone into the matter enough to have any opinion about whether he has got the best way of analysing the results of other systematic designs. It is a matter of stating explicitly what terms in the fertility are to be considered, and working out the maximum likelihood way of eliminating them. Neither 'Student's' paper nor that of E.S. Pearson is sufficiently explicit on this point to convince me that they have got to the root of the matter. I should say that any design will give a valid estimate of error if the results are analysed correctly.

When I sent 'Student' a copy of my paper 'The relation between direct and inverse methods', I got a short note in reply, calling attention to the fact that in the title of his table the words 'a unique sample' occurred. I should like to acknowledge this, because it shows that the condition that I said he had assumed without mentioning it, that the sample constituted the only relevant information, was in fact in his mind. If there are several samples from the same population everything in the direct argument still holds but nobody would use such a set separately, so that the extra condition comes in in the transition from the argument to the use that is made of it. . . .

[1] *CP* 165. See Fisher's letter of 19 September 1938 (p.169).
[2] Jeffreys, H. (1931). *Scientific inference*. Cambridge University Press.

Fisher to H. Jeffreys: 26 *September* 1938

So far as I can judge, 'Student' and I would have differed quite inappreciably on randomisation if we had seen enough of each other to know exactly what the other meant, and if he had not felt in duty bound, not only to extol the merits, but also to deny the defects of Beaven's half drill strip system.

For example, randomisation was never intended from the first moment it was advocated to exclude the elimination from the error of components of error which could be completely eliminated, as in the case of differences between blocks in a randomised block system. It only requires that these components shall equally be eliminated from the estimation of error, as is now usual, though it was not appreciated at the time I first wrote on the subject. I often put this by saying that it is only the components which contribute to the actual error of the experiment which need to be randomised to provide an estimate of that error. In the Latin Square, for example, these are the components of variation that would remain after elimination of the best formal additive in rows and columns.

The second point on which there has been some misapprehension is that I should take for granted that the experimenter can choose anything he likes to be regarded as a single plot. For example, in sampling agricultural crops, a sampling unit, as it is there called, may consist of 50 roots scattered over the whole sampling area, with a rigid structural relationship *inter se*, so that the whole unit is determined by a single act of randomisation. For example, one might choose roots 7, 27, 47, etc. walking up and down a line, until the whole sampling area has been covered. The number 7 has been chosen at random from the numbers from 1 to 20, and to provide an estimate of error, a second sampling unit of the same kind is always taken from the same sampling area, e.g. the series based on number 12. Comparison of the results from numerous such pairs of sampling units from the different sampling areas will then provide a valid estimate of the sampling error of the samples used.

This is actually parallel to putting down a systematic comparison of two barley varieties at a number of centres and using the discrepancies among the comparisons made at different centres to estimate the precision of the aggregate comparison. In doing this, of course, we should make no pretence of knowing how precisely the comparison has been made at each particular point, i.e. in relation to local circumstances, which might favour one variety rather than another. The assertions made are valid only for the aggregate of all such tests, and if the difference between the two varieties is different on different farms, we shall have no valid estimate of error on which to judge the significance of such differences. One of the practical points on which, so far as I can judge, I differed from 'Student', is that he was willing to ignore such local differences, possibly because, in any case, Guinness wanted Ireland to grow only one barley variety, whereas I felt that we owed it to the farmer to encourage him to grow whatever variety seems most profitable to *him*.

I fancy also that Gosset never realised that a fertility gradient when, as in my experience is not very frequent, it is important enough to bother about, can easily be eliminated from a randomised experiment. It is, I think, my fault that I have not made this clear earlier, but until the last two years I had really thought that 'Student' accepted all that I had put forward on behalf of randomisation. In the next edition of *Statistical Methods* I am exhibiting the procedure of eliminating a fertility gradient, as one lot of uniformity trial data used in that book happens to show a gradient which it is profitable to eliminate in this way.

Thanks for sending me your paper on sampling a mixed population. I am glad to see how close to my position you come in the parts referred to. I suppose you will send it to Cambridge Phil. Soc., or somewhere of the kind.

Fisher to D.A. Kislovsky: 16 *February* 1929

I was much interested in your reprint from the *Journal of Heredity*[1] and especially in your conclusions respecting the effects of mass selection on extensive livestock. The method of Heincke appears to be sounder that that of Pearson in that the coefficient of racial likeness depends essentially on the sizes of the samples examined, and is therefore a crude test of significance of racial difference, rather than a measure of it.

Both methods I believe ignore the correlations between characters, and this becomes especially important where they are numerous, partly because it is reasonable to attach the greatest weight to distinctions in characters which are within each group least variable, partly because if the number of measurements is greatly increased the prediction of any one from the others becomes so accurate that they differ only by errors of measurement, and then a number of the apparent variates will represent not real biometrical variates, but errors of measurement only.

Logically, though this might be tedious, I can see no more proper way than to analyse the sum of squares of each variate, and the series of products of each pair of variates, into the two portions, *within races, between races*, taking exact account of the degrees of freedom. The examination of the relationships between the two quadratic forms so obtained would be quite new work, but I think a straightforward process could be evolved, and if you would care to send over your data, either after or before the preliminary analysis, I should be pleased to try to see what could be done with it.

[1] Kislovsky, D. (1927). Types in animal breeding and their analytical study. *J. Hered.* **18**, 447–55.

Fisher to P.A. MacMahon: 3 *July* 1924

Many thanks for your letter. I have sent to the R.A.S. library for your book, but have not yet seen it. It is possible that your method of solution will solve for me at least the first of the outstanding questions.

(i) What experimental technique of filling up the square will give each solution an equal chance of appearing?

One might fill the first two at random in $n!$ ways, and then fill the first column at random in $(n-1)!$ ways. Then things become more difficult. Possibly the following would suffice. Choose one of the remaining rows at random, and proceed to fill up the spaces in order rejecting any entry which conflicts with the first row. For example, if there are five spaces, one might shuffle 100 numbered cards thoroughly and take a card numbered $5m + p$ $(p = 0, 1, \ldots 4)$ to indicate the particular letter assigned to p; the first card dealt will then specify which letter is to occupy the first vacant space, unless this conflicts with the first row, in which case the first card is rejected, and the second card examined. In this way each space may be filled in succession. Questions which arise are:

(a) is it necessary to choose at random the row or column to be filled in next, or can they be filled in in order as they stand?

(b) is it necessary to choose at random the particular space to be filled in next, or can the spaces in a given row or column be filled in in order as they stand?

(c) on rejection of a card should the pack be reshuffled?

(ii) Is such a technique necessary or sufficient for the statistical validity of an agricultural experiment? This question involves points which it is not easy to reduce to a purely mathematical form.

An example will show the kind of lines along which I have been working.

The following are the actual yields of 25 areas in a uniformity trial (Mercer and Hall, 1910)

2713	2738	2673	2698	2822
2734	2657	2614	2559	2651
2753	2574	2623	2762	2765
2656	2562	2490	2605	2504
2586	2532	2489	2496	2459

The deviations of these from uniformity may be taken as giving the errors to which an experiment of this sort is subject. The sum of the squares of the 25 deviations from the mean is 253 106; dividing this by 24 we have a variance of 10 546. As an estimate of the standard error of the mean of 5 such plots taken at random, we divide this by 5 giving 2109.2 and take out the square root, 45.92.

If we had used five different treatments or varieties, each replicated five times, we should not actually reproduce this estimate; for example, assign plots to treatments wholly at random, as follows:

		Total deviation	Mean deviation
A C C E D	*A*	− 68	−13.6
C E C A A	*B*	−166	−33.2
B B E D E	*C*	+105	+21.0
A D D B B	*D*	− 3	− 0.6
B E C A D	*E*	+132	+26.4

The arrangement has evidently happened to make the errors less than usual; in consequence of this the estimate of error found by comparing plots treated alike will be more than usual. The relation between the two things is best shown by what I call an analysis of variance:

Variance between	Degrees of freedom	Sum of Squares	Mean Square
Different treatments	4	12 127.6	3 031.90
Plots treated alike	20	240 978.4	12 048.92
Total	24	253 106	10 546

The 24 degrees of freedom between the 25 plots may be divided into (i) 4 degrees of freedom between the five different treatments, and (ii) 20 degrees of freedom representing 4 comparisons between the 5 parallels for each of the 5 treatments. Corresponding to this division the sum of the squares of the deviations, 253 106, may also be divided into two parts, the first of which may be found by squaring the total deviations for *A,B,C,D,E,* summing and dividing by 5, and the second (in practice found by subtraction) is the sum of the 25 squares of the deviations of the individual values from the mean values for the corresponding treatments. The 'Mean Square' in the last column is found by dividing the Sum of Squares by the degrees of freedom. On the average of a large number of trials it is obvious that the mean square will take the same value in each line, apart from any real difference between the treatments. In fact the significance of any apparent difference between the treatments must be judged by the frequency with which a ratio between the 'Mean Square' values, as high as or higher than that observed, will occur under uniform treatment for given numbers (4,20) of degrees of freedom.

This distribution I have been able to obtain and sufficiently to tabulate assuming the original deviations to have been a sample from a normal distribution; this assumption is not, I think, a cause of trouble, since I find that moderate deviations from normality make wonderfully little difference.

The Latin square comes in at the next stage when we want to improve the experiment by eliminating part of the variability due to heterogeneity of different parts of the field. Treating the rows and columns as I have treated the 'treatments' above, we have an analysis of variance as follows:

Variance between	Degrees of freedom	Sum of Squares	Mean Square	S.D. of mean of 5
Rows	4	162 315.6		
Columns	4	32 842		
Remainder	16	57 948.4	3 621.775	26.92
Total	24	253 106	10 546	45.92

The variance is cut down to nearly a third of its previous value, if we can eliminate from the errors of the comparisons between treatments, differences due to the different fertility of the rows and columns. That is what a solution of the Latin Square does; if we take an arrangement such as

		Total deviation	Mean deviation
D E C B A	*A*	−112	−22.4
B D E A C	*B*	− 67	−13.4
C A B D E	*C*	+ 71	+14.2
E B A C D	*D*	− 18	− 3.6
A C D E B	*E*	+126	+25.2

the deviations are only a little smaller than before, though again the arrangement is rather a fortunate one, none of the deviations exceeding the S.D. The full analysis of the experiment is now:

Variance between	Degrees of freedom	Sum of Squares	Mean Square
Rows	4	162 315.6	
Columns	4	32 842	
Treatments	4	7 654.8	1913.7
Remainder	12	50 293.6	4191.1
Total	24	253 106	

The advantage of the arrangement lies not only in the fact that the errors will generally be somewhat smaller, but in the fact that the estimate of error, obtained from the 'remainder' will generally be smaller. The advantage is somewhat counter-balanced by the fact that the estimate is based on fewer degrees of freedom, but from all that I have seen from uniformity trials, it is very well worth while. Everything depends on the validity of comparison of the mean square for the 'treatments' and the 'remainder', and I am fairly sure that it is valid provided every solution of the Latin square has an equal chance of occurrence.

P.A. MacMahon to Fisher: 30 *July* 1924

The question of the number of arrangements of letters of n different kinds in a square $n \times n$ so that each of the letters shall appear precisely once in each row and in each column is known as the problem of the Latin Square. I have given the mathematical solution and you will find it in my *Combinatory Analysis*, Vol. 1, p.250.

For $n = 2$, no. of arrangements is	2	1[1]
3 " " "	12	1[1]
4 " " "	576	4[1]
5 " " "	149 760	52[1]

and I have not calculated the numbers any further.

I have had some difficulty in reading your handwriting and also in understanding what your further point is to which you allude in the last paragraph of your letter.[2]

If you will put a definite question to me in connection with it I will do my best.

[1] The corresponding numbers of reduced squares, as given in MacMahon's book, were written in by Fisher.

[2] Fisher's letters at this time were usually handwritten but the earliest of his letters to MacMahon of which we have a copy — bearing the date 3 July 1924 and reproduced above — was typewritten. MacMahon had evidently received an earlier handwritten letter from Fisher. To judge from the content of the Fisher–MacMahon correspondence as shown here from July 1924, it seems likely that the date given on one of the letters was incorrect, with Fisher's typed letter dated 3 July 1924 being his *reply* to MacMahon's letter dated 30 July 1924 and possibly incorporating material from the earlier handwritten letter that MacMahon had difficulty in reading.

P.A. MacMahon to Fisher: 19 *September* 1924

I have lost my Toronto note. Is the correct number of Latin Squares (reduced) now 56 instead of my number 52?[1]

[1] See *CP* 110, p.493.

Fisher to P.A. MacMahon: 21 *January* 1926

I think I have enumerated the 6 × 6 Latin Squares, making a total of 6008, of which 456 are symmetrical.[1]

The method of enumeration is to write down all possible diagonals, (types of diagonal), as on the enclosed sheet. The 10 arrangements in the right hand column give, on trial, no solution, the remaining 24 all give solutions, usually only a few each. For example, the diagonal *ACDBAB* gives two conjugate pairs of solutions. In this case all the letters are distinctly characterised,

B is the letter that occurs twice,
C " " " " takes *B*'s place,
D " " " " " *C*'s ",
E " " " whose place is taken by *A*,

consequently by permuting *BCDEF* in any order and arranging rows and columns to make the first row and column come in the standard order, each solution gives 120 solutions or 480 in all.

With the more degenerate groups I have no resources but to make a number of trial permutations, until I am more or less satisfied as to the groups to which they belong, but this should not, I fancy, affect the number of solutions in the total. For example there are 60 different diagonals of the type *AAABBC*, and one of them on trial is found to give one pair of conjugate solutions; the total is therefore 120, irrespective of the fact that the substitution $D-E$ changes one solution into the other.

I hope you will think the method a valid one, although it is scarcely applicable to 7×7 as there must be about 10 000 groups in that case.

[P.S.] You will be interested to know that the Latin square has been a great success agriculturally.

[1] See Fisher, R.A. and Yates, F. (1934). The 6 × 6 Latin Squares. *Proc. Camb. Phil. Soc.* **30**, 492–507, (*CP* 110), where the correct enumeration was finally obtained.

P.A. MacMahon to Fisher: 24 *January* 1926

I congratulate [you] on a 'tour de force' in enumerating (actually) the Latin Squares of order 6. You obtain 5552 unsymmetrical and 456 symmetrical, total 6008. If you halve the number 5552 you get 2776 squares to be read both by rows and by columns. Thus $2776 + 456 = 3232 = 2^5 \cdot 101$ are necessary to exhibit the whole of the 6008.

In some recent work I have had to examine certain square formations connected with determinant theory and found that the grouping proceeded by diagonals — so that I am quite prepared to find that the diagonal method suits your enumeration best. I have no reason to doubt the validity of your method. As you say, the number increases with the order with great rapidity.

Fisher to K. Mather: 8 *May* 1950

After looking at some recent American books on statistics I have felt constrained to draft the enclosed as an addition to Section 65 of *The Design of Experiments*. I do not know whether you have been troubled by finding otherwise suitable textbooks showing the kind of misapprehension at which I aim. If not, of course, I don't see why you should be, but you may already have the question on your mind.

K. Mather to Fisher: 18 *May* 1950

Thank you for letting me see your draft for inclusion in *The Design of Experiments*. The distinction between definite and indefinite interactions is most illuminating and I am sure that it will be most valuable in clarifying what to most people is an obscure and difficult subject.

I suppose that most biological interaction will be indefinite; but the plant physiologist, for example, might well run up against the sort of situation you illustrate from industry. . . .

Fisher to K. Mather: 20 *May* 1950

I think categories such as established varieties and inbred lines will be definite enough, because reproducible at will in future work. On the other hand, genotypic differences among F_2 plants as shown by F_3 progenies would be fairly typical indefinite categories.

What had shocked me, however, was a number of recent American books with practical aims specialising in industrial experimentation getting themselves into great difficulties through assuming that main effects were always to be compared with interactions with whatever factors were included in the experiment. Logically this makes them think that an experiment without e.g. variation of temperature gave definite information of a kind not available if the whole experiment were repeated at five different temperatures. This is manifest nonsense, and my new paragraphs aim at clarifying the confusion of such paradoxes. . . .

D. Michie to Fisher: 30 *October* 1956

. . . I would also be glad of your opinion on a statistical problem which my friend Dr. Murdoch Mitchison has brought to me.

He has been measuring the dry mass of individual living yeast cells by an ingenious optical method which allows successive observations to be made on the same cell without disturbing it. He is interested in the form which the increase of dry mass takes during the period between fissions. Is the increase linear with time, or is there some degree of curvature?

He has a number of independent series of measurements, each series on a separate cell (perhaps 20 series in all). The various series differ among themselves in the number of measurements and the time intervals between them. He is not able to assume that the different series agree with each other either in the magnitude of observer error or in the culture conditions — either their initial state or their rate of subsequent deterioration. . . .

Neither the slope nor the degree of curvature can be assumed the same from one series to the next. Dr. Mitchison is quite prepared to envisage cases where an intrinsic tendency of the cell to give a concave-upwards curve is actually reversed by rapid deterioration of the conditions in a particular culture so that curvature of opposite sign results.

He asks the question 'Is there a significant overall tendency towards curvilinearity as such, without regard either to the degree or to the sign of curvature from one series to the next?' I rephrase his question as follows: 'After fitting a linear regression line to each series, is a significantly large fraction of the residual variation between observations within series attributable to some degree of quadratic curvature in some of the series?'

The recipe which I have given him is to set out an analysis of variance for each series independently, after fitting to each a regression equation of the form $Y = a+bx+cx^2$, as follows:

Source of variation	*S.S*	*D.F*	*M.S.*
Linear term of regression equation		1	A
Quadratic term		1	B
Residual		$n - 3$	C
Total		$n - 1$	

n = no. of observations in the given series.

Tabulate the mean square ratio B/C for each series, together with the corresponding value of P derived by interpolation in a table of the Variance Ratio. On the set of P values obtained in this way, perform a combined test of significance according to the method described in section 21.1 of *Statistical Methods*.

Of course, if the combined test gave a significant answer he would still be faced with a problem of interpretation. But if it did not he would be able to say that his data did not contradict the idea that the dry mass of a yeast cell increases linearly with time.

I would very much value your comments on the approach which I have suggested to him. He had already calculated the required analysis of variance before consulting me, but since on crude inspection they showed great heterogeneity between series in almost every respect he was uncertain as to how to bring the relevant information from all series to bear upon the specific hypothesis of linearity.

Fisher to D. Michie: 31 *October* 1956

. . . On Mitchison's problem your solution is highly ingenious, but if I understand it correctly it does not take account of the *sign* of each quadratic term, whereas probably what Mitchison is interested in is just whether they are largely of the same sign representing curvature upward or downward.

You could take the square root of B and give it a weight inversely to its variance. This would be the classical method of combination, which is mildly defective in not taking account of the precision of these weights.

Simplicissime, and as an additional check, the regression line gives an expected midway value, and the observed points if odd in number give one realized, and if even in number I should not hesitate to take the mean of the middle two; so you have a number of deviations from expectation, some positive and some negative; and ignoring their variable precision, on which the data tell one rather little, you could give them a simple t test. If the two methods do not agree substantially, do not trust either of them.

Fisher to J.A Nelder: 6 *November* 1956

I have just seen your review (*Journal of the Royal Statistical Society*, volume 119, page 340) of Federer's book on *Experimental Design, Theory and Application*. I notice that you reprove the author for quoting without comment my expression for the loss of information in experimental work due to the use of estimated, in place of true, variances. Your suggestion, if I read it aright, is that Federer ought to have used some 'other intuitively reasonable' measure of this effect.

I am, of course, aware that a number of people have found difficulty in understanding the cogency of the fiducial argument, though a good deal of progress seems to have been made already in this respect. I cannot expect it to be obvious, therefore, that the amount of information supplied by a test with an empirically estimated variance actually is the amount which I have calculated. I should, however, be glad to know what method is in your mind as an alternative, remembering that the planner of an experiment does not know in advance what value of t may emerge for any particular comparison, and would be foolish to suppose, as has sometimes been urged, that the purposes of his experiment are fulfilled only by deciding whether t does, or does not, exceed some value chosen in advance. Generally, a number of different comparisons are tested simultaneously in the same design.

Of course I have no objection to others discussing problems different from those which I have at some time set myself, but you will perhaps see reason in an objection to their solutions being claimed as solutions of the same problem as that which I have treated.

J.A. Nelder to Fisher: 8 *November* 1956

In reply to your letter of 6 Nov. about a point I raised in a review of Federer's book, I think I can make clear my attitude fairly briefly. It seems to me that there are two ways in which a text book can be written. It can be written from a strongly 'personal' view-point, in which the ideas of the author form a unifying background to the discussion. I would regard your books as coming into this category; I personally prefer books of this kind, though I may not always understand or agree with all that is in them. Alternatively an author can attempt to construct an encyclopaedia of his subject in which all the diverse work on related problems is set out, but in which no strongly expressed opinions of the author's own are expressed. My remark in the review was prompted by the remembrance of Section 2.3 of Cochran and Cox's *Experimental Designs*. I imagine that you would argue that the alternative criteria considered there were all irrelevant to the problem. Personally I could not accept that, and my quarrel with the author was that since he does not appear to adopt the fiducial standpoint in his book, and appears to be attempting the encyclopaedic type of work, he should consider other approaches. It may be that these other approaches are faulty, the future alone will show that, but in my opinion if the ideas have been held by reputable workers then they should be discussed, particularly in a book containing such an immense amount of detail in other directions.

I hope these remarks of mine will make it clear that it was not my intention to suggest that your measure of relative accuracy was faulty, but merely to maintain that an encyclopaedia should be an encyclopaedia.

Fisher to J.A. Nelder: 10 *November* 1956

Thank you for your courteous letter. Looking at Cochran and Cox I suppose Table 2.3 is the relevant point. What I meant by other authors having tried to solve a different problem, namely one in which the level of significance, or the *t* value, is known in advance, is illustrated by Cochran and Cox's Table 2.3. What I wanted you to see was that this is a different problem, which indeed had already been solved by 'Student', from the one with which I was concerned, namely that to a man planning an experiment, who has a choice of the numbers of degrees of freedom which will be available but no knowledge of how large the several deviations may be, it is useful to know how much information is lost by the circumstance that the number of degrees of freedom will be finite, and how much more is lost when it is small than when it is large. Of course there are plenty of people to act as gramophone records of Neyman's views. I suppose if Federer had been confronted in the literature with alternative solutions of the same problem, it would be proper to expect him to mention both and perhaps to adjudicate between them. What perhaps you have overlooked is that the problem, when the experimenter has no knowledge of what levels of significance he may expect to find, and of course many different levels may appear for the same estimate of error even in the same experiment, and especially if he does not (and this is still the view of the majority of experimenters) think that his experiment exists solely to make an acceptance decision, is a different problem from any that Neyman and his followers have discussed.

On the following pages indeed, 28 and 29, the authors advise, 'For general purposes it is suggested that this table' (namely that of Fisher's correction) 'be used to take account of the differences in degrees of freedom for error in two designs that are being compared'. It would seem, therefore, that after considering Neyman's contribution these authors do not suggest that it supplies an alternative method.

J.A. Nelder to Fisher: 28 November 1956

Thank you for your letter of 10 November. I agree with you that Cochran and Cox suggest the use of the quantity $(n+1)/(n+3)s^2$ for the comparison of designs with different nos. of degrees of freedom. I take this to mean that they believe that if the experimenter is offered a choice of two designs, one having error variance 100 and a large no. of d.f. and the other 50 and 1 d.f. he is to regard them as equivalent. There are two difficulties here for me; first, as an experimenter I would intuitively reject the equivalence, and secondly I am not sure that it is a correct interpretation of your original derivation in Sec. 74 of *The Design of Experiments*. My own belief in the matter is that this is a problem with no unique solution, so that the discrepancies in Cochran and Cox's Table 2.3 are to be expected and are not resolvable without further detail in the statement of the problem. Thus in the kind of work we do here, where we have a general long-term aim to improve the practice of vegetable growing, an experimenter will always be finding in the course of his work certain indications, as a side-product of his experiments, that might be followed up. He will often do a rather small experiment to see whether such a possible line of work would be worth following up. His decision on the results of such an experiment will approximate rather closely to an acceptance procedure. In deciding, he will have to assess the sort of treatment difference that would be practically interesting and to assess a reasonable significance level. In practice, I admit, this will be difficult to do rationally, but nevertheless different assessments would lead to different assessments of relative accuracy after the manner of Table 2.3. There is one point concerning Sec. 74 of *The Design of Experiments* which I should be glad if you would clarify for me. In the other examples of the chapter the information is derived from the likelihood function and expressed in terms of the parameters of the population, but in section 74, it is derived not from the likelihood function based on $(x-\mu)/\sigma$ but from the t-distribution $(x-\mu)/s$. This latter may be regarded as derived from the likelihood function by integrating out σ according to its fiducial distribution. There are thus two processes, the elimination of the unwanted parameter σ and the formation of the information function. If, however, we carry out these processes in the reverse order, the information function gives us $1/\sigma^2$, and its integration over the fiducial distribution of σ gives $1/s^2$. The point on which I am not clear is the logical basis for preferring one order of operations to the other. I should be most grateful if you could elucidate this point for me.

Fisher to J.A. Nelder: 3 December 1956

. . . I suppose your misapprehension as to what Cochran and Cox said about my formula $(n+1)/(n+3)[s^2]$ was due either to your misunderstanding one of the teachers here,[1] or to his own misapprehension on this point. . . .

In the superfluous little table provided by Cochran and Cox on page 28, they say 'equal amounts of information', and perhaps you have read into this the more elaborate concept of equivalence. In the table on the previous page, the author, perhaps Welch, has fallen between two stools by calculating the *average* limits of error at various levels of significance, whereas 'Student' had given the actual limits at each level, and this is what the experimenter needs.

If I understand your letter aright, Federer is to be scolded for quoting my formula not because the latter had been shown to be inexact, and not because any effective alternative has been offered, but because you hope at some future time to offer such an alternative, not being yourself satisfied with the matter as I have left it.

With respect to the calculation of the amount of information, this has been the same and examples of it have been available every year, at least of the last 35; and again it must be the peculiarities of your teaching at Cambridge which led you to think that some other method is more authentic. Common forms are

$$\Sigma \frac{1}{m}\left(\frac{\partial m}{\partial \theta}\right)^2, \quad \int \frac{1}{y}\left(\frac{\partial y}{\partial \theta}\right)^2 dx.$$

Probably, however, you were not taught to regard the fiducial distribution of μ as a frequency distribution at all.

[1] University of Cambridge.

J.A. Prescott to Fisher: 4 *October* 1930

For some time I have been intending to write to you about one or two problems of sampling and experimental method. The first deals with the matter of systematic sampling over a large area of, say, several thousand, or even million, acres. Two come to my mind particularly, one the selection of a number of sites in Egypt for the daily counting of the flowers of cotton plants, and the second relating to the determination of the probable distribution of salt in a large suspected area where it would be almost hopeless to map accurately the actual distribution within a reasonable time. My recollection of the former case is that a grid was recommended to be laid down over a map of the Nile delta and that localities for observation were to be selected say at points of intersections. I am no longer particularly interested in this matter, but the second one is very vital at the moment in Western Australia. . . . In this case an examination of soil samples from 650 sites on the so-called forest soils in an area of about 2000 square miles indicated that about one-half of the samples contained dangerous quantities of salt.

A similar instance is the systematic examination of soil profiles from an area of about 500 000 acres in Victoria commanded by the waters of the Murray impounded by the Hume dam. In this latter case the soils have been sampled on a grid basis, while in the former the sampling was at regular intervals along and on each side of prospectors' tracks through the virgin 'forest'.

The problem is in effect what system of systematic or random sampling should be adopted in order to obtain a reasonably accurate estimate of the probable distribution of certain soil types or soil conditions without actually mapping the area.

. . . It is interesting to note that, having selected an area as suitable for survey, we frequently meet all the major soil types in the course of a few days' work and that thereafter for several weeks no new types are met.

Another problem of systematic as against random sampling has been referred to me. . . . In this case the examination of records from areas showing definite yield gradient across the field has given . . . some difficulty. This appears to apply particularly to pasture work. . . .

Am I not right in assuming that your system of randomized blocks more or less takes care of the above problem?

There is one final point which has appealed to me personally and which may be already covered by your general theory of sampling and that is that sometimes one gets an impression that the older statistical method underestimates the reliability of the final result. . . . In a latin square of four treatments, there is actually one-fourth of the area under any given treatment, and with the four plots distributed at random the estimate of the yield from the whole area should be very near the truth. I can quite see that it is possible to treat this problem rather differently from the older method of taking say 1000 small plots and grouping them into different combinations until the reliability is sufficiently increased. Would it not be better to group these results into as many possible areas of say one-twentieth of an acre, scattered over the whole area? . . . The size of plot is not only a matter for determination by statistical analysis, but there is also the important question of field technique, the implements and labour available, weighing appliances, and so on, which all have a decided bearing on the final size of the plot. In Egypt we found this to be 1/40th acre for wheat, 1/20th acre for maize, and 1/10th acre for cotton.

I should very much appreciate any remarks you may have to make on the above-mentioned problems.

Fisher to J.A. Prescott: 10 *November* 1930

The problem you mention of the large scale survey is one of those which are dealt with very satisfactorily by multiple regression. The standard example of this treatment of an analogous problem is in Section 29 of my book *Statistical Methods for Research Workers* where I treat of the relationship of rainfall to Latitude, Longitude and Altitude, from the point of view of one wishing to estimate the probable rainfall at a given point, or within a given area, in view of the values observed at a number of stations.

When there is an appreciable gradient in any quantity (say) from North to South, one must distinguish between the probable error of an estimate at a known Latitude from that at a point of undetermined Latitude. The first will evidently be the smaller. In field experiments, as you say, the method of randomised blocks eliminates the greater part of any such gradient effect; it should be noticed that it also eliminates much of the heterogeneity which is not simple enough to be represented by a gradient, and what is most important, it provides a valid estimate of the errors not eliminated. In this it differs from all systematic arrangements.

I quite agree with you that higher accuracy can always be attained by increasing an experiment from one to several Latin squares; usually it can also be done without increasing the area by sub-dividing the experimental area available. However, parallel Latin squares at different stations have the great advantage of broadening the inductive basis of the result.

Personally, I always advise that size and shape of the plot should be determined by practical agricultural convenience, while the structure and extent of the experiment must be governed by the accuracy aimed at. The advantage of using the easiest and most expeditious field methods is thus applied to make further replication possible.

C.R. Rao to Fisher: 18 *September* 1952

I am writing this to refer to you some difficulty which I had in reading Fairfield Smith's paper on discriminant function for plant selection.[1] In the problem of predicting the genetic component $\psi = a_1\psi_1 + a_2\psi_2 + \ldots$ by a linear combination of the observations $l_1\bar{x}_1 + l_2\bar{x}_2 + \ldots$ I find the constants $l_1, l_2, \ldots$ depend on the sample size on which $\bar{x}_1, \bar{x}_2, \ldots$ are based so that no single set is ideally suited to predict genotypic values of various lines providing different numbers of observations. So the least square estimates weighted by n_r

$$\sum_r (\psi - l_1\bar{x}_1 + \ldots)^2 \, n_r$$

do not provide the best functions for individual cases except when $n_r = n$ for all r as in Fairfield Smith's problem. But in Panse's problem of poultry n_r are all different. But of course one can ask for a common linear function with which the genotypic values can be ordered. In this case the least square solution weighted by n_r is a good one. Having estimated the covariance matrix of the genotypic values $\psi_1, \psi_2, \ldots$ and also of within lines from covariance analysis it is possible to calculate the appropriate constants for each sample size within the lines and obtain the genotypic value for the ith line as

$$l_{1i}\bar{x}_1 + \ldots + l_{pi}\bar{x}_p$$

and then arrange the lines in decreasing order for selection. Am I correct? I am also developing tests for the adequacy of an assigned discriminant function. If you think I am on right lines I propose to send the manuscript for publication in *Annals of Eugenics* or *Heredity*. I will send you a copy in advance.

[1] Smith H.F. (1936). A discriminant function for plant selection. *Ann.Eugen*. **7**, 240–50.

Fisher to C.R. Rao: 23 *September* 1952

Thanks for your letter letting me know that you propose to send a manuscript for publication in the *Annals of Eugenics* or *Heredity*. It is only of the latter that I am now editor. I think it was intrinsic of Fairfield Smith's approach that he had a single multivariate sample to work on so that the number of cases was the same for all variates. I do not think I have heard of Panse's problem with poultry. I only know of his work with cotton. Another thing to remember about Fairfield Smith's paper, which perhaps does not concern you, is that he was developing a method supposedly based on the covariation of yield factors in one year with measurements available for selection in the previous year. Unfortunately the only material he had at hand to illustrate this method consisted in wheat data for a single year and this has led Mather and others somewhat to misunderstand the nature and purpose of the discriminant function.

Fisher to P.R. Rider: 19 *January* 1932

I was much interested in your long letter on B. Peirce's method for the rejection of observations;[1] the final step is certainly Maximum Likelihood in which Peirce was presumably following Gauss. I am not clear, however, whether or not Peirce is trying to do more than is possible.

Any set of residuals from a mean (and the same applies to residuals from more complicated regressions) may be represented by a point on a many dimensional sphere, and any criterion for rejection must consist in choosing certain portions of this sphere which shall be deemed inadmissible. If the criterion also determines how many observations are to be rejected, it must divide the sphere into portions

(i) in which no observation is rejected,
(ii) in which one is rejected,
(iii) in which two are rejected and so on.

Necessarily, the volumes of these portions must be fixed *a priori*; i.e. the probability of rejecting 1, 2, 3, . . . observations out of n, when no gross errors have really occurred. We might for example decide to reject one or more observations in one case out of 20, two or more in one case out of $(20)^2$ and so on; and we shall obtain such a series if we decide to reject observations one at a time by the repeated application of the same criterion. The question is what particular values does Peirce reject, and how do the frequencies run?

This puzzles me a good deal, as it is only when the rejected area can be mapped out that one can judge of the plausibility of the method.

[1] Rider had asked if the final step in the paper by B. Peirce (1852), Criterion for the rejection of doubtful observations, *Astron J.* **2**, 161–3, was not 'just the principle of maximum likelihood'.

Fisher to H.L. Rietz: 25 *February* 1930[1]

The approach to sampling problems by way of the geometry of generalised Euclidean space is an exceedingly fruitful one, but one which at present has to rely to some extent on the mathematical intuition of the reader. At least I know of no source to which one can be referred for the elementary general properties of such space; I was looking only recently at Baker's big work to see if he treated of the matter, and though there is much about cubic and biquadratic loci, the simple properties of the generalised Euclidean space are not explicitly developed.

The solution of the 1915 paper [*CP* 4], while obtained geometrically, may be easily verified analytically as in 'Sur la solution de l'équation intégrale de M.V. Romanovsky' (1925), (*Comptes Rendus de l'Académie des Sciences*, vol. 181, pp. 88–89) [*CP* 46]. I have also given in 'Applications of 'Student's' distribution', (*Metron*, vol. V, pp. 90–104, 1925) [*CP* 43], a general analytic argument equivalent to the very widespread use of hypergeometry in considering degrees of freedom. I fancy the only result of mine which now rests wholly upon the geometrical argument is that for the 'Distribution of the partial correlation coefficient' (*Metron,* vol. III, pp. 329–332), (1924) [*CP* 35], where it is beautifully direct; however, in this case also, the thing can be reduced to the analytic dissection of a sum of squares.

[1] Rietz had written seeking references to the properties of *n*-dimensional space used by Fisher in studying the distribution of correlation coefficients.

Fisher to H.L. Rietz: 4 *March* 1937

My attention has been called to a statement ascribed to you by one, S. Kullback,[1] which is either misleading, or, if not so in its original connection, appears to have misled Kullback.

This is to the effect that the analysis of variance is obtained largely by inferences based on the number of degrees of freedom of the variates, rather than upon formal mathematical proofs. Mr. Kullback seems to take this to mean, *inter alia*, that the distribution of the multiple correlation coefficient has not, in the course of my work, ever been demonstrated.

I need scarcely remind you that this and a great many other similar problems were solved independently prior to 1925, before the particular arithmetical arrangement known as 'the analysis of variance' had been developed. A reference for the correlation ratio and the multiple correlation in the null case is *J.R.S.S. 85*: 595–612 [*CP* 20]. It was about 1923 that I recognised the unity of these different solutions, giving the general form of the z distribution at the Toronto Conference in 1924 [*CP* 36], and an analytical demonstration, of which all analyses of variance are more or less direct applications, in *Metron 5*, part 3, pp. 90–104 [*CP* 43].

The analysis of variance merely presents the properties of the quadratic forms involved in, I think, a very simple manner, and it is too much like deliberately ignoring the body of work on which it was based to suggest to students that no adequate proofs have been given. Whether you have overlooked this previous work, which was, of course, deliberately ignored by my distinguished predecessor, Professor Karl Pearson, or whether you think that the proofs therein given are not complete, I do not, of course, know. I have, however, never been inaccessible to correspondents who wish for a more explicit statement of the steps of my proof.

[1] Kullback, S. (1935). A note on the analysis of variance. *Ann. Math. Stat.* **6**, 76–7.

H.L. Rietz to Fisher: 22 *March* 1937

In reply to your letter of March fourth, I regret the matter very much if I have misled Kullback or any one else by my remark about arguments based on degrees of freedom without formal proof. Those arguments have been difficult for me to follow and I feel sure the same difficulty has been experienced by many of your interested readers.

Perhaps I should say first that Kullback has never been a student of mine. I met him in Washington on the occasion of his examination for the doctorate. That is the only personal contact I have had with him.

The introduction to Kullback's note is unfortunate. So far as I see, he was merely attempting to present, by the use of characteristic functions, an alternate demonstration of the distribution of the multiple correlation coefficient from uncorrelated normally distributed material.

With respect to the reader's difficulties in following arguments based on degrees of freedom, I venture to say it would be very useful to your interested readers to be shown in considerable detail the meaning of degrees of freedom in relation to properties of the quadratic forms under consideration.

I shall be glad to avail myself of the kind suggestion in the last sentence of your letter.

It will be much appreciated if you will include my name on the list to whom you send your current reprints.

Fisher to H.L. Rietz: 1 *April* 1937

Thanks for you courteous letter. I think the most general approach to the analysis of quadratic forms is that given, in 1925 or 6, in Applications of 'Student's' distribution, *Metron* V: 90–104 [*CP* 43]. In Section 6 on the significance of regression coefficients, I see that the printer has put ζ for ξ, but I think it is sufficiently correct in other ways, and you will readily recognise the generality of the demonstration. I am afraid I have now no spare copies of this paper. I have added your name to my mailing list.

Fisher to V. Satakopan: 30 *January* 1939

Thanks for your letter and kind congratulations.

For my own part I should never have pitted orthogonal polynomials and harmonic components against each other as rival methods,[1] since the cases to which they are appropriate seem to me quite distinct. However, Whipple has evidently for years harboured a feeling that polynomials have no place in meteorology, and in this I think he is wrong, as soon as meteorologists interest themselves, not merely in the average of a number of years, but in the differences between these years as individuals.

The difference arises in the fact that, whereas the average of any element at different times of the year through a number of years is necessarily a periodic function, yet the same element for any one individual year may have quite a different value at the end from what it had at the beginning. This is true of ordinary meteorological elements, though it was brought to my mind most forcibly when the element I was studying was the average effect of a given amount of rainfall on a crop harvested at the end of August, when it was obviously more than ridiculous to suppose that the average effect on August 31st in the year in which the crop was harvested approximated to the average effect on September 1st, 6 weeks before the crop was sown.

Otherwise the parallelism is close, apart from the fact that users of harmonic analysis have not always confined themselves to the truly orthogonal components, namely, those with period n/s where n is the number of observations in the record and s is a whole number from 1 to $(n-1)/2$ in case n is odd, and $(n-2)/2$ when n is even. Periodogram values for periods other than this series are complicated compounds, involving the rejection of varying amounts of the data, and I do not think anything useful can be done with them, but the series of coefficients of sines and cosines having these periods are strictly comparable to the coefficients of the orthogonal polynomials, save that they come in pairs instead of in a single series. . . .

[1] Satakopan had said in his letter that he had been asked to initiate a discussion on 'harmonic functions vs. orthogonal polynomials in meteorology and geophysics'.

Fisher to W.H. Sayers: 20 *August* 1941

Thanks for your long letter, which, if I have followed it correctly, turns largely on the meaning of the word *experiment*, and perhaps partly on that of the word *design*.[1] Perhaps I can make my meaning clear by a discussion of this point.

The experiments with which I was concerned are observations, or sets of observations, intended and adapted to the intention — that is to say *designed* — to increase knowledge, or, in other words, to remove doubt. Whatever degree of confidence we may feel in various assertions which can be made about the subject matter, there are, in these cases, at least some which we are willing to revise, or reject, in consequence of our observations. They may, of course, be very easy to reject, i.e. very crude, or few, or inaccurate observations may be sufficient to disprove them, or again they may be difficult to disprove in that they are compatible with common or crude observations, while a doubt still remains as to their truth. Further, it may be noted that only such assertions as influence our future expectations, beyond observations already made, have scientific relevance; e.g. the statement that Napoleon died at some exact age, though it may be a verifiable fact, is a contribution rather to history than to science. The statements which experiment is designed to test are to this extent all in the nature of generalisations, i.e. statements concerning some aggregates of possible future observations. For example, the statement that, on killing a particular sheep, we shall find in the intestine some particular species of parasitic worm can be proved true or false; but, except for its implication that this parasite may be found in other sheep which we have not killed, its verification has no scientific value. The statement that between x and y per cent of the sheep of the country contain the worm, or that some particular drug diminishes its incidence, are, on the other hand, both within the field of scientific experimental enquiry. Logically I suggest that the contribution to knowledge of any experiment lies in its capacity to disprove more or less conclusively some statement or generalisation respecting future observations which previously was tenable. If this is clearly borne in mind, a certain amount of supposed experimentation may be set aside as non-functional or irrelevant.

I do not clearly follow on what grounds you feel that doubt as to the precision of units plays any determining part in the nature of experimentation, nor really do I think that the amount of statistical work required in interpretation plays any central part. It is not very long since statistical methods have been brought into any relation at all with experimental design, and clearly throughout they are only one aspect of the subject, though co-extensive with it. Obviously the best designed experiments are those of which the interpretation is most plain, and these often require so little statistical investigation that we are scarcely aware there is any at all.

[1] Sayers, a physicist, had questioned Fisher's choice of title for his book *The design of experiments*, claiming that not all experiments come within its scope and that the role of statistics in the field of experiment is 'merely that of providing a special type of metrology'.

W.H. Sayers to Fisher: 24 August 1941

Many thanks for your letter of August 20th. I agree that our difference ultimately is in respect of the definition of 'experiment'. I think your interpretation is a result of long association with experimental work of a special type and that it is unduly narrow.

One of your fellow biologists (T.H. Morgan) says that experiment is 'The method by which Science separates the wheat from the chaff — namely the use of working hypotheses controlled by quantitative measurement'.

I do not regard this as a completely satisfactory definition but it does stress the essential characteristics of experiment — there must be a hypothesis, which is to some extent in doubt, and there must be an experimental result which can be expressed in the form of a measure of the validity of that hypothesis. It may be accepted that the term 'experiment' should be restricted to cases where the hypothesis to be tested is not trivial.

I imagine that you will agree so far.

Now on any such basis it is not permissible to dismiss, as you appear to do, the test of a single sheep for infestation by a particular parasite as 'not an experiment'.

I understand exactly why you have so dismissed this particular test — it is because you have tacitly made certain assumptions both as to the nature of the hypothesis which might be tested in this way, and as to the 'significance' of the result so obtainable on hypotheses of this kind. These assumptions would as a matter of fact usually be valid in your everyday practice. But suppose that the parasite in question is one which infests sheep in say Australia, but has never been known to occur in Britain. Suppose further the life history of the parasite to be unknown, but that you have formed the hypothesis that it enters the sheep with its food in a larval form, this larval form being known, but hitherto assumed to be a separate species. The suspect larval form is, like the parasite, unknown in Britain. You obtain a culture of the suspect organism from Australia, you introduce this into the food of a British sheep, and after a suitable interval, examine the sheep and find the adult parasite. Would this not be an 'experiment' and a highly significant one at that?

This example is a very interesting one from my point of view. Your hypothesis is that the suspect larva, α say, develops into the adult parasite β. From this hypothesis you deduce a secondary hypothesis from the first, that if α is introduced into a British sheep, β will subsequently appear in that sheep. If the test confirms the secondary hypothesis it confirms that the primary hypothesis is probably valid, the probability under the particular circumstances being obviously very high indeed.

But a more direct test of the primary hypothesis is conceivable. You may be able to find a suitable culture medium in which you can observe that α does in fact develop into β. If you can do this there is no question of confirming the primary hypothesis only to a high order of probability — you have confirmed the initial hypothesis absolutely.

I do not suppose you will deny that both methods of testing the initial hypothesis are deserving of the title of experiment. But there is a fundamental difference between the character of the two tests. One confirms — or otherwise — by indirect inference what is to some degree uncertain — the significance of the result obtained is measurable only in terms of the probability that the indirect inference is valid. In the other there is no indirect inference with its attendant uncertainty. And for this reason I still maintain that there is a field of experiment which does not call for the use of statistical concepts in the analysis of results.

This field is fairly easily recognisable. You pose your hypothesis in the form of a question — for example, 'Does α develop into β?' and you answer it by saying, 'I believe (estimate) the answer is Yes'.

Now if your experiment, or test, gives you an answer to the hypothetical question in precisely the same terms as those of the question — if the answer, that is, is 'Yes, α does develop into β', no indirect or uncertain inference is involved and the experiment lies outside the field in which statistical methods are required in its interpretation. But if the answer given by your experiment is not in this form, if it is merely that certain results have been observed which are consistent with the hypothesis that α develops into β, you have then to examine what are the probabilities that this observed consistency is not the consequence of the validity of your assumption. And this probability is — under such conditions — never zero. In such cases, statistics — the metrology of probability — must play a part in interpreting the result.

I am quite sure you will not attempt to obscure the argument by pointing out that in fact there is always some measure of indirect inference in every real experiment, since your measurements are of uncertain precision. If the uncertainty of the measurements are within limits too small significantly to affect the final result, they do not introduce any practical element of uncertain inference into the interpretation of the result.

Fisher to W.H. Sayers: 1 *September* 1941

Thanks for your long letter. I see that there is a type of work which may involve experiment, the object of which is merely identification. In the framework of ideas every adult has a unique larval form, and *vice versa*, and if we exclude such real possibilities as that those of polymorphism in adults or larvae, or the possibility that critical differences in larvae may have been overlooked, then rearing a single larva, e.g. from marine plankton in an experimental tank up to a recognisable adult, would, in ordinary language, be an experiment of this logically simple type. It is analogous to the position of a palaeo-botanist who has recognised certain leaf impressions and fruiting bodies separately at a given horizon, that is, searching for a plant fragment that should prove the two to belong to the same species. Confirmation or refutation of such identification is usually supplied by totally independent workers, but is certainly felt to be desirable.

Personally I do not much mind if the word 'experiment' is commonly applied to such rather ideal special cases. For the worker with any future applications in view, frequency, e.g. of successive infestation, will always be of essential importance.

W.H. Sayers to Fisher: 8 September 1941

Thanks for your letter of September 1st. I am not quite clear that I have yet made my point quite clear to you. If an observed experimental result is expressible in precisely the same terms as the hypothesis which that result is to be used to check, then you may classify the experiment as pure 'observation'. It is this factor of the commensurate relation between experimental answer and hypothetical question which determines whether statistical analysis has any part to play in interpreting the result.

In a very large number of experiments in the realm of physics, you can so design your experiment that your result is in such commensurate terms — apart from the question of the precision of the instruments you are using.

In general the determination of this precision is a matter of calibrating instruments — and the calibration is an experiment on the instruments separate from the experiment for which you use the instruments. Whether practically speaking you need to apply statistical methods to the analysis of the calibration experiment depends on circumstances, but ultimately calibration must, to be of any real value, have a statistical basis.

Your type of biological experiment is rather like trying to carry out chemical analyses using reagents of unknown and variable purity and concentration. You can 'calibrate' chemical reagents, but you can't 'calibrate' a plot of soil for fertility, or a spell of weather for its influence on plant growth, etc. Therefore you have so to design your experiments that each experiment provides in itself the data from which you can derive some equivalent to a calibration of the reagents you are using in that particular experiment.

It is I suggest this experience with forms of experiment in which 'calibration' of the experimental equipment and measurement of the experimental result are necessarily part of one and the same experiment that has led you to make somewhat too extensive claims for the rôle of statistical analyses in experiment generally.

Fisher to W.H. Sayers: 29 September 1941

Of course I agree with the generality of the distinction you make between physical experiments in which the necessary calibration of the instruments is carried on on a different occasion, often at a different time, and in a different institution from their experimental use, owing to the legitimacy in much physical material of the assumption that the calibration remains valid, whereas for biological material this assumption is often so palpably untenable that a good biological experiment has to contain within itself the controls needed to give confidence in the conclusions to be drawn. It is, I imagine, this common simplification of physical research which has led to the problem of experimentation being much more thoroughly elaborated in biological and psychological than in physical research. A good biological experiment, for example in agriculture, is in fact from the logical standpoint a far more perfect whole than a physical experiment need ordinarily be.

Fisher to G.W. Snedecor: 25 *April* 1935

I was glad to see a paper of yours in the *Journal of the* [*American*] Statistical Association,[1] and that you have been trying out the analysis of covariance in its application to the educational material at Ames. I am, however, entirely puzzled by your references to my work. You say, at the end, that in my fifth edition [*SMRW*] I do not 'accept all Bartlett's conclusions.' Perhaps not, if the phrase is taken quite literally, but it seems to refer to the method you had just given of isolating, in the analysis of variance, the components due to differences between regressions, either the regressions between different lots or that among lots. It would be strange if I did not agree with this, as I taught him the method while he was at University College, and communicated his paper[2] to the Cambridge Philosophical Society. The secretary, Whyte, must have forgotten that I am a Fellow, for he put the paper in as communicated by himself. Bartlett does not suggest that his procedure is other than mine, though the acknowledgement of its origin (p.333) is a bit ambiguous.

The fact is that Wishart seems to have jumped to the conclusion, some time, I should think, about 1933, that because the sum of squares of deviations of the adjusted means of groups may be derived simply from the sum of squares and products in the covariance table, in the way I set out in my fourth edition, that no allowance for the sampling error of the estimated regression coefficient was necessary in making an exact z test. As an approximation this is harmless enough in experimental data where the material has been properly randomised, for in that case, the independent variate will have been randomised also. So that the differences between the means of the different groups will be small compared with the differences within the groups. Anyhow, in the summer of 1933, several people working with Wishart, Wilks, Bartlett and Cochran, among others, were all at work applying the method wrongly to material in which the approximation was very bad indeed; and it was not until their papers began to come in to me, as referee, in the autumn and winter of that year, that I knew what they had been doing. Indeed, by that time some of them certainly had a suspicion that the method was wrong, though without seeing how to put it right.

In Wilks' case the effect was most unfortunate. He is by nature extremely academic, and his training has not made him familiar with the research aspect of statistical problems. If his time had been better spent in this country, he might really have widened his grasp of what mathematics are good for in this connection. As it was, he got hold of the idea that the whole question of analysis of variance needed a rigorous overhauling, and no one seems to have taken the trouble to refer him to the original papers in which the method was built up. At one time, I understand, he thought he had proved that the Latin square was invalid, but I understand he gave that up when Cochran showed him the simple algebra of the kind which Irwin has from time to time published, by which the unbiassed character of the test is demonstrable merely from the use of randomisation. Finally Wishart persuaded Yule to submit a paper of his to the Royal Society, but owing to his lack of background, it was so pointless that it had to be turned down.[3]

The mistake was evidently partly my fault, for it had happened that neither in my first paper on the analysis of covariance, with Eden, in 1927 (*J.A.S.* 17. 548) [*CP* 57] which Tippett had reproduced in his book,[4] nor in the section which I had added to the fourth edition of my book, had I had occasion to apply the z test, and Tippett had not done so, either. I do think, however, that Wishart would have shown better sense if he had consulted me as soon as he had found there was any difficulty, instead of stolidly teaching the wrong method, at least up to the time of his address to the Statistical Society (Jan. 1934)[5] when I happened to be Chairman. If you look up the *J.R.S.S. Supplement*, Vol.1 p.43, you will see his statement, and his bewildered footnote, added later. I took the next opportunity, at a meeting of the Section, to give the exact test (Vol.1. p. 198),[6] as Pearson and Miss James had produced a wildly erroneous application.

It may interest you that an exactly similar situation arises when correction is made for a missing observation. If you put x for such a missing observation, all the entries for sums of squares will be simple quadratic functions of x, and, if that for error is $(a + 2bx + cx^2)$, then obviously its minimum value is $a - b^2/c$.

The value of x which gives this minimum value is the best estimate that can be made of the missing observations, but if this value is inserted in the analysis, or, as Wishart says (*J.A.S.* Vol. 20 p.406),[7] we 'proceed as usual with the analysis of variance, using the estimated figure for the missing yield,' we shall ignore the fact that the precision of our comparison is lowered whenever an observation is lost. To make an exact z test, it is only necessary to minimise equally the sum of treatments plus error, and to ascribe the difference between this minimum and that for error to the treatment effects. The sum of squares for treatment, obtained in this way, will always be less than that using the best estimated treatment means, for it allows for the greater inaccuracy with which they were estimated. In fact, the particular component of which the error is enhanced has been scaled down until it has the same precision as the others. In Wishart's example, p.405, the sum of squares for treatment should be reduced from 22148 to 20892, if I have done the arithmetic right.

The analysis of covariance is useful for such a variety of purposes that I took a great deal of trouble, the winter before last, to demonstrate its proper use and to prove its exactitude when properly used. Both points needed a good deal of hammering at. I believe the exactitude of the test is no longer disputed, but I think you may be glad to have the facts before you for any future reference you may make to the subject.

[1] Snedecor, G.W. (1935). Analysis of covariance of statistically controlled grades. *J. Am. Statist. Ass.* **30**, 263–8.

[2] Bartlett, M.S. (1934). The vector representation of a sample. *Proc. Camb. Phil. Soc.* **30**, 327–40.

[3] See Fisher's letter of 27 December 1933 to Wilks (p.299).

[4] Tippett, L.H.C. (1931). *The methods of statistics*. Williams and Norgate, London.

[5] Wishart, J. (1934). Statistics in agricultural research. *J. R. Statist. Soc., Suppl.* **1**, 26–51.

[6] See Discussion following Wilsdon, B.H. (1934). Discrimination by specification statistically considered and illustrated by the standard specification for Portland cement. (with Appendices by E.S. Pearson and F.E. James) *J. R. Statist. Soc., Suppl.* **1**, 152–92.

[7] Allan, F.E. and Wishart, J. (1930). A method of estimating the yield of a missing plot in field experimental work. *J. Agric. Sci.* **20**, 399–406.

Fisher to R. Summerby: 25 April 1932

It is a great pleasure to hear from you, and to discuss again some of the problems of experimental design and interpretation. . . .

As to when you should use the *P* tabled in my book, and when the half value, is a point which I might well have discussed more fully.[1] Let me take an example. A man comes and declares that if we would only use a rotary cultivator we should get much better germination with our mangolds. We might try an unreplicated experiment, with single areas under the two treatments, then if the germination really does better we shall think there is something in the assertion, but if not we will think no more about it. In circumstances like these when we come to replication and estimates of error, it is clear that the new method scores a significant success only if it differs from the old in the positive direction; consequently if we decide to use the 5 per cent point we ought in fairness to give him 5 per cent in the right hand side of the curve, i.e. everything beyond $t = +1.645$, if n is large, because he is not going to claim any success if the difference is negative, however large it is.

On the other hand with a pair of varieties the fact that one beats the other in a single trial would not contribute anything to our decision, unless we already had prior evidence that one was better than the other, and are just testing it further. One or other is bound to win; and so in the test of significance we shall count both tails as significant, and for the 5 per cent point allot 2.5 per cent to each, so counting anything outside the range $t = \pm 1.960$.

. . .

[1] Summerby had asked Fisher to elaborate on his discussion of this point in Section 12 (p.45) of *SMRW*.

D.S. Villars to Fisher: 26 September 1938

. . . I have recently been interested in using a new test for homogeneity of a set of variances and I wonder if I could again presume upon your kindness to clear up a point which I am not sure I fully understand. . . . Thus, suppose we have the set of variance estimates s_1^2, s_2^2, . . . s_n^2. Compute $z_1' = \frac{1}{2}\ln s_1^2$, $z_2' = \frac{1}{2}\ln s_2^2$, . . . $z_n' = \frac{1}{2}\ln s_n^2$ (where ln means *natural* logarithm). Next compute the sum of squares of deviations of the z''s from their mean. This sum of squares of deviations must equal the sum of squares of deviations of legitimate z''s which might have been computed by forming ratios of each s_i^2 to any new independent variance estimate, s_{n+1}^2. So, form the ratio of this sum of squares of deviations to the true variance of z which you state in your book (*Statistical Methods*, 6 [edn.], 231)[1] to be

$$\sigma^2 = \tfrac{1}{2}\left\{\frac{1}{f_1} + \frac{1}{f_{n+1}}\right\}$$

(where the f's are the number of degrees of freedom) and look up in the χ^2 table for $n - 1$ degrees of freedom. This test is valid for the case where we have a set of variances of equal numbers of degrees of freedom.

Now what I am not quite certain about is this. What should one use for f_{n+1} in the formula for the variance of z? . . . Also I would like to ask if you see any fundamental objection to this method. . . .

[1] *SMRW*, p. 228.

Fisher to D.S. Villars: 4 October 1938

The method you propose for testing for significant variation among a group of estimated variances is quite a reliable one, which I have sometimes used, though I do not know that it has anywhere been published.

As you know, if σ^2 is the true variance and s^2 an estimate of it based on n degrees of freedom, then ns^2/σ^2 is distributed as is χ^2; consequently the sampling variance of half the natural logarithm of s^2 is the sampling variance of log χ.

This may be expressed exactly as the sum of an infinite series

$$\frac{1}{n^2} + \frac{1}{(n+2)^2} + \frac{1}{(n+4)^2} + \ldots$$

This is a function which has been tabulated under the name of the Trigamma Function (B.A. *Mathematical Tables*, Vol. I), actually what you want being 1/4 trigamma in $(n-2)/2$. For the first few values of n the actual values are as shown below, where I have put the approximate value $1/2(n - 1)$ for comparison:

1	1.2337055	
2	0.4112335	0.5

3	0.2337006	0.25
4	0.1612335	0.167
5	0.1225944	0.125
6	0.0987335	0.1000
7	0.0825944	0.0833
8	0.0709557	0.0714

You will see that, except for very small samples, you would need an enormous number of them for the difference between the exact and approximate variance to matter appreciably.

Knowing the true variance of the natural logarithm of s, which, of course, is half the natural logarithm of s^2, you can, as you suggest, make a very precise test of heterogeneity. Using the method of Section 21.03, Ex. 14, in my book [*SMRW*], making a comparison between the observed and the expected sum of squares of deviations from the mean of the logarithm, you will see that there is no need to introduce the total number of degrees of freedom, as the sum of squares you use is itself an empirical measure of the variance of the quantities among which significant variation is being tested. . . .

H.M. Walker to Fisher: 15 *March* 1940

. . . A problem which arises often in educational research, with which I do not know how to deal, is this. Suppose that the means of two samples have been compared on k different traits, and suppose that p is the probability that on a single trait a given value of t, say t_0, would occur by chance if in the population the two means are equal. Then we should expect pk differences to be, by chance, as large as t. If more than pk differences are of this size, how do we interpret the significance of the several differences? This is not a problem for multivariate analysis. We do not want to know whether the two groups can be reliably differentiated on a group of traits, but we want to know on *which individual* traits they can be differentiated. This search for traits which will differentiate two criterion groups occurs in *many* of our studies. Recently I read a study in which the author studied 2500 items on a group of 700 persons to see which ones were reliably related to a criterion. On how many of these items would he be justified in drawing conclusions? If this has been discussed in any paper to which you can give me a reference, I shall be most grateful, for I can find nothing on the subject.

Fisher to H.M. Walker: 22 *April* 1940

. . . I have rather a dread of the procrustean process of forcing a problem to fit a favourite method; for, quite early in life, I was impressed by the fatal effects of this in Karl Pearson's work; but I am not sure that your problem is not one for multivariate analysis. To begin with, unless the best multivariate discriminant is significant, I imagine no one would wish to go further, or to draw conclusions about individual traits; but, if it is significant, this means that it is significantly different from a kind of null criterion having all its coefficients zero. Now, one can equally test significance from less severely restricted criteria, e.g. from criteria having all coefficients of variables with suffix greater than r [equal to] zero. An insignificant difference here would surely allow one to say that only x_1 to x_r taken jointly are needed, and failure to obtain an insignificant difference when any of these is omitted would indicate that all were necessary. . . .

Fisher to S.S. Wilks: 27 December 1933

My wife and I were very glad to have your Christmas card. I hope you have had the best of good times this winter.

I have been looking through the paper submitted for you to the Royal, on the analysis of variance, and am much puzzled as to why you (and apparently Wishart also, if he suggested the problem) should feel that such an elementary point as that you discuss should need a new and very elaborate discussion.[1]

As regards the proofs I have given, I think from 1923–25, I was content with the partly intuitive treatment of linear forms as restrictions in multiple space, but in 1925 I gave an explicit analytical treatment, very simple in character, which enables you to prove all the cases you deal with in a page or two, and what is more important to perceive exactly what other cases it must apply to. In addition, cases in which Blocks (say) are not homogeneous are readily seen to present no difficulty, such as you suggest in your summary.

As the method seems not to have come to your notice, and as considerable harm is being done by the idea that the analysis of variance contains unproved assumptions, instead of being merely a convenient arithmetical arrangement arrived at some years after the sampling problems had been solved, I may as well outline it now.

If $x_1, \ldots, x_a$ are independent variates each normally distributed about zero with unit variance, then

$$\xi_r = p_{1r}x_1 + p_{2r}x_2 + \ldots + p_{ar}x_a$$

is normally distributed about zero, and its variance is unity if

$$\sum_{\alpha=1}^{a} (p_{\alpha r}^2) = 1. \qquad \text{(A)}$$

Also ξ_r and ξ_s are independent if

$$\sum_{\alpha=1}^{a} (p_{\alpha r} p_{\alpha s}) = 0 \tag{B}$$

If r variates, ξ, have already been constructed, there are r linear homogeneous equations for $p_{\alpha s}$ to satisfy, and this can be done up to $r = a - 1$, but no further. The condition (B) at once excludes the possibility that any set of coefficients can be linear functions of any selection of the other sets. Thus a new variates, ξ, may be constructed, in an infinite variety of ways, such that they are all normally and independently distributed with unit variance.

Note further that the rows of the p-matrix are the coefficients of ξ expressed in terms of x, and the columns are the coefficients of x expressed in terms of ξ. Also that

$$S(x^2) = S(\xi^2).$$

Hence

$$S(x^2) - \overset{r}{\underset{1}{S}}(\xi^2) = \overset{a}{\underset{r+1}{S}}(\xi^2)$$

which is distributed in samples as the sum of the squares of $a - r$ independent variates, wholly independent of each of $\xi_1, \ldots, \xi_r$, or any combination of them.

The validity of the z-distribution for any analysis of variance thus depends only on satisfying the condition (B) of orthogonality. It was the ignoring of this condition which spoilt Wishart's paper[2] in the German journal, where he failed to see that the treatments did not occur in all rows and columns, and in consequence obtained much too low a sum of squares in the residue. This is the only case I know of the analysis being applied wrongly.

For example, take deviations from the mean, let

$$\xi_1 = (x_1 + \ldots + x_a)/\sqrt{a}\,;$$

this satisfies (A), consequently

$$S(x^2) - S^2(x)/a = S(x - \bar{x})^2$$

is distributed as is the sum of $(a - 1)$ squares of independent variates having the same variance. Had Helmert taken this step he would have anticipated 'Student's' result by 32 years. These $(a - 1)$ variates may be specified in a number of ways, such as

$$\begin{aligned}
\xi_2 &= (x_1 - x_2)/\sqrt{2} \\
\xi_3 &= (x_1 + x_2 - 2x_3)/\sqrt{6} \\
\xi_4 &= (x_1 + x_2 + x_3 - 3x_4)/\sqrt{12} \\
\xi_a &= \{x_1 + \ldots + x_{a-1} - (a-1)x_a\}/\sqrt{a(a-1)}
\end{aligned}$$

which are easily seen to be orthogonal.

Now suppose you have *ab* variates, *x*, arranged in *a* columns and *b* rows. From each row you can put down $\xi_p (p = 2, \ldots, a)$, and from the *b* values of ξ_p from the different rows you can similarly calculate $\xi_{pq} (q = 2, \ldots, b)$, giving $(a - 1)(b - 1)$ independent variates, all mutually orthogonal, and all independent of the sums of the rows and columns. There is, of course, no need to specify the $(a - 1)(b - 1)$ particular linear functions in this way; all that is needed is the general proof that it can always be done; and that the sum of their squares may be obtained by subtraction from the total.

For a 4 × 4 Latin square

A	B	C	D
B	D	A	C
C	A	D	B
D	C	B	A

you can take coefficients

	1 1 1 1	1 1 1 1	1 1 1 1	1 1 1 1	Grand Total
	1 1 1 1	−1 −1 −1 −1	0 0 0 0	0 0 0 0	
	1 1 1 1	1 1 1 1	−2 −2 −2 −2	0 0 0 0	Rows 3
	1 1 1 1	1 1 1 1	1 1 1 1	−3 −3 −3 −3	
	1 −1 0 0	1 −1 0 0	1 −1 0 0	1 −1 0 0	
	1 1 −2 0	1 1 −2 0	1 1 −2 0	1 1 −2 0	Columns 3
	1 1 1 −3	1 1 1 −3	1 1 1 −3	1 1 1 −3	
A−B	1 −1 0 0	−1 0 1 0	0 1 0 −1	0 0 −1 1	
A+B−2C	1 1 −2 0	1 0 1 −2	−2 1 0 1	0 −2 1 1	Treat-
A+B+C−3D	1 1 1 −3	1 −3 1 1	1 1 −3 1	−3 1 1 1	ments 3

All necessarily orthogonal if each treatment comes once in each row and once in each column. Whence it follows that the remaining 6 degrees of freedom are distributed as required independently of these 9.

It is possible that your attention has not been called to the 1925 paper, 'Applications of 'Student's' distribution' [*CP* 43], where this method is set out with especial reference to its application to regression, to which all the other applications can be reduced as special cases.

E.g. If *u* is a variate having values 1 in the first column, −1 in the second, and zero in the others, *v* has 1 in the first two, and −2 in the third, *w* has 1 in the first three and −3 in the fourth, then *u*, *v* and *w* are uncorrelated independent variates, and their elimination jointly is the same as the elimination of variation between columns. Since all that is assumed in the general case is normality and homogeneity of deviations from the regression surface, the differences among the columns may be as great as you please without affecting the validity of the comparisons among the other components. The limitation of the proof of the validity of the test of significance of treatments to the case when rows and columns have no significant effect is quite unnecessary. The general case is established by the same argument as the special case you have chosen for discussion.

If you find on consideration that you agree that the whole idea that complete and adequate proofs have not been given of the validity of the *z*-test in the analysis of variance is a mare's nest, and that cases to which such analysis can be applied are already satisfactorily defined, the question arises as to whether your paper can be amended so as to avoid the criticisms to which in its present form it is exposed. Probably you will think it wisest to withdraw it for the present, on the ground that it was written under a misapprehension as to what had already been demonstrated (for which you can in no way be held to blame).[3]

The ground would then be clear for any new paper you cared to publish on the characteristic functions of the distribution of the quadratic forms occurring in the analysis.

[1] Wilks's paper was entitled, 'On the independence of sums of squares in the analysis of variance'.
[2] Wishart, J. (1931). The analysis of variance illustrated in its application to a complex agricultural experiment in sugar beet. *Arch. f. Pflanz.* **5**, 561–84.
[3] See also Fisher's letter of 25 April 1935 to Snedecor (p. 294).

Fisher to S.S. Wilks: 6 *February* 1934

Thanks for your letter of January the 20th, which has just reached me.[1] It has done much to clear up your attitude, and I see, at least, that though you say you were aware of the existence of the 1925 *Metron* paper, you have evidently, even now, not grasped the simplicity and generality of the proof there given, of the distributional problem needed for tests of significance in the analysis of variance. I say this (i) because you speak of it as containing an analytical method which *could* be applied, followed with 'but in as much as it had not been applied in the literature', as if something more needed to be done. Now, of course, a method of proof constitutes the proof in all cases to which the method is shown to be applicable, just as Euclid's fifth proposition is a method of proof shown by him to be applicable to any given isosceles triangle, and thereby constitutes a proof of this property for all such triangles. The class of cases to which I showed the method to be applicable was that of equivariant variation about any regression line or surface. And this includes, as no doubt you will readily appreciate, the experimental treatments which you have discussed.

In his paper of, I think, 1931, Irwin expressly disclaimed any novelty, in the result to be proved, and referred the reader to my paper of 1925 for a complete discussion.[2] In your letter you say 'The paper is definitely one of technique and does not purport to be a presentation of new discoveries'. I think, however, that every reader would judge that it did purport to supply proofs on points on which no adequate proof had yet been given. My second reason for thinking that you have not apprehended the process of proof given in 1925, lies in your discussion on page 2 of your letter, where, after 'it is necessary, as I understand it to show' you discuss a procedure of most superfluous complexity.[3]

The example I had given in my letter of the natural method of applying the general process to any particular case was a Latin square, which I chose because that arrangement was dealt with in your paper as though it were one involving particular difficulty. Your comment 'A general proof, of course, would be one that holds for any Latin square whatever, without having to consider any particular one' seems to overlook the fact that what I illustrated was a method of writing down the coefficients of the linear components, applicable to any Latin square whatever. I cannot suppose that you want me to think that you do not perceive that it is so applicable, since its applicability rests merely on the mutual orthogonality of all differences among rows, all differences among columns, and all differences among treatments. If you were in doubt, perhaps the best method of satisfying yourself as to its general applicability would be to try and set up a Latin square to which it could not be immediately applied. This should lead you at least to perceive very clearly the simple reasons that make it generally applicable. You will perceive, too, that, apart from the arbitrary subdivision of the sets of 3 degrees of freedom into unitary parts, the sets of components are those picked out in the ordinary arithmetical procedure of the analysis, as is quite generally the case in the analysis of other experimental arrangements. That you have not grasped the point of the method, perhaps because it is too easy for you to start thinking about it, is shown by your statement 'The case of proving the mutual independence of sums of squares of deviations from means of rows, columns and treatments and residuals . . . is a little more difficult.' Perhaps you have not noticed that the sum of the squares of the three components assigned to rows *is* the sum of squares of deviations of rows from their mean, and that its independence of other sums of squares is assured by the mutual orthogonality of the linear components. With regard to residuals the simplest general treatment is that given in *Metron*.

I had hoped that on reading my previous letter you would have perceived that, in view of your unawareness of the very direct and complete proof already available, your paper, submitted to the Royal Society, is not one that would add to your reputation as a mathematician, however interesting you yourself may have found it, to develop the characteristic functions of the quadratic forms involved. You refer to a number of novel points, which come up as by-products, but you do not say what these are. Any such should, I suppose, receive publication in a short paper, in an appropriate technical journal. The Royal Society confines itself, as far as possible, to papers that are not only good of their kind, but also of more than specialist interest, and I should judge that I should be doing both the Society and yourself an ill service if I were to recommend your paper, as it now stands, for publication.[4]

If you are in any doubt it might be wise to ask the opinion of Professor Hotelling or some other American friend of standing. I need hardly say that I shall always be willing to assist you, or any other mathematician of merit, to gain adequate recognition for your contributions to knowledge. In the present case I judge that the whole work was undertaken under a most regrettable misapprehension, and that you may need some little time to familiarise yourself with methods of reasoning and of demonstration other than that in which you have so far specialised.

[1] Wilks asked Fisher to reconsider his paper, the purpose of which was, he said, to give by means of characteristic functions the proofs of independence of the sums of squares entering into the analysis of variance.

[2] In his letter to Fisher, Wilks had referred to Irwin's paper (Irwin, J.O. (1931). Mathematical theorems involved in the analysis of variance. *J.R. Statist. Soc.* **94**, 285–300) saying that it 'did not go into the question of independence — thus omitting a very important part of the theory'. Cf. Irwin, J.O. (1934). On the independence of the constituent items in the analysis of variance. *J. R. Statist. Soc., Suppl.* **1**, 236–51.

[3] The complete sentence in Wilks's letter is as follows:
'In order to carry out an analytical proof of the independence of a number of sums of squares by the method you give, it is necessary, as I understand it, to show that you can reduce each sum of squares to a sum of independently distributed squares diminished in number by the number of linear restrictions, and further, that the squares entering into each sum are independent of those entering any other.'

[4] Fisher's report on Wilks's paper, which he sent to the Royal Society on 26 March 1934, concludes as follows:
'I cannot help feeling, of the paper as a whole, that its aim has been so much misdirected from the first that the author would, in a few years' time, be glad that it should not be now published, since, as far as it goes, the proofs it provides amount to no more than that certain widely used experimental arrangements possess the properties that they were originally designed to possess, and which have never been, save by temporary and accidental confusion, in the least doubt. However, if other Fellows of the Society greatly wish for its publication, in spite of this opinion, I would raise no further objection, provided it is immediately followed by a demonstration that the cases here considered fall into the general class of simple regressions, and are demonstrable individually by a very simple application of the general proof.'
Wilks's paper was not, in fact, published by the Royal Society.

Fisher to E.B. Wilson: 25 *January* 1940

I am glad to have your nice long letter of January 3rd, which I have only just received. It is really a long while since I heard from you. I have heard from Fréchet, and hope to see him when he visits England, as I understand he has some such mission in view. So far, I am afraid, the Galton Laboratory has not contributed towards winning the war. We have been reluctantly evicted from our former premises, on the ground that the College was 'closed, and no infringement of its state of evacuation could be allowed'. Consequently our former habitation stands empty. The Serological section of my Department was taken over by the Medical Research Council, who, thank Goodness, have had the sense to maintain them in existence as a unit, and, after a struggle, the rest of us are permitted to continue our researches in rooms which Rothamsted has been good enough to lend us.

I have been doing nothing more useful than tabulating blood group frequencies, which show a very pleasing gradient from the North to the South of Great Britain. However, the *Annals of Eugenics* is not at present threatened, and we have had so far a good supply of first class material. . . .

I followed Shewhart's earlier work on Quality Control with great interest, but have not seen much of it later, and so do not know whether it has at all developed. It seemed to me that what he had to say he said very well. The best sense, I think, that can be made of Frisch's notions[1] was made by a Dutchman, Koopmans, in his book called *Linear Regression Analysis of Economic Time Series*, which might interest you if you have not seen it.

One of the central confusions in this subject seems to be that regression coefficients are thought of, or used, by economists in two discrepant senses: 1) as coefficients in abstract economic laws connecting a number of variables either exactly or with random errors superimposed, and 2) as the best coefficients to use for predicting one quantity not to be observed from others of which we have the values. The second which, I fancy, is the more frequent, in practical use, is straightforward regression work without modification. Even if we knew the values in the first type of problem, we should prefer to use those of the second type for its own purpose of making predictions. However, when the first type is wanted, or is under discussion, then there is no avoiding the disturbance caused by errors of measurement of the independent variates, which seem to be inevitably indeterminate in economics, and those not indeterminate, never actually determined in experimental physics.

[1] Wilson had asked if Fisher had made much of Ragnar Frisch's monograph on Confluence Analysis.

Fisher to E.C. Wood: 9 December 1947

Thanks for you letter.[1] For many years before the table was computed I had come to the conclusion that the only proper treatment of ranked data, looking at the matter from the point of view of estimation theory and on the assumption that the ranked series was based on a sample from a bivariate normal distribution, lay in using the mean deviate corresponding with each observed rank. The immense simplification in the scoring of ties was an accidental by-product. I had studied the integrals obtained and in outline the appropriate method of numerical computing, having calculated some of the simpler values.

It was the stimulus of getting out a definitive table collection with Yates which led me to get the table completed, partly with the help of Stevens at the Galton Laboratory. . . .

[1] Wood had asked about the origin of Table XX in *Statistical tables* for converting ranked data into normally distributed scores.

Fisher to A. Zeller: 2 February 1957

Thanks for your letter of 1st February. . . . With respect to your question on the analysis of variance,[1] I have from time to time over the last thirty years expressed the opinion, arising out of the earlier practice of agronomists, that to calculate a standard error for an individual variety, based on say three degrees of freedom, is less accurate, even for that variety, than to use thirty degrees of freedom found by pooling ten different varieties, even though these may not have accurately exactly the same variability. In such cases the judgement may be confident that the true variability between varieties is rather trifling compared with the sampling errors of their determination.

There are, of course, numerous cases in experimentation in which degrees of freedom for high order interactions may reasonably be judged on much the same grounds to be effectively pure error. I think it rather a dangerous tendency among some mathematical writers on the design of experiments to carry this assumption so far as to dispense with replication altogether, or to use so-called fractional replication, except in cases where the factors in question have been rather fully explored in previous experiments. In fact I always advise, personally, that it is well worth while in a design to have a number of degrees of freedom providing *nothing but confirmation* of the general ideas on which an experiment is planned. This, I think, is important in real research, but of course much less important in routine technological determinations, though even there, if I were directing such work, I should like to insert checks on the technological perfection of the processes actually employed. . . .

[1] Zeller asked Fisher to say to what extent one may pool with the error sum of squares other sums of squares (e.g. for blocks or treatments) having non-significant variance ratios.

3
History of Statistics

Fisher to W.R. Buckland: 20 *July* 1954

. . . You will know that many of these ideas[1] and the definitions of the corresponding technical terms originated in my work which, as these ideas developed, contains sufficiently clear statements as to the meanings of the words introduced, e.g. I never used the word efficiency of estimation in reference to the mean square deviation of estimates from finite samples, still less did I ever introduce this practice which had been indeed latent, though not explicit, in all books on the theory of errors since the time of Gauss. My definition depends on amount of information and is aimed at affording an invariant comparison for transforms of the parameter such as a mean square deviation cannot give.

[1] Fisher had agreed to act as a consultant on terms relating to estimation in the preparation of M.G. Kendall and W.R. Buckland's *Dictionary of statistical terms* and then found himself in difficulties over the definitions presented to him. The *Dictionary* was published in 1957 by Oliver and Boyd, Edinburgh.

Fisher to H.D. Dufrénoy: 12 *May* 1936

Allow me to congratulate you on your fine paper, 'Méthodes statistiques appliquées à la pathologie végétale', which will, I am sure, be most valuable to all engaged in agronomic and phyto-pathological work.

Perhaps you will permit me to make clear a certain confusion, originating in America, as to the work developed in this country, which has, I see, influenced your mode of expression in some places.

The mathematical problem solved by 'Student' in 1908 constituted a most important refinement of the theory of errors, the possibility of which had been previously overlooked by mathematical writers on this subject. He supplied for the first time a rigorously exact test of significance for the mean of a finite sample drawn from a normally distributed population.

Perceiving the importance of what he had done, and being myself interested in the mathematical problems presented by the exact treatment of finite samples, I later extended his procedure to cover such experimentally important cases as a comparison of the means of two samples, the test of significance of a regression coefficient, and for the difference between the means of two regression coefficients, all of which problems may be reduced to the same distribution as that found by 'Student'.

Further, in a number of other problems, such as the comparison of the means of more than two samples, the significance of an interclass correlation, that of a multiple correlation, and the test of significance of deviations from regression formulae of any assigned form, I found that the principles of 'Student's' approach might be applied, but in these cases gave rise to a more general distribution, the variate of which I have denoted by z.

In respect to notation, 'Student' at first expressed his distribution in terms of the symbol Z, standing for the ratio of the mean value of the sample to the standard deviation as estimated by the followers of Professor Pearson by dividing the sum of the squares of deviations from the mean by the number of observations. Later he appreciated the advantage of using the number of degrees of freedom as a divisor in estimating for variance, and at my suggestion adopted the further modification of comparing the mean (or its deviation from its expected value) with its own standard error, rather than with that of a single observation; the joint effect of these modifications being to use the quantity t connected with Z by the relation $t^2 = nZ^2$ where n is the number of degrees of freedom.

By doing this he overcame the difficulty that for increasing values of n the value at which Z becomes significant is diminished without limit, whereas that at which t becomes significant tends to a finite limit determined by the normal distribution. In 1925 'Student' and I published a series of four short papers in *Metron*, giving the new Tables which 'Student' had calculated, with an explanation of their applications, method of calculation, notation, etc.. I did not decide to use the symbol z for my more general distribution until I was assured that 'Student' was convinced of the desirability of changing his notation and the form of presentation of his Tables. Consequently the use of Z for 'Student's' distribution is an occasion of unnecessary confusion to workers in this subject.

With respect to the 'method of pairing', since in 1908 'Student' had solved only the problem of the significance of the mean of the unique sample, the only practical applications to experimentation which he could exhibit were those in which pairing had been used. He has himself expressly disclaimed the invention of the method of pairing, 'which', he says, 'must be as old as Noah', and to which the improvement he has made in the *theory of errors* is not now confined in this application. His protest was, of course, aroused by the conviction of Professor Love that pairing was an essential part of 'Student's' method.

Pray excuse this long digression on a point which you may think unimportant, and which is certainly principally of academic and historical interest.

Fisher to C. Eisenhart: 28 *October* 1938

. . . I think your idea of a book on 'Readings in Statistical Methods' on the lines of a source book is really a brilliant one. I think mathematics has suffered more than most subjects from the professional teacher acting as rédacteur and with the best intentions — usually the elimination of difficulties and complications — often presenting the ideas of his originals very inadequately. This is especially so when, as is too often the case, he is himself only familiar with their work at third hand.

Naturally, the selection of such an anthology will be greatly a matter of judgment, and what is worth choosing will depend greatly on the interests of the supposed readers. The field chosen might be very wide, or considerably more narrow, and this is a point which it will be necessary to think about carefully, I should judge, at all stages during the growth of the collection.

As regards my own work, I have tried to look at it from the point of view of such a compiler, and I send you a bibliography in which the various strains of ideas have been followed by classifying the different papers.[1] My early work in mathematical statistics was principally concerned with giving the exact solution of a number of problems of distribution which, until then, had been almost totally neglected. These lead to the exact tests of significance. Initially these problems all appear under different disguises, although later the whole system could be telescoped, and can now be taught quite briefly in a course on the *Analysis of Variance*. I have marked the whole series of which this is the central idea with an α, putting it in brackets where the new solutions were given, incidentally, as problems arose in special lines of work. The α series really culminates in 1928–9 with such problems as the *general sampling distribution of the multiple correlation coefficient*, the *tests of significance in harmonic analysis*, and the general formulae for *moments and product moments of sampling distributions*.

Closely associated with α was the interest in the more abstract and philosophical problem of estimation. I have marked these papers β. The practical results of these two lines of research have been best made known in *Statistical Methods*. β leads naturally to γ, a series of papers on inductive inference dealing with inverse probability, likelihood and fiducial probability. It is easy to see what totally different emphasis these series deserve, according to the interests of the reader. The group of applications (δ) concerned with heredity, evolution and eugenics commences almost as early, and inevitably overlaps on theoretical points; thus the 1918 paper [*CP* 9] has the first hint of the analysis of variance, which was not fully developed until the 1923 paper on crop variation [*CP* 32]. The 1919 paper [*CP* 11] also contains a pretty solution of the distribution problem. The later papers on human genetics are also very largely methodological. I have marked ϵ papers much concerned with the interpretation of experimental data, and these lead naturally to η, concerned primarily with the design of experiments, always in conjunction with their interpretation.

I am afraid that numerical illustrations of statistical methods not previously used occur practically throughout the series, so that a classification for which this feature was particularly important would cut across the one I have attempted. I have also marked ζ for the beginning of a series on the interpretation of multiple measurements, which, from the point of view of tests of significance, fits into the analysis of variance, but involves ideas with very different applications. . . .

[1] The following list showing Fisher's classification of his publications was kindly provided by Dr. Eisenhart.

α: *CP* 4, (11), 12, 14, 19, 20, (22), 30, 31, (32), 34, 35, 36, (38), 43, 44, (46), 49, 50, 56, 61, 62, 63, 74, 75, 83, 90, 91, 123, 141; *SMRW*.
β: *CP* 1, 12, 18, 42, 62, 71, 108; *SMRW*.
γ: *CP* 84, 95, 102, 109, 124, 125, 137.
δ: *CP* 9, 11, 15, 24, 25, 47, 55, 59, 65, 68, 69, 70, 71, 72, 81, 86, 87, 93, 96, 105, 106, 113, 114, 115, 116, 118, 119, 130, 131, 132, 133, 135, 141, 142, 143, 144, 153; *GTNS*.
ε: *CP* 16, 21, 22, 32, 37, 39, 51, 58, 64, 77, 104.
η: *CP* 48, 57, 78, 79, 85, 113, 114, 115, 131, 132, 139.
ζ: *CP* 138, 141.

A. Fisher to Fisher: 5 May 1931

From an announcement by the Agricultural School of Iowa, I notice that you are to give a series of lectures in Iowa City during the coming summer. Someone also told me you expect to arrive in New York at the beginning of June. When you do come, be sure to look me up, so that we can have a chat together. Both of us are — if I so may say — somewhat unorthodox fellows, and although we may not agree on some questions, nevertheless we ought to have a jolly time in talking about other things.

Moreover, another important question should be given some attention. For unless I am totally mistaken, you are not exactly what might be described as a teetotaller. The strain of being continually pure may therefore eventually pall upon you, and you probably will have to unbend as it were so as to bring about the proper reactions in the way of slight dissipations. Permit me therefore to be a guide to you in the exploration of various places of sin such as our American 'speakeasies', which, by the way, offer a most excellent, nay almost perfect, sample of a typical cross-section of American society. At any rate, be sure to look me up.

Now for an entirely different matter. I know perfectly well that you — as what one might call a civilized person — do not take yourself too seriously. But this is not always the case among your adoring disciples, who, in their effort to expand their own egos at the expense of their environment, are apt to overdo a good thing. Take our friend Hotelling, for instance. According to him the adherents of the classical theory of probability must have sunk into a state of ease, almost approaching apathetic luxury, so that the time has become ripe for the Messiah, and lo, according to Hotelling, you are the saviour to lead the statisticians out of the wilderness. As pointed out by Norman Douglas, renunciation and hope, these spiritually correlated Siamese twins, have always exercised an irresistible charm to polite society. It makes us all feel so comfortable to be told on the one hand that we are going to the devil and on the other hand that the coming Messiah is in the near offing. But what do you think of this new role which has been assigned to you?

Some of your younger disciples, Dr. Shewhart for instance, are like the early Christians imbued with an almost religious fervor. But youth oftentimes speaks from the heart rather than from the brain. . . .

As to the whole theory of small samples, I am myself rather an old hand at the game. In the first decennium of the present century, when I was not yet out of my teens, I became interested in the problem of small samples in connection with the yeast physiology of sparkling wines at a time when some of my near relatives were engaged in the wine industry and in the export of high-grade wines from California. I originally consulted some of my friends . . . in Copenhagen . . . and they told me that the problem, apart from its purely physiological and bacteriological aspects, was mainly one involving mathematical statistics and advised me to consult the late Mr. Gram. I then found that considerable work and that practically a finished theory had been developed both by Gram and by Professor Thiele, some of which, although not nearly all of it, had been described in Thiele's Danish 1889 text on *Almindelig Iagttagelseslaere*.

Bearing in mind these early facts, I am therefore rather surprised, in reading an article by your assistant, Dr. Wishart, in the July 1930 *Biometrika*, of which you were good enough to send me a reprint. Wishart refers to Thiele's work as a paper. This is wrong. The book is a complete and self-sufficient text. Its full title, translated in English, is: *Lectures on the General Doctrine of Observations*. Moreover, Wishart's mention of the symmetric functions by Cayley, and other writers, and that 'later workers were probably unaware that the tables of symmetric functions had already been published in works on pure mathematics', seems to me to be a statement of such radiant improbability that little credence can be placed thereon. For the facts surely must have been known to English readers since 1903, because the tables by Cayley and by the German, Fiedler, as well as Thiele's own tables of 1889 were specifically mentioned in the English text on *Theory of Observations* by Thiele, published in 1903. . . .

There are a lot of other things in your colleague Wishart's paper, which, at least from a historical standpoint, is open to criticism, and we may talk it over when you arrive in this country. Perhaps you may also be able to elucidate me on what your famous 'maximum of likelihood' really implies. . . .

But why go further? The quarrel over inverse probability and over Bayes' Rule is bound to go on between the mathematicians. To me it seems that we take them altogether too seriously. . . .

I am more inclined to quarrel with you over the introduction by you in statistical method of some outlandish and barbarous technical terms. They stand out like quills upon the porcupine, ready to impale the sceptical critic.

Where, for instance, did you get that atrocity, *a statistic*? . . .

Why is it that your disciples always are so serious? Whenever I see them ascend the rostrum to expound — *ex cathedra* as it were — their pet dogmas in a solemn and gravelike manner, I cannot help thinking that they take themselves altogether too seriously and that humour could be used far more advantageously for administering a wholesome dose of truth.

But be this as it may, however. In spite of my scepticism towards some of your work, I cannot help having a sneaking fondness for R.A. Fisher. He is so refreshingly unorthodox, a cooling breeze as it were in the sultry air of our academic circles. Therefore by all means look me up whenever you happen to be in New York. We need a breeze.

Fisher to A. Fisher: 15 *May* 1931

Many thanks for your long, grave, frivolous, breezy and generally Rabelaisian letter. I must certainly try and hunt you up as I pass through New York, probably on Monday, June 1st; but I shall be in the devil of a hustle and dry as a cinder, for, though I have no doubts about Copenhagen, even your eloquence has not convinced me that New York is the place for a binge.

You ask me a lot of fine enigmatical questions, of which I have found one that I can half answer. I do not know in the least how seriously I take myself, and not being an alienist I never met anyone who did, but I certainly don't take myself seriously enough to feel any responsibility for justifying, explaining, or excusing anything written by anybody else. Stuff from my own laboratory, the publications of which I officially 'approve', is passed out on the comprehensive ground that I approve of anyone who has thought about a subject expressing his own opinion over his own name. This probably needs explaining, as a good many continental institutions and faculties take a very different view of the matter, and even Denmark may have been infected by this characteristically Alpine view, prevalent in Italy, France and Germany, of intellectual organisation. I think my attitude is fairly general in England, though Pearson's laboratory is an extreme exception, and, even if I had not believed it to be a salutary national tradition, I should have adopted it in view of the 'awful warning' which his laboratory provides.

You must have observed for longer than I have, one remarkably interesting characteristic of your adopted country, I mean its susceptibility to the influence of vogue. Pondering upon it, I have often thought that it brought to American Science quite as many advantages as disadvantages. It is a real advantage that new ideas should sometimes be tried out, quickly and eagerly, up to the limits of their usefulness, even if this involves many blunders, inflated reputations, and misapplications of research funds. In my own country a man's reputation is normally at its highest just 20 years after his death, at which date his copyrights expire, but we are sometimes so impatient, in the case of the very long lived, as to applaud their opinions when, apart from the effects of senility, they are about 50 years out of date. In fact, I have felt that, if an eye were kept on its weaknesses, the reign of Vogue was rather beneficial. It was not until I saw Mr. Grove's review of my book on Statistical Methods in the *American Mathematical Monthly*[1] that the humourous and other possibilities of *rival vogues* occurred to me, and I wondered if you, of whom, among others, Mr. Grove evidently regarded himself as the spokesman, were prepared to stage Danish and English statistics, like rival mannequins, jealous for the privilege of setting the fashion. For myself I found the scene rather disgusting, though admittedly extremely funny.

Personally, I am rather annoyed; partly because five to ten years ago I had quite an uphill job to free English opinion from the authoritarian opinion that all enlightenment was to be found in the immaculate gospel of Pearson, and I see no need or profit in undertaking the same thankless task for Thiele; partly because I have always been a particular admirer of the Danish work; and yet it looks as though, for the next five years or so, I should have to add to every explanation of a modern result some sort of codicil as — 'this is how Thiele blundered over this' — 'This is how Gram handled that' — 'This is how Arne Fisher or Frisch or Steffenson still approach the matter'. Intellectually it will be easy enough. 40 years do not pass without some progress, and progress in the last 10 has been particularly rapid. A result that is worth publishing is an improvement on the methods of *all* previous writers, and the more active they are the more surely will they be found fumbling at it ineffectually. But that is not the way I should prefer to treat either the great writers of the past or the ablest contemporary workers. I should prefer to write for their appreciation, as I should talk over a dinner table.

On the whole, then, I would rather give my beard a Messianic cut, if this would encourage Shewhart and Hotelling to go ahead and do their job in a workmanlike manner. In any case you will not expect me to be too lenient with the doctrine of the infallibility of the Holy Writ of the nineteenth century, whether in Danish or in English.

P.S. I had forgotten your point about vocabulary. I use special words as the best way of expressing special meanings. Thiele and Pearson were quite content to use the same words for what they were estimating and for their estimates of it. Hence the chaos in which they left the problem of estimation. Those of us who wish to distinguish the two ideas prefer to use different words, hence 'parameter' and 'statistic'. No one who does not feel this need is under any obligation to use them. Also, to Hell with pedantry!

I see I have coupled Pearson's name with Thiele's too often. I do *not* mean that his work is rotten with mathematical errors, only that he had no more glimmer than Pearson of some of the ideas we now use.

[1] Grove, C.C. (1930). *Am. Math. Month.* **37**, 547–50.

Fisher to R. Frisch: 6 March 1931

I am much obliged for the copy you send me of your letter to Prof. Hotelling on the question of the terminology you advocate for the coefficients of the logarithm of the characteristic function. It is perhaps a pity that Laplace did not once and for all give a name to these coefficients, as well as to the function in the expansion of which they occur, which function is of course analytically of much the greater importance. Perhaps we may take it that he left it to the good sense of subsequent mathematicians to say explicitly what they meant, without quarrelling about points of terminology.

The concurrent use of alternative terms for the same objects does not in practice give rise to any inconvenience, and has the great advantage that each person may suit his own convenience; by this process moreover, the more generally useful term, in respect of brevity, of its descriptive quality and of the facility with which it associates itself with a wider terminological system, will in time inevitably come into the more general use.

This, of course, is what we should all desire, in a philosophical spirit, even if our own special associations have given us a strong preference for one particular usage; and it is perhaps all to the good, that willy-nilly this is the process which slowly perhaps, but inevitably, determines the use of words in the English language.

Apart from its associations, which perhaps ought to weigh with a Scandinavian, I have, personally, always felt a certain awkwardness, both logical and linguistic, about the use of the term half-invariants and semi-invariants, and should be glad to see them replaced by more suitable terms, which should of course include a term for the function in the expansion of which they appear. Mathematically these coefficients are not in any sense invariant; and it is not very clear by what process of calculation the fraction *half* is arrived at! Linguistically, 'half-invariant', the use of which by Thiele gives it the stronger claim to consideration of the two, is what is called a bastard word, and for this reason will not be readily admitted into good usage. The term 'semi-invariant' is an attempt to repair this defect, but is scarcely so euphonious as to make up for its other deficiencies. There is an added inconvenience if one is led to speak of the semi-invariantive function, which would seem to be the proper cognate term.

You do not develop in your letter any objections to the terms cumulant and cumulative function, though one is led to feel that you have some strong objections which you do not mention. I must conclude, therefore, that your objection is based solely upon the undesirability of making a change in a usage with which you are already familiar. I should advise that whenever such a change of usage seems inconvenient, it should not be made, and then the newer terms will be used only by those who feel that they have real advantages. You will not, however, and I am glad to see that your letter does not, question the right of mathematicians to make use of whatever terms they find most suitable for the expression of their meaning.

R. Frisch to Fisher: 7 April 1931

Thank you very much for your letter of March 6. Of course, I do not question the right of mathematicians to make use of whatever terms they find most suitable for the expression of their meanings. I have only expressed my personal opinion in the matter of the term 'semi-invariants'. I believe that no pressing need exists for introducing another term than this and that the mere fact that the term 'semi-invariants' has been used for a long time is an asset with this term which weighs very heavily.

As to the term 'cumulants', the term is not self-explanatory and I was not present at the meeting when Hotelling explained it at Cleveland, so I probably do not understand quite what it is meant to cognate. Does it refer to some properties of semi-invariants of a *function* of observations? Or does it refer to the fact that these parameters are defined by integrals, that is, by 'cumulation'? In the latter case it seems that the term cumulation is just as proper in connection with the original moments as with anything else.

Whatever the situation may be in this respect, I must admit that I have still some difficulty in seeing the necessity of a change in the present case.

I have recently received a letter from Dr. Craig who is of the same opinion as me in this matter.

Fisher to R. Frisch: 17 April 1931

I wonder if I shall have an opportunity of discussing these questions of terminology or what will be more interesting, the subject matter of them, when later in the Summer I shall be at Minneapolis. I had, of course, fully understood from your previous letter that you would prefer to retain Thiele's term, and I had hoped to make quite clear that I saw no reason why you should not.

The cumulative property has nothing to do with the use of integrals in the definition of the population moments and thence of any parameters functionally related to the moments. It is the property, of which both Laplace and Gauss fully saw the importance, in studying the cumulative effects of independent causes, namely that certain parameters, such as the mean, the variance, etc., of the resultant distribution were simple sums of the corresponding parameters of the component distributions. It is this property of what I call the cumulants that has attracted the attention of mathematicians from the first. In the higher theory, of course, the importance of the property is seen to lie in its application to the cumulative function, the logarithm of the characteristic function in Levy's sense, since this is not always to be expanded in a power series.

For my own part I am perfectly content to leave it, as in the end it must in any case be left, to the convenience of teachers and users of mathematics to decide what terms they shall apply. There is no need, therefore, for me to reiterate the reasons for my own preference.

Fisher to R. Frisch: 19 *March* 1934

I have looked through Arne Fisher's paper and, as you asked me, am sending you my comments. The paper seems to have two objects, which might perhaps with advantage be kept distinct; one is to establish the merit and priority of certain work by Thiele and subsequent work as influenced by him, especially in recent times Bertelsen and Arne Fisher, and the other is to deplore the influence in contemporary America of some recent English work with which the critic is much less familiar. The confusion of the two subjects weakens the paper through the author's anxiety to say at the same time (i) The formulae of which modern writers are so proud are to be found identically in Thiele's writings and (ii) The modern work is erroneous in comparison to Thiele's. The confusion is also, I suggest, generally undesirable in making it difficult to refer with ordinary generosity to a distinguished earlier writer from whom one may occasionally differ, if in such references one is forced to dissociate himself from the kind of well-meaning but indefensible claims which a modern writer like Arne Fisher has put forward on his behalf.

For example, it seems proper to give Thiele credit for first giving the formulae corresponding to the partitions (2^2), (2^3), (3^2) and (4^2) but rather childish to make a song about the formulae for 1^r which must have been familiar at least since Laplace introduced the characteristic function, or about the mean values of $\mu_2, \ldots, \mu_6$, i.e. of the primary series of statistics with which he chose to work. But the set of results given by Thiele can scarcely be claimed as giving the analytical expression in general for partitions of the type r^i, still less for the more general partitions in which the parts are unequal. When I wrote my paper on the solution of the general problem by partitional methods I was surprised in looking through the works of Tchouproff and Church to find how few seemed to have been given by previous writers, and, had I known of it, should have been glad to cite Thiele amongst the others as the earliest writer who had realised that exact formulae of this kind could be derived by direct algebra.

Again I think it is now clear that Thiele's system of presumptive values is not the same as my series of k-statistics and the implication that it is the same does no credit to Thiele, for his intention was clearly different from mine. His principle in making presumptive values was to choose those values of the parameter for which the observed values of his h statistics would be equal to their expectations and this is not the same as the definition of a presumptive value given as Tchouproff's on p. 6. I imagine that Thiele knew what he was about and did what he wanted to do whereas Bertelsen seems to have done as I did and chosen statistics having K as their mean values. Why Bertelsen did so is not explained, and should be of interest, but the implication seems to be that he was aiming at an improved method of estimation, and in consequence fell into the grievous error, of which I alone am accused in the numerical example in the attached sheet. Bertelsen's aim must evidently have been different from mine since he was a good boy to do in 1927 what was very naughty of me in 1929.

As far as I am concerned anyone who troubled to read my papers would know that I was not concerned with estimation at all, first because, in the paper on the theory of estimation [*CP* 42], I am particular to dismiss the question of bias at an early stage as of no interest to the theory and in the paper on moments of moments [*CP* 74] I go out of my way to point out that moment functions only provide statistical estimates of high efficiency for a special type of distribution (p.200) with a reference to the paper in which I first demonstrated their great inefficiency for all of the Pearsonian types of curve which depart at all widely from the normal [*CP* 18]. My interest in the formulae arose solely from the facts that they can be made exact and general for all distributions, that very few of them had been previously given, and these in very complicated forms, and that no general notions seem to have been developed as to their classification, method of derivation in the more complicated cases, or extension to multivariate distributions. In fact what I had done was just what Craig in a paper published nearly contemporaneously suggested needed to be done if the formulae were to be brought within manageable compass.

The quotation given by Arne Fisher in translation from Thiele's paper 1889, bottom of p. 5, is of interest as showing that Thiele also was only partly concerned with estimation and that he was quite aware that in the current state of the subject he was unable to give any satisfactory account of the principles on which methods of estimation should be based. It seems to me much more to the credit of a man of science that he should be aware of the limitations of his own knowledge and of that of his age than that he should be represented as opposed to the very advances of which, as an active thinker, he must have felt the need. . . .

Fisher to R. Grant: 17 *May* 1950

I am enclosing a letter from Chicago[1] which may interest you, as they are taking the quite extraordinary step of celebrating the 25th anniversary of *Statistical Methods*. Of course, this is partly coincidental, in that its first edition just struck the quarter century, and it will be the half century that they will be particularly concerned to memorialize.

I should not be able without great difficulty to meet their request for a personal account of the origin of the book, because, partly by reason of the book's effect, the whole body of possible readers will be now taking for granted points of view unknown in 1925. However, I will think about it, and may be able to do something. Perhaps you would also consider whether some account of the book as a publication would be worth supplying.

[1] See W.A. Wallis's letter to Fisher (p.325).

R. Grant to Fisher: 19 *May* 1950

It was a real pleasure to receive your letter of 17th May with the letter from W. Allen Wallis, Committee on Statistics, University of Chicago, U.S.A. which you enclosed. A letter such as this surely goes far to make work worth while!

It all takes my mind back to that day when Frank Crew called relative to your manuscript, how he spoke of its quality, the formative work that it contained, and urged publication if only on the grounds that statistics in future would and must form part of research work in every science.

Some publication particulars have been extracted from the copyright ledger and they are enclosed.[1]

Now I feel there is no one who could write a more acceptable note, such as is required, than Frank Crew. Accordingly, a copy of these publication notes has been sent to him with the request that he gives this suggestion his consideration.

[1] See *J. Am. Statist. Ass.* **46**, 31 (1951).

Fisher to R. Grant: 23 *May* 1950

Thanks for your letter of 19th May. I shall be very much interested to see what Crew makes of your little assignment. It was [D.W.] Cutler who approached me, probably after consulting Crew, and certainly he came at the right moment, for I did not have to do any mathematical research *ad hoc*, but only had to select and work out in expository detail the examples of the different methods proposed. It was often quite a job to find a good example. Indeed, some were not so good and have been omitted in later editions. . . .

M. Greenwood to Fisher: 10 *June* 1933

I heard yesterday unofficially but I hope correctly that you are to be the next Galton Professor and should like to be the first to congratulate you and wish you many years of happiness and successful work in the university. There is, in my opinion, no other man alive worthy to sit in old K.P.'s chair; his mantle has descended on your shoulders and a double portion of his spirit is yours. Like all of us, you owe him much. If K.P. had never lived, you would have assuredly been one of our foremost men of science but very likely your field of work would not have been statistics, but perhaps, pure mathematics. You will repay that debt, not by silly adulation of the 'illustrious predecessor'—as so often happens when a great personality leaves the stage—but by carrying the work further. This appointment really makes me happy for—although K.P. has often enough wounded my feelings and insulted my friends—I still love the man and venerate his genius; I should have been sad if his kingdom had fallen into the hands of a second-rater. I hope that when you are my neighbour[1] we shall often meet and that if there is any way in which I can help you you will not fail to let me do so.

[1] Greenwood was Professor of Epidemiology and Vital Statistics in the School of Hygiene and Tropical Medicine in the University of London.

Fisher to M. Greenwood: 13 *June* 1933

Thank you immensely for your very kind letter, which expresses better than I could do just how I should like to carry the work forward. I understand that great circumspection and some restraint will be needed to avoid hurting feelings unnecessarily, and this is my chief concern in accepting the post. It will necessarily be some time before the department can be a coherent unit, but I look forward to establishing helpful relations with your School, as well as with the other departments at University College.

Fisher to H.W. Heckstall-Smith: 9 *February* 1956

Thanks for your letter. I have always regarded the *F*-test and the *z*-test as the same, and indeed in 1924 I calculated the variance-ratio as a means of getting their natural logarithms.[1] Of course if everyone had a computing machine at hand on their desk the variance ratios are the quicker, but those without computing machines, and who can use four-figure tables, are about five times more numerous.

Owing to the misapprehensions that have been spread abroad, especially in the United States, I have added in the 12th edition of *Statistical Methods* (page 226)[2] a short historical note which I hope will prevent expositors from representing the *F*-test, or as the French have got it 'test de Snedecor', with the *z*-test, or 'test de Fisher'. I think it was only an afterthought that led Snedecor to say that the capital *F* he had used was intended as a compliment to myself.[3] In any case you might find it useful to bear in mind that some of your readers are not equipped with calculating machinery, and are used either to slide rules or to logarithms. Of course the differences between logarithms to different bases are quite trivial in practice. . . .

[1] Heckstall-Smith had suggested that in an article he was writing for a medical journal he would need to use the *F*-test rather than the *z*-test because natural logarithms would not be tolerated.
[2] *SMRW*, p. 229.
[3] See also Fisher's letter of 16 February 1938 to Snedecor (p. 323).

H. Hotelling to Fisher: 21 *October* 1930

. . . C.C. Craig, who is here for the year, tells me that the distribution in samples of the standard deviation was published by F.R. Helmert in the *Zeitschrift für Mathematik und Physik* in 1876. His proof is copied in Czuber's *Theorie der Beobachtungsfehler*. . . .

Fisher to H. Hotelling: 14 *November* 1930

I have just been looking at Helmert, and unless I have misunderstood him he certainly has not anticipated 'Student' in giving the exact distribution of the variance as estimated from squared deviations.

He seems to be only concerned with the distribution of quantities such as

$$(\epsilon_1^m + \ldots \epsilon_n^m)/n$$

when the distribution of ϵ is known; not with the distribution of sums of powers of deviations *from the mean of the sample.*[1]

The last three sections seem to be only a crude kind of discussion of the theory of estimation from such sums of powers as he is discussing, but I cannot find that any account is taken of the fact that the mean from which the deviations are measured is also subject to errors of estimation.

Will you look at it and confirm or correct my views?

[1] Compare with Fisher's letter of 24 July 1934 to E.B. Wilson (p. 326).

H. Hotelling to Fisher: 26 *November* 1930

. . . Today I read Helmert's article in the *Zeitschrift für Math. u. Physik* for 1876, and agree with you that it is pretty crude. The distribution he reaches on p.203 is the same as that of χ^2, or s^2/σ^2, with a number of degrees of freedom equal to the number in the sample. As you remark, he makes no allowance for the errors of estimate in the mean, though it would have been easy to do so. Of course he missed the vital idea of the distribution of $(\bar{x}-m)/s$. . . .

Fisher to H. Hotelling: 1 *May* 1931

. . . Is it not a curious fact historically that the first appearance of the normal distribution was as an approximation to the binomial series, whereas if Bernoulli and de Moivre had been more advanced analysts and had got the expression for the true sum of a broken binomial series, they would have found the integral of the *z*-distribution?

About semi-invariants I find that Thiele actually defines them as the coefficients of $t^r/r!$ in the expansion of

$$S(e^{tx})/n$$

where x is the value observed in the sample. They are therefore in his usage primarily statistics, and only by the current contemporary confusion between statistics and parameters were used to designate the population values. If the point is to be governed by historical accuracy, therefore, now that the distinction has been made and is regarded as essential by statisticians, the term 'semi-invariants' should be confined to the series of statistics

$$m_2, m_3, m_4-3m_2^2, \ldots$$

while 'cumulants' is the first term to be given specifically to the population parameters, of which Thiele's semi-invariants may be regarded as estimates. As such they are consistent, but as Craig's work has shown, introduce serious algebraic difficulties. Still it is all to the good to have a distinctive name for them.

Fisher to M.G. Kendall: 19 *August* 1940

Many thanks for your letter and the paper enclosed.[1] I am very pleased to have the latter for the *Annals*, as it makes a worthy addition to the two valuable papers you have written on the subject previously.

I am glad also you find the term 'cumulants' satisfactory for purposes for which I use it. My own reasons, as usual in such cases, were somewhat complex, but I was influenced by the fact that Thiele introduced his term 'half invariants' for a system of statistics calculable from the data, and only later, in the rather casual manner of his time, came to use them for the parametric moment functions to which French and American writers especially later came to apply the term semi-invariants. I have noted Thiele's usage in *Statistical Methods*, pp. 75–78 of the 7th edition.[2] Of course I also wanted a term rather more distinctive than the general term, introduced I believe by Cayley, for the wide class of symmetric functions. I fancy the usage you suggest now corresponds with Cayley's original meaning. Thirdly, I felt some need to avoid such a cumbrous phrase as 'the semi-invariant generating function' for what I call the cumulative function. . . .

[1] Published eventually as Kendall, M.G. (1940). Some properties of *k*-statistics. *Ann. Eugen.* **10**, 106–11.
[2] *SMRW*, p. 72.

Fisher to F.J. McGuigan: 8 *April* 1958

Thanks for your interesting letter.[1]

I chose the term 'null hypothesis' without particular regard for its etymological justification but by analogy with a usage, formerly and perhaps still current among physicists, of speaking of a null experiment, or a null method of measurement, to refer to a case in which a proposed value is inserted experimentally in the apparatus and the value is corrected, adjusted, and finally verified, when the correct value has been found; because the set-up is such, as in the Wheatstone Bridge, that a very sensitive galvanometer shows no deflection when exactly the right value has been inserted.

The governing consideration physically is that an instrument made for direct measurement is usually much less sensitive than one which can be made to kick one way or the other according to whether too large or too small a value has been inserted.

Without reference to the history of this usage in physics, I think one can say that your second interpretation is very nearly right. One might put it by saying that if the hypothesis is exactly true no amount of experimentation will easily give a significant discrepancy, or, that the discrepancy is null apart from errors of random sampling.

Thanks for raising an interesting point.

[1] McGuigan asked what Fisher had in mind when he introduced the term 'null hypothesis', saying that he knew two possible interpretations: (i) that it was an hypothesis asserting that the difference between two groups is null; (ii) that the term came from the word 'nullify' and a null hypothesis is any hypothesis set up for the purpose of being nullifieid.

Fisher to W.L.B. Nixon: 13 *January* 1956

I was extremely interested in your long letter.[1] . . .

The sources of data given in *Statistical Methods* refer to J.A. Harris (1913) 'On the calculation of intraclass and interclass coefficients of correlation from class moments when the number of possible combinations is large', *Biometrika*, ix, 446–472. This may not be the earliest use of the term intraclass correlation, but it was to Arthur Harris's credit that he distinguished correlations of the interclass and intraclass kinds which, of course, had been confused in the literature of the previous twenty years. Indeed it was largely to give Harris the credit of having made this advance and to show how far it led, i.e. to the analysis of variance applicable to many other problems, that I gave the space to intraclass correlation. . . .

[1] Nixon asked about the origin of the term 'intraclass correlation'.

V.I. Romanovsky to Fisher: 4 *December* 1935

I would be very obliged to you, if you indicated me how are established two approximative formulae, p.221 of your *Statistical Methods for Research Workers*.

I am also puzzled why you use, in the analysis of variance, $z = \frac{1}{2}\ln(s_1^2/s_2^2)$ instead of s_1^2/s_2^2. . . .

Fisher to V.I. Romanovsky: 20 *December* 1935

I am glad to have your letter.

So far as I remember I obtained the approximations[1] for the test of the significance of z where both n_1 and n_2 are large by obtaining the moments of the distribution of z, or rather its cumulants, from its characteristic function. I forget the details, but clearly the factor $(1/n_1)-(1/n_2)$ is a simple allowance for the third moment, while the first term is derived from the normal distribution.

I had a good many reasons for using *z* instead of some function of it in the test of significance in the analysis of variance. One important reason was that in order to make a compact table it is necessary that the test value should be well interpolated by what I call asymptotic interpolation using the reciprocals of the numbers of degrees of freedom, and this is more true of *z* than of any other simple function. A second point is that half the tabulation is saved by the fact that reversing the sign of *z* and interchanging n_1 and n_2 we have the 5% and 1% points at the opposite ends of the distribution. Finally, the close analogy between interclass and intraclass correlations is paralleled by that of the values of *z* obtained from *r* by the same transformation. The advantages of this transformation I have set out in the book. . . .

[1] *SMRW*, p. 234.

Fisher to G.W. Snedecor: 16 *February* 1938

. . . I am afraid I have a difficulty which you may not know of in assigning the particular symbol *F* to the variance ratio,[1] and that is, that a table of this value derived from my table of *z* was published in India, I think in *Sankhyā*,[2] before yours.[3] The author, if I remember right, used a different symbol — probably *x*. I feel it would be too complicated to make a sub-heading using the different designations.[4] . . .

[1] Snedecor had suggested to Fisher that 'it would be clarifying to associate your variance ratio table in some manner with *F*.'
[2] See Mahalanobis, P.C. (1932). Auxiliary tables for Fisher's *z* test in the analysis of variance. *Ind. J. Agric. Sci.* **2**, 679–93.
[3] Snedecor, G.W. (1934). *Analysis of variance and covariance*. Collegiate Press, Ames.
[4] See also Fisher's letter of 9 February 1956 to Heckstall-Smith (p.319).

Fisher to L.H.C. Tippett: 12 *November* 1931

Many thanks for sending me the copy of your book *The methods of statistics* which arrived here this morning. I had read some of it before and was very glad indeed to see how much use you were making of the methods which from time to time I have put forward. This is a real encouragement to me, and since it has been obvious for some time that most of the books on statistics are much out of date, I hope and think it will help your book, though of course it will expose you to the attack of the more recalcitrant of the Pearson group. . . .

You have evidently found Woo's table[1] useful, which surprises me a little as this involves subtracting $\overline{\eta^2}$ (written $\bar{\eta}^2$), and dividing by σ_η. This makes you use the term correlation ratio in a much wider sense than is usual, for example in place of *R* the multiple correlation coefficient. Of course, once the analysis of variance is grasped the equivalence of the test of significance for the whole group of problems including η and R is apparent. But it is the analysis of variance itself which keeps the arithmetic straight and easy, and not the possibility of regarding the ratio of two sums of squares as a sort of η^2. I wonder they did not make Woo work the actual 1 per cent and perhaps 5 per cent points for η^2, as even using interpolation it would have been easy to get the values accurate.

I think your book will do a lot of good and help a good many people. The examples seem to me especially well-chosen.

[1] Woo, T.L. (1929). Tables for ascertaining the significance of association measured by the correlation ratio. *Biometrika* **21**, 1–66.

L.H.C. Tippett to Fisher: 15 *November* 1931

Thank you very much for your letter of the 12th and for the comments about the book. . . .

I personally never use η^2 as a test of association, but it *is* used, and therefore seemed desirable that it should be linked up with the other tests. I think that one of the most hopeful ways of getting the Pearsonians to use the Analysis of Variance is by showing its equivalence to η^2. That the former is much neater and easier to calculate should be obvious. . . .

Fisher to L.H.C. Tippett: 17 *November* 1931

Thanks for your letter. . . .

I am interested in what you say about the Biometric Laboratory and the analysis of variance. In the highest quarters I fear a presentation of the newer methods is acceptable, if at all, only if their origin is disguised; witness Mr. Woo's being introduced to the true distribution of η^2 through Hotelling's paper which Hotelling wrote, as he tells me, in a state of forgetfulness, after having read my 'Goodness of fit of regression formulae' [*CP* 20], in which the distribution is first given. In this circle my work has now passed through the stages of being regarded as

i) criticism so impudent and ignorant that it may be ignored;

ii) applicable only to such special cases as to be without interest;

iii) the stage of containing some points of value, which are, however, better obtained from other sources.

But apart from this very small knot of incorrigibles I should advise the use of the analysis of variance, with which, as it is only a convenient arithmetical arrangement, no author need be associated and to forget the fancy symbols which from time to time and in various particular cases have been attached to the ratios which may be deduced from the analysis. Of course η^2 is the limiting value of R^2 when the number of adjustable constants is made equal to the number of arrays and it may be useful occasionally to know that R^2 can be made no higher. But even in the case of R^2 the arithmetical presentation gives a much clearer idea of what it means than that it is the square of a generalised sort of correlation coefficient.

I am very glad to hear about the lectures. I am sure they will do no end of good. I find myself that students knowing little mathematics become extremely keen on these 'advanced' methods simply because, I think, on working through the arithmetic (and certainly not without) they find them readily intelligible, and they take no harm from knowing that they are far more exact mathematically than what is taught to astronomers and biometricians.

W.A. Wallis to Fisher: 10 *May* 1950

This year marks the twenty-fifth anniversary of the publication of *Statistical Methods for Research Workers*. The *Journal of the American Statistical Association*, of which I have recently become Editor, hopes to mark this event by one or two articles on the character and consequences of that volume. In that connection, I would appreciate your help on two points.

First, can you suggest two or three persons whom I might invite to prepare articles? I naturally want people who have a thorough comprehension of, and sympathy with, your work, so that we will get a treatment that is both deep and broad, and in more than a formal sense. The two names that come first to my mind are Hotelling and Cochran. . . .

Second, would it be possible for you to prepare a paper for us on the history of *SMRW*? I have in mind something on how you came to write it, on any special problems you encountered, on the revisions, deletions, and additions of the various editions, etc. If the publishers have no commercial objections, I believe that statistics on the sales, by year, country, and edition, might be of considerable interest. In general my hope is that you could without embarrassment give a sort of biography of the volume. But anything you see fit to write will be more than welcome. Maybe you would find the next twenty-five years a more stimulating topic. . . .

Fisher to W.A. Wallis: 17 *May* 1950

Thank you for your two letters of May 10th. I am sure that the choice of Hotelling and Cochran is an excellent one, and hope that they will both have something to say. They are, however, both professed mathematicians, and the book was, of course, written not for mathematicians but for practitioners, who, in so far as they understand their fields of application, are good judges of the kind of statistics which aids them in their work. They constitute, I believe, the real and ultimate judges of such a book as mine, though, of course, the majority of them are rather inarticulate in mathematical circles.

I should suggest, therefore, and have already suggested to the programme committee, that others from one or more fields of application should be invited to show what the book or the ideas which it was intended to express have done for their own subjects. . . .

As to your suggestion of my preparing a paper on the history of the book, I am not quite sure what can be done, but I will at once consult my publishers and see whether they are interested in putting together an account from their point of view.[1] In preparing recently notes on Messrs. Wiley's republication of some of my old papers I have been greatly impressed by the extent to which thought has changed and by the consequent difficulty of reconstructing for modern readers the state of statistical science in 1925. So at the moment I am not sure whether I could do anything at all adequate. However, I will think about it.[2]

[1] See p.317 for Fisher's correspondence on this with Robert Grant, a director of the publishers, Oliver and Boyd.

[2] The American Statistical Association devoted a session at its meeting in December 1950 to 'The influence of *SMRW* on the development of the science of statistics' and papers under this heading by F. Yates, H. Hotelling, W.J. Youden, and K. Mather were published in *J. Am. Statist. Ass.* **46**, 19-54 (1951).

On his copy of Hotelling's paper 'The impact of R.A. Fisher on statistics', Fisher has written several comments.

On p.35 alongside Hotelling's statement that 'Karl Pearson began as a mathematical physicist, specializing in the theory of elasticity, but with the enthusiastic rise of biology that followed Darwin was drawn into his series of "Mathematical Contributions to Evolution"', Fisher has written 'Response to Galton's endowment'.

On p.36 where Hotelling says '. . . in considering the development of statistics in the first quarter of the twentieth century it is well to bear in mind the British tradition of mathematical physics, which sought primarily for useful formulae and could ignore some incompleteness in the details of their deductions, some over-condensation and abridgement of formal rigor . . .', Fisher has underlined 'useful formulae' and has written in the margin 'New quantitative operations. Heaviside. Matrix algebra'.

On p.37 where Hotelling says Gosset received his B.A. degree in astronomy from Cambridge, Fisher has written 'mathematics'.

On p.42 where Hotelling says Sir John Russell 'persuaded Fisher to leave the classroom for a temporary stay at Rothamsted in 1919', Fisher has written that 'he was unemployed, but had recently been invited by Pearson to come to U.C.' [University College].

On p.46 alongside Hotelling's statement that 'The Continental mathematical tradition of explicit and complete argument has now taken a firm hold both in this country and in England, and should help to minimize errors, misunderstandings, and controversies.', Fisher has scrawled 'Hasn't it!'

Fisher to E.B. Wilson: 24 *July* 1934

. . . A great many statistical discussions in astronomy could be made very much clearer and easier to read, and, I suspect, also to write, if the arithmetical arrangement, which I call the analysis of variance, . . . were familiar to the authors.

Historically, it is interesting that the distribution of s^2 was discovered by an astronomer, Helmert, as early as 1875,[1] but its importance was obviously overlooked, owing to the widespread use of the mean error (Peters' formula) in astronomical work. Perhaps the more so, as Peters was the editor of the journal in which Helmert published his results, which appeared as insignificant excrescences of papers relating principally to the sampling errors of Peters' formula. . . .

[1] Compare with Fisher's letter of 14 November 1930 to Hotelling (p. 320) and also *CP* 165, p. 3.

4
Teaching of Statistics

Fisher to D.J. Finney: 27 June 1957

For further guidance I am wanting to say distinctly that opportunities for research are the attraction.[1]

As you say, I should have no difficulty at Caius or at Rothamsted of going on with the work, in which more and more extensive fields continually open up, of pointing out faulty and misleading methods now taught to students. There is surely a lifetime's work in front of me there, but it is less agreeable than doing a bit towards clarifying the situation in one of the natural sciences, where the application of statistical methods is not bogged down with all the pedantic verbiage that seems to be necessary in mathematical departments, where it seems to amount to no more than finding excuses for replacing known competent methods by less adequate ones. If you want to forget that a test of significance is liable to be invalid if based on inefficient estimation, it seems to be sufficient to speak of it as a non-parametric test.

[1] Fisher was considering where and with whom he might work after his retirement from the Chair of Genetics, University of Cambridge, in 1957.

Fisher to H. Hotelling: 22 February 1946

. . . One side of your problem,[1] which I have no doubt has not escaped your attention, is the difficulty of combining in the same man, mathematical skill with the kind of width of logical outlook only, I think, easily acquired by long contact with the atmosphere of experimental research. In statistical methods, as it seems to me, we are developing the mathematics appropriate to a logical attitude which has not in formal pure mathematics ever found coherent expression, and personally I have often felt exasperated at the obtuseness of perfectly good mathematicians when faced with any logical situation involving gaining new knowledge by experience or experimental observation. I am sure that this is simply because the logical presuppositions with which they have for years been familiar in deductive analysis are by this time taken for granted, not only as axiomatic, but as universally sufficient. . . .

[1] Hotelling had sought Fisher's help in obtaining suitable personnel for the new Department of Mathematical Statistics at the University of North Carolina.

Fisher to T. Koopmans: 14 *January* 1936

I have at last been able to look through your interesting lectures on Modern Sampling Theory. I have annotated the typescript with a few suggested amendments, mostly quite trifling.

The main point which I think deserves consideration is that of order of development and presentation. After an introduction to the general notions of population and sample, parameter and statistic, specification, estimation and distribution, making the point that we must judge of the value of different possible methods of estimation by means of the distribution of the estimates, I believe it would be wise to give some considerable amount of time to problems of distribution, which, though full of mathematical interest, have the advantage of being purely deductive, and of involving only the classical concept of probability representing an ideal frequency.

I think at this stage I should illustrate three methods by which exact sampling distributions may be obtained: the method of generalized space, of the use of which the solution of the distribution of the correlation coefficient is a good example, with, possibly, extension to the multiple correlation coefficient. The procedure here developed includes, as by-products, the simultaneous distribution of the mean and variance of a single normal variate. Another good example of the geometrical method is that of the test of significance for the largest Fourier component in harmonic analysis. The two other methods which should be illustrated are the use of mathematical induction for samples of increasing size, and the use of the characteristic function. Of these, the former works well for demonstrating the χ^2 distribution for the sums of squares of independent normal deviates having the same variance, and leads naturally to the three principal distributions, χ^2, t and z, used in tests of significance. The use of the characteristic function, though beautiful in its generality, has not often led to new distributions which have not been found in other ways. There are, I think, some examples in my paper 'Two new properties' [*CP* 108]. A special case of importance, with a very direct method of solution, is that of an estimated percentile, such as a median of an odd number of observations. Under the heading of 'problems of distribution' also should be included the approximate large sample formulae for the variance of any chosen statistic, best developed perhaps from the general multinomial distribution.

On this groundwork of knowledge of the distributions of estimates the criteria which I have developed are easily intelligible. As an illustration of an inconsistent statistic I should not use one which tends to no limit at all, but rather one which tends to the wrong limit, as, for example, in using Sheppard's correction with grouped normal data. If a statistic tends to no limit at all, I should not know whether it were consistent or not. The notion of efficiency can be introduced with easy examples, such as the median of a normal sample, and is consolidated by proving that there is a lower limit to the limiting value of nV as the sample tends to infinity, i.e. that such a limit exists whenever the distribution of the statistic tends to normality. This proof needs supplementing by the proof that it is possible to find a statistic having this limiting value for nV, as can be satisfactorily done by considering all statistics which are solutions of equations linear in the frequencies and showing that, when the variance is minimised, we are led to the equation of maximum likelihood with a limiting variance as small as possible. Consequently, the equation of maximum likelihood leads to one efficient estimate. The maximum precision obtainable in large sample estimation leads, by one of the most important steps of the whole theory, to the concept of the intrinsic accuracy of an error curve definable for all curves for which it has a finite value, e.g. for 'Student's' distribution when $n=1$, and to the recognition of the intrinsic accuracy of the error curve of a statistic as the amount of information which that statistic extracts from the data. The proof of the existence of a maximum precision is then easily adapted to show that estimation may involve a loss, but cannot introduce a gain of information, and that the loss can only be zero when the sets of sample for which $d(\log L)/d\theta$ is constant are the same for all values of θ. This is the essential property of sufficient statistics, and from it the fact that no additional information can be supplied by other statistics can be demonstrated. It should also be shown at this stage that, in general, a finite amount of information is lost by maximum likelihood estimates, but that, in the limit for large samples, the loss is less for such estimates than for other efficient statistics. At this stage also I should show that the loss of information may be further reduced by the use, in conjunction with the estimate, of one or more ancillary statistics which themselves provide no information, and that in special and easy cases, as in location and scaling, the whole of the information may be recovered in this way.

Turning now to tests of significance, I should show that when there is a variable parameter any test of significance, whatever level of significance be chosen, will generally reject some of the possible values. If the parametric values form a continuum, a portion may be defined for each test of significance, within which the values are not rejected, but outside which all values are rejected, the probability of rejecting the right value being known with exactitude. If, therefore, the test is based upon statistics which contain the whole of the information supplied by the sample, we may make a probability statement respecting the unknown parameter which is entirely well defined. Its logical content differs entirely from a statement of inverse probability. It is independent of all prior knowledge respecting the frequency distribution of the parameter. To distinguish it, it is called a statement of fiducial probability. The aggregate of such statements possible define the fiducial frequency distribution of the parameter as inferred from the observations. Such statements derived from a portion only of the observations would necessarily be erroneous, as would be, for the same reason, statements of fiducial probability based on non-sufficient estimates. No such fiducial distribution is available from discontinuous observations, though the observational programme may sometimes be modified deliberately to lead to conclusions of this type. It is, in my view, most important that the limited applicability of the notion of fiducial probability be stressed.

I am returning your typescript herewith.

Fisher to J. Maclean: 3 March 1930

. . . I ought to say that it is undoubtedly the practitioners rather than the teachers who are absorbing the book [*SMRW*]. I can understand this as since I wrote it I have had to do some lecturing and examining in the subject, and am now convinced that, apart from really first class mathematicians, statistics should be taught by the direct handling of numerical data in the laboratory.

Fisher to P.C. Mahalanobis: 9 October 1954

. . . I think the arrangement for the seminar on statistical inference would suit me perfectly, but though I may have to appear sometimes before a large crowd, I have no illusions as to how little teaching can be effected in this way. In a large crowd with important names present, young and unknown men who might really have something to gain are completely inhibited. If, however, you have anyone with a keen mind among your younger students, I hope you will not hesitate to encourage him to set his wits to work and make of my seminar a genuine discussion rather than a monologue by a super-Pundit whose word may not be doubted.

The world has enough dictators anyway. . . .

Fisher to K. Mather: 12 March 1943

. . . I should like to impose as a necessary condition for being accredited as competent in business statistics the ability to test the significance of a 2 × 2 table. After all, when business men can do that, it may spread anywhere — into the Insurance Offices or even into the higher ranks of the Civil Service. . . .

Fisher to H.F. Smith: 29 May 1946

. . . About Indian and indeed most foreign mathematicians, I seem to feel much as you do that the choice between what has to be elaborately explained and what can be assumed to be spontaneously evident differs a great deal in different schools of mathematical thought. I recently received a very highbrow treatise of a first class Swede, namely Cramer, purporting to deal with mathematical statistics, of which I think a full half was a comprehensive introduction to a theory of point sets, which I should not feel spontaneously had any real bearing in reasoning from frequencies or the inter-relation of distribution functions, but I suppose in view of the didactic manner in which the theory of probability is approached in Russia, for example, and in France, and in recent years in the United States, this sort of thing must seem a quite essential clarification of the subject.

In Paris recently I found an interesting and perhaps useful distinction being made between *statisticians* and *probabilists*, broadly speaking putting me in the first class and Cramer in the second. I also learned that a certain, I believe very learned, Russian called Kolmogoroff has postulated an axiom, which purports to justify axiomatically certain types of argument for which I am responsible. It runs something like this. If the probability of an event dependent on a number of parameters, $\theta_1, \theta_2, \ldots \theta_s$, has one and the same value ω for all possible combinations of the parametric values, then the absolute probability of the event exists and is equal to ω. So now when I am in any difficulty I can insert with confidence 'by Kolmogoroff's axiom'. . . .

Fisher to H.C.S. Thom: 8 *October* 1941

I have just received your note of August 30th about the new 8th edition of *Statistical Methods*. The edition had, however, been completed before I saw your letter, so that my consideration of its suggestion has been what is called 'academic'.

I think there are advantages in the idea,[1] and really only one serious disadvantage. Your reader wants, perhaps, a concise little elucidation of the mathematical connection stated and assumed in the text. Then he wants, and may expect to be referred to, a technically competent source for this particular mathematical result, from which he can easily pick out exactly what he wants. His interests, in fact, are centripetal. Having the haystick he wants to find the needle.

Now the chief use for a student of looking up papers, either in my own bibliography or in the general list of sources, is centrifugal. When a man is ready to widen his association with a particular method, to consider more fully some of the difficulties of its particular application, and the marginal applications in which its appropriateness is doubtful, or if he is inclined to browse in the history of the subject and detect how this idea led to that, I think that foundation papers may make useful reading; but they are nearly all mixed in respect of theory and practice, theoretical statements being put in *ad hoc,* when they are immediately needed, perhaps on discovering that the point had not previously been made explicitly clear, and by no means in the orderly succession of a connected theoretical treatise.

In my view a statistician ought to strive above all to acquire versatility and resourcefulness, based on a repertoire of tried procedures, always keeping aware that the next case he wants to deal with seriously may not fit any particular recipe. Of course I know that my book has been described as though it supplied such cut and dried recipes, and has been criticised for not supplying the mathematical background by which they can be better understood; but you will see from the above that, from my point of view, this is a misapprehension, based on the belief that the understanding required can be obtained from the mathematical background rather than, as I think, from the particular peculiarities of the actual body of data to be examined. This kind of understanding seems to be ignored to a frightful extent in mathematical discussions.

[1] Thom had suggested that if Fisher's papers listed in the bibliography of the end of *SMRW* had specific references to them inserted at relevant places in the book, mathematically inclined readers would be greatly helped and the book would then be a useful reference for the applied mathematician interested in theoretical developments.

Fisher to G.T. Walker: 27 February 1942

. . . I need not say how glad I am that you have been looking at some of my own work. I have resisted all temptation so far to overload my book on Statistical Methods with mathematical demonstrations for I should far rather that most workers should give their minds to thinking of the nature and origin of the data before them in relation to the appropriateness of any contemplated procedure than to what is usually rather irrelevant, the algebraic manipulation by which results are formally demonstrated. Consequently I have usually given formal analysis, not in this book but in other papers, such as the enclosed, written from time to time as the occasion seemed to require. . . .

5
History and Philosophy of Science

Fisher to W.E. Le G. Clark: 17 *November* 1952

. . . The point and policy of this little group[1] is indeed to make such real contact between practitioners, e.g. ecologists, pharmacologists, etc., and more theoretical types so that the methodological work done shall really be fit to cohere with problems arising in the real world. It would be no bad thing if people like yourself, concerned largely with forming just conclusions on morphological grounds involving some element of subjective judgement, should come to think of biometricians not so much as 'natural enemies', but rather as auxiliary specialists having had some practice in the difficult art of weighing numerical evidence. . . .

[1] The Biometric Society. Fisher's letter led to Clark joining the Society.

Fisher to R.F. Harrod: 17 *July* 1942

I have just read your paper on 'Population and the future', which I thought put the true state of the case quite excellently. I can only excuse your depreciation of it by my own frequent experience. A year or two after I have finished a piece of work I find myself remembering my efforts with a most uncomfortable distaste. But then, if a few years later I have occasion to reread what I actually wrote, I am usually astonished to find my own bad opinion to be very largely ill-founded. Try rereading your own article as if it were someone else's new contribution to the subject. It is a most encouraging experience.

The worst of trying to move the world is that you have to stick at it so long to move it even just a little. But I suppose I ought to have thought of that before starting.

Fisher to H.W. Heckstall-Smith: 30 *July* 1957

I like your quotation from Wittgenstein,[1] and to me the phrase 'because we prefer our hypotheses simple' is unanswerable, but 'we prefer to believe that the truth is simple' is a highly metaphysical statement. However, we must use what words we have with even less pedantry than we use what ideas or hypotheses we can form, so I am against a general ban on the words 'truth', 'proof', and 'real', but in favour of exposing their ineptitude in particular cases.

[1] Heckstall-Smith had written quoting a colleague's justification for choosing a particular hypothesis: 'Because we prefer our hypotheses simple. We prefer to believe that the truth is simple.' and added that he disagreed with this and would like to exclude the words *true, proof,* and *real* from scientific and statistical theory. He said he was 'deeply impressed' by the following quotation from L. Wittgenstein's *Remarks on the foundations of mathematics* (IV, 48, page 156): ' "*Mathematical logic*" has completely deformed the thinking of mathematicians and of philosophers, by setting up a superficial interpretation of the forms of our everyday language as an analysis of the structures of facts. Of course in this it has only continued to build on the Aristotelian logic.'

D. McKie to Fisher: 2 *June* 1943

I have been meaning to send the enclosed to you for some time — but you know how things are in these days. . . . I expect you have read Brewster's *Newton*: I find it a very irritating book. The new biography by L.T. More [1934] is nearly as bad. My own reaction, for instance, to the calculus controversy — purely a historian's reaction — is that Newton fares badly at the hands of both biographers. This reaction, of course, is due to what I hold to be the proper plying of my trade of historian — due heed to dates and documents and contacts, not too much stress on men's foibles and touchiness. However, what I feel is really needed is for a clear opinion by a mathematician on whether certain mathematical results used have been independently derived — no one seems to have put that right, not even Rouse Ball. I do get the very clear impression (as a historian/not a mathematician) that Newton's complaint against Leibnitz was just (altho' his method of dealing with it was unfortunate).

Fisher to D. McKie: 4 *June* 1943

Many thanks for sending me your notes about Newton's Chemical Philosophy which seem to me a much more worthy tercentenary notice than some of those I listened to at the Royal Institution and the subsequent dinner of the Royal Society Dining Club. Lord Keynes tried hard on that occasion to make all our fleshes creep with his stories of the mysterious contents of the big black box which Newton packed up when he moved to the Mint. To me this was very unconvincing.

There is one point often overlooked which of course only comes into question if on occasion Newton, like some other mathematicians, ever did show himself to be credulous, and that is that some degree of this weakness would seem naturally to be associated with any remarkable precocity of intellectual development, for the simple reason that the child in such cases is perfectly able to understand abstract distinctions and to follow long trains of reasoning, so satisfying his curiosity with apparent finality on a number of problems, using as unquestioned data the kind of over-simplified statements thought suitable for immature minds, which in later life are reinterpreted with a great deal of qualification. I mean such statements as that the Bible is the 'Word of God', which to any simple-minded child, not suspecting the adult world of a conspiracy of hypocrisy, seems to mean the devil of a lot more than it does to any bishop.

I think it very probable that Newton showed a credulity of this kind in that he had thought out, or at least satisfied himself when quite young as to the importance of different kinds of human endeavour, and would not easily be persuaded in later life that what he could do in mathematics was really more important than clearing up obscurities in the chronology of the Bible.

Thanks again for sending your note.

Fisher to J. Maclean: 11 *October* 1940

Thanks for your letter of September 1st which has just reached me with your enclosure.[1] Have you ever thought how eccentric mankind is in the application of his *attention* or *interest*? On almost any point on which public interest is lively and sustained, mankind, with or without using the organisations we call States, can do apparently almost what it likes. By some geometrical accident, for 50 years or so, most of mankind was interested in the discovery of, or rather access to, the poles of the earth's rotation. The difficulties were enormous and organised States did not do very much to help, but one shot followed another until what had seemed impossible has become quite easy, and there may be Cook's tours to the North Pole when our aviators and machines are looking for jobs after the next demobilisation.

When something difficult, remote and unprofitable like this can be done, it would seem easier to reorganise our internal affairs to bring about any possible clearly conceived and generally desired object. The difficulty I feel is that since the French Revolution (I mean the one about 150 years ago) the programme has been (1) abuse, (2) destroy, and then, perhaps, (3) reform, that is to say that distrust, hostility and hatred towards existing institutions and persons is quite taken for granted as desirable, honourable and effectual in most groups which are at all impatiently concerned with getting anything put right.

In fact, I suggest that this agitator's mentality is by far the most serious obstacle to the internal reform of Society and, in fact, to the very reforms which the agitator and his supporters would like to bring about.

Devotion to France is genuinely and effectively traditional among modern Frenchmen; yet this year the majority of Frenchmen acquiesced in a fairly obvious betrayal of their own country. The reason, of course, lies in the internal dissensions of French politics. The France to which that loyalty was pledged was threatened both from Berlin and from Moscow. The Communist danger developed first, and the *front populaire* with the Spanish Civil war showed that it was really dangerous, and gave not only die-hards but all law-abiding Frenchmen a feeling that there were worse things than Fascism. Consequently, on the collapse of the French army, it was alarmed extremists of the type of Laval who were able to seize power and, as they had perhaps some reason to believe, to save France from revolution.

I merely mention this as an example of how completely human effort can be frustrated where opinion is divided as to social aims. Hitherto, in the past, unity as to the fundamentals of social aims seems only to have come from religion, through religious control of education. I think it is noteworthy that the secularisation of education in Europe has only lasted two or three generations before collapsing, especially where it had been logically and thoroughly carried out, as in Italy and Germany, where the totalitarian parties have attempted to find an artificial substitute for the previous influence of the Church; the first example of this was, of course, provided by Russia. So far as I can judge, the substitute is of a very inferior brand in all three countries but its immediate aim is quite clearly to give to a people something of the unity of social aims and interests which I have been talking of.

I am afraid this is not very much to your point. You must take it merely as my reaction to your stimulating letter and enclosure.

[1] Maclean had written to Fisher from Wilson College, Bombay, referring to the world situation and saying he believed there was an urgent and challenging intellectual problem in economic relations — 'one well worthy of your powers' — and that 'it might make all the difference to mankind if you could arrange to put your unique strength into straightening things out in this respect'.

Fisher to J. Needham: 3 *November* 1956

I am returning the galley proof,[1] and I think your correspondent, or whoever wrote the marginal notes in red ink, has made some useful comments. Of course, in this sort of field the difficulty of a conscientious historian is to know what was taken for granted as common knowledge, as certain widely taught arithmetical processes may be, and what was novel and for some purposes effectual. I imagine one of the outstanding qualities of Chinese science is that they kept their eye on specific purposes, and though they may have used general or abstract considerations in ordering their thoughts, did not so much elaborate these in theoretical disquisitions, as the Greeks often did very elegantly and modern mathematicians often with intolerable prolixity, as write down a specific formulation for a particular job. I do not think that there is a higher and a lower in all this so much as a traditional difference in mode and style. The difficulty I feel must bother the historian is that to the modern European mathematician the theory of finite differences can only be said to be introduced or used from the time of the first generalized and systematic treatises, and not from the time of such effectual private usage as undoubtedly Newton was doing in his early twenties. I understand that in fact most of the named theorems in this subject up to about 1850 should be called 'Newton's theorem', though only one of them bears his name.

In this case, and because it was Newton, I believe it would be admitted that finite differences were fully developed in the sixties of the seventeenth century, but if a nameless mathematical schoolmaster had used the same methods very adroitly for astronomical observations it might be 200 years before printed material and textbooks showed an acquaintance with the methods mastered long before.

[1] Presumably the proof of an extract from Needham's *Science and civilization in China*.

R.N. Salaman to Fisher: 20 *March* 1944

Probably you have seen in *Nature*, March 11th, p. 298, a short article by Dingwall on Telepathic Phenomena.

I would like to know whether, in your opinion, it is to be taken as serious evidence that such occur, because as far as I understand it the man with the cards is not actively wishing the guesser the answer, nor is the guesser making any wishful effort to learn what the other sees, but would simply seem to be making guesses which for some reason are influenced by the fact that another man at a distance has the cards in his hands.

I find it very difficult to attune one's mind to accept the interpretation that either party is influencing the other, but I gather the statistical evidence almost proves that such occurs.

If you have a spare moment, I would like very much to know what you think.

Fisher to R.N. Salaman: 22 *March* 1944

The question you ask me is one which appeals to me a great deal, and though I have constantly seen these claims to immensely high significance made on behalf of work with which I personally have had nothing to do, yet in all of the good many cases in which I have been asked to look at data actually secured with a view to demonstrating extra-sensory perception there has never been any such decisive evidence.

I suppose what gives rise to these confident statements is something like this case within my own experience, and of which I remember the details clearly:

A very rational and entirely honest girl who had become interested in supposed telepathy had formed the conclusion that one could not investigate evidence of supposed telepathy unless one had first a clear basis of knowledge as to the possibility or impossibility of clairvoyance. She therefore designed a test to see whether any considerable number of ordinary people possessed a clairvoyant faculty, at least in a slight degree. I think she grasped, and this also seemed to me enlightened, that for evidential value a slight but constant measure of successes among a large number of subjects was much more important than evidence of a knock-out star conjurer's performance done once under special conditions and not in any sense reproducible.

What she asked people to do was to turn up five cards in succession from a well-shuffled pack, noting in each case what they thought the card was going to be, and after turning it up, what it actually was. She thought that five cards at a time was the right amount, but asked each person to do this five times so as to provide the results of five guesses or attempted clairvues. I made out for her a system of scoring partial successes, as when you guess the King of Clubs and the Knave of Clubs turns up, framing the system so that the average score was zero, and the standard deviation for random guessing, I think, ten points, so that the total of 25 guesses had a standard error of 50, and the mean of 25 scores had standard error of 2.

She got to work and persuaded no less than 240 people to co-operate, finally producing 6000 records for examination. Before I saw them she had scored the whole lot and found positive scores well in excess of negative, and in the aggregate very significantly so.

Of course, some of her subjects may have cheated, and possibly those who reported the most improbable successes should be discounted, but even if one sets these aside the preponderance of small positive scores among the remainder requires an explanation. I was led to think by my own experience that though all but one or two of the collaborators may have been entirely honest, yet they may have vitiated the results just as much as if they had sent in false returns. I had myself received a set of forms to fill in, and when first I had them, sat down, did 5 guesses, and filled in one form. I was not successful, no interest was evoked, the forms went into a drawer and were forgotten. Later, further correspondence induced me to fill in a second form, but I never completed the five. What I think is that if a chance success had come my way in the first five trials, I should have been interested, completed the set (unless later experience was too disappointing), and sent in my return to be included with the others.

I was able to show that in fact the large body of data collected had been vitiated in just this sort of way, and the confirmation is important, for of course it is just as easy to make a hypothetical and unfounded objection to an experimental result as it is to make a false experimental claim. It happens that when people guess cards in numbers, certain [general?] preferences begin to show themselves quite strongly; red cards are guessed more often than black, odd numbers are guessed more often than even, and so on. In 6000 guesses, each card of the pack should have appeared about 120 times, but actually the frequencies had a very wide range, from about half of this amount nearly to 200. This of course proves nothing except that you could make money in the right company by offering slightly better than the calculated odds to anyone who will guess the card in your hand, provided you make sure that, on the whole, unpopular cards are the ones you hold. What is really informative about the card-guessing data that I'm speaking of is that, after tabulating frequency of choice of the cards *guessed*, one could quite independently tabulate the frequency with which each of these same cards was *drawn*. Here at least one might expect the frequencies to be in accordance with chance, namely equal numbers of Black and Red, equal numbers of Odd and Even, and so on; but the frequencies of cards drawn were also disturbed by some factor other than chance, and what I think was enlightening was that these frequencies were in fact a pale reflection of the frequencies appropriate to cards chosen. The differences were not so much as half as great, but they were regularly in the same direction. Consequently I think that one must infer that a large number of the cases in which unpopular cards were drawn have somehow been eliminated from the record in just such a way as would be brought about the suppression, by non-completion, of trials started inauspiciously.

It seems to be one of the ways in which faith moves mountains!

Fisher to E.B. Wilson: 5 February 1957

I see that I have not answered yet, as I had hoped to do, your letter of November 24th, in which you refer to Bridgman's contribution in *Science*[1] to the discussion on 'Science and the Supernatural' which that journal organized.

I believe a great deal is done towards clarifying this tangled situation by the realization that however complicated some of the cases may be, in which uncertain inductive inference is useful, there is yet an underlying simplicity recognizable in the simple disjunction: Either the hypothesis is false (including the milder faults of being incomplete, or imperfectly true), Or a very extraordinary coincidence has taken place. I do not believe that the practical use of this simple type of disjunction is misunderstood by practical experimenters, unless, and until, they come under the influence of attempts, originating in mathematical economists, to reinterpret this simple disjunction as an 'acceptance procedure' or 'decision function'.

In relation to parapsychology, I have never had the least hesitation in admitting that the data as reported do present the dilemma: Either there is something here which you do not understand, or a very remarkable coincidence has occurred, but it only presents this dilemma to those who are already satisfied (a) that they are not being hoaxed, (b) that the observation and recording is of scientific standard, and (c) that the reporting is unbiassed, i.e. that there has been no exaggeration, or the omission of mitigating circumstances. I do not at all believe, in the light of the attitude exhibited by those who report parapsychological work, that anyone has reason to be satisfied with these three preliminaries.

It is, to my mind, a very ominous sign that a parapsychologist should exhibit himself as insulted, and throw off the gestures of moral indignation, at the thought that he is not strictly and habitually truthful, even when his reputation and career is largely at stake. I think the psalmist was more nearly right than the parapsychologist. In genetics I do not think that touchy dignity should prevent one from saying, 'Here is some material which I have bred, and which I can hand over to you so that you can verify the remarkable phenomena which I believe it will exhibit'. Without being able to take some such attitude to a critic, it seems to me sheer impudence for Soal and others to claim that they have provided anything like a scientific demonstration. Years ago in my book on *The Design of Experiments*, in Section 7 headed 'The Test of Significance', I tried to express this idea by saying:

'In order to assert that a natural phenomen is experimentally demonstrable we need, not an isolated record, but a reliable method of procedure. In relation to the test of significance, we may say that a phenomenon is experimentally demonstrable when we know how to conduct an experiment which will rarely fail to give us a statistically significant result.'

I believe experimenters in the Natural Sciences will take that as not too hard a saying, but it appears to be fatal to the parapsychologists. I suggest, therefore, that the weakness lies not with tests of significance as applied to their own purposes in the Natural Sciences, as Bridgman seems to suggest, but with the peculiar use that psychologists have tried to make of them.

[1] Bridgman, P.W. (1956). Probability, logic and ESP. *Science* **123**, 15–17.

6
Scientists and Scientific Research

Fisher to J.R. Baker: 4 November 1940

Thanks for your letter on your new Group or Society on behalf of individual freedom in research.[1] I subscribe all too easily to your four propositions, too easily because I am sure they would receive adequate lip-service from the most dangerous opponents of your ideas. I want something more positive, and therefore more difficult. If a group could hammer out a set of consistent views, principles, or policy, answering the question, 'How should research be *organised*?', it would have far more practical influence. This, of course, would only be possible by frequent meetings over a period of a year or two.

The first answer, that research cannot, or should not, be organised at all, on the ground that all the outstanding workers in the past found the means to do their work, and, in fact, achieved their discoveries without organisation, is not a little unsatisfying to those who suppose

(a) that the great majority of whole-time research men now and in the future live on salaries from the State, Universities and Research Institutions, and have their expenses defrayed from the same sources;

(b) that the professional administration responsible for these funds will be largely ignorant of, and often indifferent or antagonistic to, the advance of knowledge.

(c) In most lines of research one relies intimately on past work of living contemporaries, and often consciously does a good deal with their problem specifically in view.

(d) Many workers hold strongly the belief that their work could be greatly aided by the technical co-operation of other specialists, e.g. in the production of new apparatus for their special requirements, and think, perhaps optimistically, that this need can be met by collaboration within large research institutions.

(e) Finally, I personally submit that when I have solved a problem to my own satisfaction, I have still made no contribution to Science until my ideas and methods have been grasped by at least some other minds, and that the contribution only really materialises when this number of minds approaches the hundreds.

It is a tragedy of the history of science that we have, I fancy, no good example of a great scientist using his penetration, knowledge, and prestige primarily to enable others to do more interesting, or more important, or more successful work.

All this implies that the word 'organisation' is not to be taken in the narrow sense of projects to be filed in triplicate with allocations approved by a series of remote pundits, but rather in the sense that we really do begin to appreciate the work of a great man when it throws new light on our own problems.

Organisation is merely a name for intention rather than inadvertence in relationships between human beings. The would-be research worker, inexperienced and fresh from a University course, has a number of quite definite needs, which anything worth calling organisation should attempt to supply. I think your group ought to survey these and discover how they can be met. Let me try to list a few here:

1. He needs experience, for his university training has certainly not introduced him to the actual difficulties of personal research, or to the attitude of mind of those who successfully make it their business. Contact with other research workers of various ages is valuable here.
2. Very frequently he needs moral support of the kind given when workers of standing think that his problems, or preliminary results, are of real importance. If he can be brought into contact as assistant with men whose work he already has reason to admire, his confidence may be very greatly strengthened.
3. Whatever the framework of his work, he must feel free to take up and give something like half his time to any problem which excites his interest, and which he feels he is fitted to grapple with. I think I have before compared the recognition by a man of his own problem with falling in love, as it seems to have almost as personal an appeal.
4. As University teaching is bound to have left him with great limitations, both on the methods and the subject-matter of research, it is often of great value to be constrained to face some problem of practical application, especially as this will confront him with a great deal of unscientific uncertainty, and with dealing with men interested only in practical results. The test is sometimes rather a tough one, but research gains immensely whenever correct principles or exact methods can be applied in new fields. Many of the best research men react most happily to a situation in which nothing can be supposed definitely known.

I think it is unfortunate, at least in mathematics, that the aristocratic prestige of the word *pure* should be applied often to ineffectual pottering of a kind which avoids the real difficulties which come in sight when applications are attempted. In your proposition 2 I should say the advancement of knowledge by scientific research is measured by the increase of power which it gives to other men to overcome their difficulties, theoretical or practical. I rather doubt if proposition 4 touches a practical issue. I know a good many scientists who do all their own work. Many of them do not want to have assistants, and would not know how to use them. They do not, in themselves, constitute any problem, except that it often seems disappointing that their earlier work of great apparent promise has not been more fully followed up.

[1] Baker's letter proposed the formation of what became known as the Society for Freedom in Science.

[The following Memorandum on Freedom in Scientific Research written by Fisher was sent to Baker in March 1941.]

Freedom in scientific research, like freedom in citizenship, is a relative term. The free man does not claim to live without prohibitions or restraints, but to be subject only to such restraints as are necessary to safeguard the just claims of others. A programme of anarchism, with the destruction of all forms of authority and organisation, is an emotional gesture, implying both impatience and laziness. It is an evasion of the problem, not an attempt to state or to solve it.

Who are the others to whom a professional research worker should pay regard, to the extent of curbing his own desires lest their just claims be infringed? First, I suppose, should be placed his obligation to civilization at large, to advance knowledge in his special field. This is, in particular, an obligation to those other scientific men having similar interests, who may be able to make use of his work. *Secondly*, one has obvious obligations to his employer, the State, a University, a Research Laboratory, or a Commercial Firm, whose salary he accepts upon a more or less clearly understood agreement as to the services which it is his duty to supply. *Thirdly*, the majority of workers have a scientific chief in whose assistance they are employed, and collateral workers with whom they may, and in most cases should, collaborate. To make claims for Liberty, without regard to such a framework of obligations, seems to be idle and unprofitable. To make claims for scientific freedom compatible and closely coherent with such a framework may well remove the grounds for legitimate grievances, and even be essential for maintaining in scientific work the intellectual integrity which it unquestionably requires. Minimum standards may be insisted on, with the whole weight of scientific opinion behind them, and used directly in aid of particular cases, provided they take account of the actual obligations of the scientific worker; whereas extravagant claims may win only a half-hearted verbal assent, since it manifestly appears that they cannot always be realized in practice.

I believe the most fruitful approach to the whole group of problems arising in this connection is to consider the requirements of the junior worker from the time he becomes engaged in scientific research, from the point of view of his personal progress and development. At this stage even those with the most brilliant academic careers will be, in certain important respects, definitely inexperienced. Confronted with differences in the apparatus, the material, and scale of the work around them, many of the less enterprising, if they had the option, and if they did not feel under an obligation to carry on the new work they had accepted, would certainly tend to gravitate to problems of a more academic nature, and more within the scope of their previous experience. I have frequently been impressed with the advantage that a worker has gained, especially in self-confidence and resourcefulness, by being confronted, *malgré lui*, with problems of the so-called applied or practical character, which in reality are problems requiring exploration and judgement rather than the application of a ready-made formula.

A second respect in which the new worker is in fact seriously handicapped by inexperience is in the practical conduct and design of experiments. In this case I believe much more could be done in post-graduate work at the Universities than is, in fact, done to prepare for the future scientific career. For the logical principles of experimental design and of reasoning from experimental results are of great interest to post-graduate students, who would appreciate definite courses in this subject. In fact, however, and at present, the majority of scientific workers enter their careers without this preparation, and learn as they go, by their own mistakes and those of their colleagues.

In these two respects it would be often far from helpful, and sometimes disastrous, if the scientific worker were to commence his career under the impression that it was his duty, thenceforth, to add to human knowledge by means of his own unaided and unguided efforts. Nevertheless, the conditions of his employment may be well or ill adapted towards fitting him step by step to exercise genuine independence of judgement; to develop the capacity for independent experimentation, and, finally, to plan a daring, original and fruitful research program. To give a basis for independence of judgement it is, I believe, of far more importance than is generally supposed that the worker should allot a considerable fraction of his working time to making himself acquainted with the published literature. This is, I suppose, in all subjects very large in volume, very diverse in scientific cogency, and varied in the ideas propounded. The student's reading may have been well directed, but it has covered almost certainly only a very small fraction of the published researches bearing on his problems. The literature is often not only very extensive, but intrinsically difficult, and requires time and deliberation, if it is to be properly assimilated. The junior worker should receive encouragement, and his duties should allow him to read, with adequate care, far beyond the limited series of papers which his chief may indicate to him as necessary for the understanding of the work of his Department. The object should be to familiarise the reader with the stages whereby current opinions have been developed, and to train him, by scrutinising the results of past experimentation, to exercise his own judgement on the value of the experimental evidence available on different disputable points. There would be more confidence that real independence of judgement was expected from all, if it were true in all laboratories, as it certainly is in some, that many assistants are better acquainted with the current literature than is the Head of the Department. Since the study of the literature is time-consuming, and many junior workers feel that their other duties have a prior claim on their time, I am inclined to lay stress on the stipulation that extensive study of the literature, including participation in scientific discussion, should be a definite part of the duties of scientific assistants, recognised as such in the terms of their appointments.

A second share of time must be deducted from that earmarked for departmental duties for independent research, initiated spontaneously by scientific assistants, within the field for which the department is equipped. That a reasonable fraction of the time available should be made free for this purpose, not as a special favour, but as a normal right attaching to research appointments is a claim which, I believe, should be made universally. We must recognise that this right will often be used unwisely, and that the investigations proposed will often be ill-considered. I do not believe these facts should be thought to justify interference. Anyone employed in scientific research has a primary right to make his own reputation, at least among his immediate colleagues, and to disregard criticism at his own risk. In respect of publication, there will, I believe, be least embarrassment if the Head of the Department is not expected, *ex officio*, either to approve or to facilitate publication by his assistants. This suggestion, I know, differs considerably from the traditions and practice of many departments, but latent grievances in respect to publication are unfortunately rather common, and if we aim, as I think we ought to do, at the frankest cooperation of independent minds within our research departments, I believe it affords the only satisfactory basis.

A word might be said here on a not uncommon situation, sometimes felt to be oppressive, in which the Head of a Department is also the Editor of an important journal and may seem, to junior workers, thereby to control the only opening for effective publication. As the result of personal experience, I believe this fear to be so much exaggerated as to be in fact illusory. Other journals which at first sight may seem less suitable channels, may, in fact, bring the work published before a more valuable audience. In my own experience, for many years the leading English Journal in Mathematical Statistics was closed to my papers, through the disapproval felt for them by the Editor at the time. In consequence, my work was published in about 30 more or less appropriate journals, and thereby came under the notice of a far wider public. Being published in relation to its applications, moreover, it came before readers whose needs it met, and who wanted it for use; a much more important class than the self-conscious theoretical experts whose reaction is merely to criticise, or to ignore anything subversive in their own field. A paper may be more quickly appreciated, appearing in one journal rather than in another, but I do not think it ever happens that the obscurity of the journal is a reason for detracting from the importance which is attached to a paper.

To sum up:— The conditions of scientific research vary enormously, according to the employer and the size of the Department in which they are carried out. We may greatly aid the personal development of a young research worker by associating him in a Department with colleagues of varying experience, and by setting him problems in which some judgement and exploration is necessary. Two safeguards are suggested, both of which, unfortunately, detract from the amount of time he will be able to give to departmental work.

(I) That it shall be a definite part of his duties to read widely in the relevant literature, and to form a critical judgement of the value of the experimental work published.

(II) That he shall be free, in respect of a reasonable fraction of his time, to undertake independent personal research on his own responsibility.

Fisher to H. Corbière: 28 *May* 1947[1]

I have earned my own living most of my life by employment as a research worker or University teacher.

As a young man I did not find it really difficult to get scientific papers published, although anything of value, that is anything leading to ideas unfamiliar to the current authorities, was liable to rejection by one journal, though publishable in others.

I found my work rather quickly appreciated in the United States, where interest in statistical methods and genetics is more widespread, and there are probably more people on the alert to notice work of potential value. This was particularly so with the first edition (1925) of my *Statistical Methods for Research Workers*, which was unfavourably received (so far as reviews are concerned) in this country, but about which I had at once several most encouraging and appreciative letters from the United States, where it evidently was felt to meet a need.

Work that has reached what is nearing completion has, of course, lost its importance. Probably the most useful aspect of my early work was to develop accurate tests of significance and methods of statistical analysis clarifying and largely replacing the deeply involved entanglement in which the subject had been left by the biometrical school of Karl Pearson. I have found people also much interested in what can be done to improve technically the design of experiments and the logic of the inductive process by which experimental data are interpreted. I have always been interested in Genetics, especially in relation to evolutionary theory.

[1] Corbière had sent Fisher a number of questions about his life as a scientist.

Fisher to H. Corbière: 2 *June* 1951[1]

I find it difficult to give an opinion on the excerpts from Jules Romains' article which seems to require some sort of semantic analysis before their meaning is quite clear. The first sentence seems to be true, merely by definition of 'major importance', for an historian of Science will pick out a few advances for emphasis merely because he judges their importance to be greater than that of numerous others, without, however, having any absolute scale on which importance may be judged.

If as a result of nuclear fission interplanetary travel became possible, it would be, I suppose, a prodigious and at the same time not confidently forseeable development, but scarcely a second rate result.

In the two fields with which I am most familiar, namely Statistics and Genetics, the scene has been so transformed during the present century that it is difficult to appreciate earlier work owing to the relatively childish nature of their language and concepts. An immense field has been opened out in each case for detailed exploration and this work might be judged by a literary man to be second rate although perhaps intellectually and economically fruitful, and in any case apt in the sense of being appropriate to our present state of intellectual advancement.

When science is, as it is tending to become, highly organised through large scale government agencies, every major advance may be regarded as an immediate set-back, since it is liable to disintegrate the basis of the existing organisation.

[1] Corbière had sought Fisher's opinion on the following excerpts which he said were taken from an article entitled, 'An epoch in science' by Jules Romains in the magazine *Science et Vie*.
'The exploration of the real knows no "pre-determined bounds"; it seems, however, that of all discoveries, those of major importance are of the same order numerically as the important discoveries of stars — "very few." The future should fall back "on prodigious and, at the present time, unforseeable developments of second-rate results and the application thereof."
Pure science has already probed into the inmost secrets of the Universe.
"Our time has little chance of being outclassed." '

Fisher to H. Corbière: 24(?) *March* 1957[1]

Morality is derived from the social tradition, including the religious tradition, interpreted in the light of experience. Such understanding of the real world as can be supplied by science, including biological science, is therefore quite relevant in the development of a morality.

[1] Corbière had asked Fisher if he thought one could draw a moral from Science and in particular from Biology.

Fisher to J.O. Irwin: 25 June 1934

I think you are Chairman of the study group which is responsible for getting Wishart to compile the Bibliography of Agricultural Statistics, published in the first number of the Society's supplement.[1]

I think you will agree with me that a bibliography ought to be rather carefully accurate, and that its value is seriously impaired if personal bias is admitted. The latter seems evident in entries 49[2] and 155[3] for Wishart does not, I understand, deny that his paper in the *Arch. f. Pflanz.* showed a complete misunderstanding of the limitations of the layout adopted and by ignoring the non-orthogonality of some of the interactions arrived at an estimate of error which was so much too low as to lead him to report a number of interactions as significant without any justification.

If the paper is still to be recommended to students as describing principles and arithmetical details of the modern layouts, the least a bibliographer should do is to point out that these principles and details have been later disputed.

Yates' paper (No. 155) reexamined this experiment in detail and showed exactly where Wishart had gone wrong. This was the main point of his paper, but Wishart's summary says nothing about it.

As regards general carelessness in descriptions, the next paper to Wishart's (No. 50)[4] affords an example of what I mean, for 'Alumnus', following Fisher is said to use multiple correlation, when in reality the method is a direct use of non-linear regressions, but there are a great number of examples of this kind of thing.

I think the idea of a descriptive or annotated bibliography is an excellent one and perhaps I might suggest to your committee the propriety of one or two safeguards which might avoid its abuse.

The bibliographer might submit his draft to a meeting of the committee in his absence, with a view to their selecting a number of papers regarded as controversial, the comments on this particular class of papers might: a) be left to the authors and have their initials appended or b) be drafted by the committee or c) by one or more experts in the subject nominated by the committee.[5]

[1] Wishart, J. (1934). Bibliography of agricultural statistics, 1931–33. *J.R. Statist. Soc., Suppl.* **1**, 94–106.

[2] '49. Wishart, J. The analysis of variance illustrated in its application to a complex agricultural experiment in sugar beet. *Arch. f. Pflanz.*, 1931, *5*, 561–584.

A particular experiment with a complex lay-out is used for describing the principles and arithmetical details of the modern lay-outs.'

[3] '155. Yates, F. The principles of orthogonality and confounding in replicated experiments. *J. Agric. Sci.*, 1933, *23*, 108–145.

A theoretical study of the lay-out of the modern field experiment. Orthogonality is shown to be at the basis of the simple accepted methods, whereas more complex lay-outs may be deliberately non-orthogonal to secure additional information within the limits of size imposed.'

[4] '50. "Alumnus". A comparison of the effect of rainfall on spring- and autumn-dressed wheat at Rothamsted Experimental Station, Harpenden. *J. Agric. Sci.*, 1932, *22*, 101–114.

The method used following Fisher is that of multiple correlation of yield on constants representing the distribution of rainfall.'

[5] Wishart wrote to Fisher after receiving a copy of this letter, saying that Paper 49 was a description of an experiment that Fisher had designed and that the paper had been read and approved by Fisher before publication. For Fisher's reply to Wishart, see p. 355.

H. Jeffreys to Fisher: 23 November 1938

I expect you are snowed under with letters of congratulations,[1] but cannot restrain myself from writing another. You have done a great work in bringing sanity into biology, and I regard it as a tribute to biologists that they have recognized it.

I am having a scrap with the C.P.S.[2] editorial committee. On the advice of their referees they have cut out about half of a paper of mine, including the part that concerned Stevens's problem, and are taking the line that it is only as a personal favour that they can contemplate publishing any of it, on the ground that their referees object to the fundamental principles. I take the line that where there are different possible modes of treatment the only fair thing is to publish both, but as far as possible, usually by suggesting personal discussion, to avoid waste of space by emphasizing apparent differences that come from sheer misunderstanding. . . .

[1] Fisher had been awarded the Royal Medal of the Royal Society.
[2] Cambridge Philosophical Society.

Fisher to H. Jeffreys: 24 November 1938

I think it is time something were done with respect to the Secretary and Editorial Committee of the C.P.S. Two and a half years ago I was forced to resign a long-standing Fellowship of the Society[1] through the attitude of one, A.H. Wilson, whom I do not know apart from this correspondence.

The situation was that M.S. Bartlett thought he had detected an error in a paper of mine dealing with a test of significance originally put forward by W.U. Behrens. Bartlett's paper[2] purporting to expose this error was published by the Society, without notification by him or them to me as the author criticised; they thereby took the ordinary risk of receiving a reply. The Referee objected to my reply on the ground that I had chosen a degenerate case, although this case was not chosen by me, but by Bartlett, and reiterated a number of facts which I had demonstrated and emphasised in my note, as though I had overlooked them and as though they were opposed to my point of view. Nevertheless, Mr. Wilson and his Committee decided to refuse publication of my answer in which I gave my grounds for dissenting from Bartlett's conclusions. This course was, I think, particularly objectionable as I had been myself responsible for first putting forward the fiducial argument, and had, therefore, to continue to shoulder the responsibility of preventing its being misunderstood, as I think Bartlett had done.

The excuse was put forward that I had misrepresented Bartlett, but when Wilson was challenged to cite the particular passage which was thought to be a misrepresentation, with a view to my modifying it if it really seemed to be such, he failed to produce any case.

I do not know who is responsible for the management of the Philosophical Society, but think it must contain a sufficient number of reasonable men to conclude that this is not a reasonable or possible way to treat Fellows of the Society. I am enclosing the correspondence, in case it may be of service in getting the Society to conduct its business in a more tolerant way.

Many thanks for your kind congratulations.

[1] Fisher's explanation of his resignation shows the situation was not as simple as is perhaps suggested by Bartlett's statement that he had heard 'from Wishart that Fisher resigned from the Cambridge Philosophical Society when they published my paper'. (in *The making of statisticians*, ed. J. Gani. Springer, New York (1982)).

[2] See Bartlett's letter of 16 April 1937 (p. 51).

H. Jeffreys to Fisher: 27 *November* 1938

There are some reasonable people on the Council; I have passed your literature on to one of them, who was very outspoken. It does seem a bit thick that they should have published Bartlett's paper without consulting you and then made it a condition of accepting your paper that you should get his permission. . . .

I had heard vaguely that you had left the Society, but had no notion of what it was about, and rather thought it might have been over my 1936 significance tests paper — in which case I should have agreed with about 3/4 of what you said. The Editorial Committee consists of Hardy, Hodge and Hall (pure mathematicians), Wilson (quantist, knows about metals), Dee and Ratcliffe (Cavendish physicists, young and enthusiastic). Except Hardy I don't think any of them is over 35; the pure mathematicians all have experience of the L.M.S.[1] council, but I should doubt whether the others have had experience on any other society. The trouble about pure mathematicians is that their decisions can be so clear cut in their own job; a result follows from the postulates or it doesn't, it is new or it isn't, and it doesn't matter what the postulates are. A paper can be short and to the point, in which case it is trivial, or long and (to a Philistine like me) not worth doing anyhow, and then it is important. It's a bad training for statistics. A mathematical logician would be more to the point; they don't believe in universal agreement. Frank Ramsey called some of *Principia Mathematica* 'sloppy' in the L.M.S. and Russell was the referee. His report was 'quite right, it is'. . . .

[1] London Mathematical Society.

Fisher to H. Jeffreys: 28 *November* 1938

. . . Many thanks for your information about the Cambridge Philosophical

Society. I will want to have my correspondence back when you have done with it; but I think it is necessary for you and your friends to have it all in order to see just how the Society's business has been conducted.

H. Jeffreys to Fisher: 1 *February* 1939

Here are your papers (by the way you didn't sign your letter but authorship was assignable by internal evidence). Appleton has had them and has been taking a keen interest in the matter, but he is leaving in a few weeks to become secretary of the D.S.I.R. and I don't expect that he will have time to do anything. I suggested a set of possible Council rulings that would meet both our problems, but I don't know whether anybody else can be found that would be both willing to take them up and have enough push to get them through. I am, meanwhile, making such efforts as a private member can.

Fisher to H. Jeffreys: 3 *February* 1939

Thanks for returning the correspondence. For the rest, more power to your elbow!

H. Jeffreys to Fisher: 13 *February* 1939

I wrote to Hodge as Editor of the C.P.S. *Proceedings*, suggesting that the Editorial Committee might climb down a bit over your row with them. The suggestion that you might rejoin if they did was my idea, as I told them. My motive is the selfish one that I like to have my mistakes pointed out before publication, and so far in the C.P.S. I have been able to get away with anything if it was wrong. I enclose Hodge's reply, which is as cordial as one could wish, though I should welcome something a bit more specific and am answering with a couple of suggestions that they might adopt. One of them has, I have been unofficially told, been adopted already, but I want to make it official. . . .

Fisher to H. Jeffreys: 15 *February* 1939

Thanks for sending me Hodge's letter, which I return herewith. It is good of you to suggest the possibility of my rejoining the Society, but I am glad you made it clear that this was your own suggestion, as I only mentioned the affair to you in corroboration of your own impression that the affairs of the Society were being pretty badly mismanaged.

So far as I am concerned, I resigned for an entirely specific reason, which I believe I made clear to the Secretary, namely, that the right of answering a misunderstanding and, as I think, a misrepresentation, of my own work in the Society's *Proceedings* was peremptorily denied. I suppose I ought, at the time, to have taken the matter up with the President, whoever he may be, but in the circumstances I was content to leave the matter as it was.

I might, indeed, sometime, like to rejoin the Society, especially if I ever came to Cambridge, but I should really hesitate to do so if the Secretary at the time thought the right way to treat Fellows whose work had been criticised in the *Proceedings* was the way in which I had been treated. . . .

Fisher to H. Jeffreys: 20 *February* 1939

Since writing to you last I find that Yates has been thinking about the problem of samples with unequal variances and has written a short paper about it. As he was one of the people who thought at first there was something in Bartlett's objection, it would afford a good opportunity for the Cambridge Phil. Soc. to put themselves right, if they cared to publish his paper. I think he would be willing, if that was their intention, to send it in for the *Proceedings*.

However, the prevailing attitude two years ago was so inexplicable that it may be that your efforts will reveal difficulties of a kind I have not suspected.

H. Jeffreys to Fisher: 16 *April* 1940

. . . By the way, now that Yates's paper[1] is out, are you ready to join the Camb. Phil. Soc. again?

[1] Yates, F. (1939). An apparent inconsistency arising from tests of significance based on fiducial distributions of unknown parameters. *Proc. Camb. Phil. Soc.* **35**, 579–91.

Fisher to H. Jeffreys: 18 *April* 1940

Thanks for your letter. Yes, I should be happy if you cared to put me up for the Camb. Phil. Soc., as, with how much reluctance I know not, they have now done what is reasonable towards correcting their previous mistaken attitude.

L.F. Richardson to Fisher: 30 *July* 1937

Thanks for two papers from the *Annals of Eugenics*. I have read the one about 'Karl Pearson and the Method of Moments' [*CP* 149] with considerable interest and a tinge of regret. With interest, because I look to you as the authority on how these things should be done. The agenda on pp. 316–317 interested me particularly. It is true that I seldom do statistics, but one never knows when one may have to. With a tinge of regret because it seems to me unnecessary to accuse K.P. of having deliberately falsified his figures. There might easily be other explanations. Anything like deception was entirely contrary to his character as I remember it. Fiercely jealous he was certainly, and I can well believe, without knowing the details, that he treated you abominably. That was in a way a compliment; for it meant that he regarded you as a formidable rival. But I think he had a relentless love of truth. Some months ago I read the obituary notice of K.P. which Udny Yule wrote for the Royal Society. I read it with delight! For Yule seemed to me to have drawn K.P.'s portrait according to the motto 'nothing extenuate nor aught set down in malice'.

Fisher to L.F. Richardson: 4 *August* 1937

Thanks for your letter of July 30. I shared your regret very heartily when I decided what I had to do about Karl Pearson's attack on Koshal. It is appalling that the very last paper that he wrote should have shown such vindictiveness, and should have required such criticism. I enjoyed Yule's obituary, as you did, and thought it most fair and well-balanced. The question whether Pearson's reputation will stand so high with a later generation as it did with Yule's is, I think, beside the point in an obituary, which should aim at being a contemporary appreciation.

As for accusing him of having deliberately falsified his figures, this I was obliged to do, as soon as I understood what he had really done. That he deliberately falsified them is a statement I base only on his own assertion. He adds the explanation that it was done in order to make his results more comparable with Koshal's, but he does not compare them with Koshal's, but with other material with which it makes his figures less comparable. Let who will accept his explanation. I, for one, cannot.

As for aspersions on the honesty of scientific colleagues in general, I do not think you can feel more strongly than I do that they are undesirable, and hitherto I have consistently avoided them. My reason for making an exception is that they seem to be less undesirable than the results of successful dishonesty. My immediate motive, of course, was to discredit Pearson's despicable attack on a comparatively helpless victim, that is on Koshal's competence, as a statistician, an attack which had already been used to threaten Koshal's scientific prospects. If such methods are allowed ever to be successful, it is certain that there are other vindictive persons who will not scruple to use them. The fact that the example was set by a name widely honoured made it particularly important to show that honest work can, with goodwill, always be defended against dishonest attack.

If you travel much, or hear much of the squabbles and disputes among foreigners, you will have realized, with much the same dismay as I have done, that what we, in this country, accept as ordinary standards of honour are not universally so accepted abroad.

Peoples are now engaged in scientific research without any long native tradition as to propriety in its conduct. In the eyes of any careful reader of Pearson's paper, once I had decided to answer it, I had to stand either as exposing a fraud, or as conniving at it.

N.M.V. Rothschild to Fisher: 6 *February* 1952

I mentioned to you that I had been thinking about the status of biometricians and statisticians in scientific publications. This problem affects the Agricultural Research Council when questions involving the grading or promotion of biometricians is under discussion. If a statistician is recommended for promotion or employment in a particular grade, it is inevitable that the A.R.C. should wish to scrutinize his publications. I don't see how this can be avoided in present circumstances, even though we may attach considerable importance to the views of Directors and referees.

Several cases have come to my notice where the candidate has published very little though his Director said he was first class and that, though he had not published much, he had often been acknowledged in other people's papers.

It seems to me that the questions at issue are:

1. What are the conditions in which a biometrician should be a co-author of a paper?
2. What are the conditions in which he should be acknowledged at the end of the paper?
3. What are the conditions in which the biometrician's work should be in an Appendix at the end of the paper?

I think one can examine these questions from at least two points of view. First, did the biometrician's contribution involve original statistical research? Secondly, did his contribution involve him in a lot of work?

In the field of agricultural research it quite often happens that the biologist does some experiments but the results are quite incomprehensible without statistical treatment; the biologist may not be able to do the necessary computation even with the advice of a statistician. In such a case the statistician, though not making much contribution from the pure statistical point of view, is involved in a lot of work. Should he be a co-author of the paper? In other cases the biologist may be able to do the computation provided he is given detailed instructions by the statistician. Should the statistician be acknowledged at the end of the paper on these occasions?

Would it be possible to make a brief classification on the above lines of the different types of contributions that statisticians can make to experiments; and to lay down general principles as to how these contributions should be acknowledged? There will always be borderline cases, I imagine.

I wonder whether you would consider submitting this letter to any of the biometrical societies that you may think fit, and let me know the results of their deliberations. I have in mind that the Agricultural Research Council might inform Directors of Institutes and other persons concerned of any 'findings' that may eventuate, and ask them to comply with these so far as is possible.

Fisher to N.M.V. Rothschild: 7 *February* 1952

Thanks for your letter of February 6th on the status of biometricians in joint work. I think it sets the ball rolling and I have sent it to Yates, now President of the Biometric Society, with an explanatory letter.[1]

[1] See p. 356.

Fisher to P.M. Sheppard: 11 *November* 1949

Thanks for your letter. I do not really see why my name should appear in this matter at all.[1] It has always been a pleasure to me to carry out any statistically detailed work which may suggest itself. If the author values this he can say so in about five gracious words, and, if he fears that his readers will not understand it, he can often add an appendix in the name of his associate. Joint papers are to be avoided for doctorate purposes, simply because university boards have such incompetent regulations with respect to them. I recently had a series of papers for a higher degree in which the author, the only interested and therefore least reliable party to the transaction, had provided a statement representing the importance of his own and the unimportance of his collaborators' contribution to each paper. As I knew some of these statements to be untrue I refused to act as referee. I do not think it is a necessity, but I do think it a fact, that university boards generally must be expected to be about as incompetent as this. So put your views before Ford, and whatever you agree to will be agreeable to me. . . .

[1] Sheppard had suggested that Fisher's name should be included with his on a paper reporting work on *Panaxia dominula*.

Fisher to W.H. Thorpe: 12 *May* 1953

Thank you for your letter and enclosure.[1] I do not know what to say about the proposed organisation, for anything like an agitation group, however well started, is almost bound to be parasitised by political agitators and to do more harm than good, as indeed I think is true of the many organisations aimed at preserving peace during the last half century. Most of us have paid very heavily for their folly and fanaticism. Yet surely one would want to help the aims you had in view.

[1] Thorpe had asked Fisher if he would be a signatory to an open letter to Cambridge scientists calling for the formation of 'a local organisation for social responsibility in science'.

Fisher to J. Wishart: 17 *September* 1934[1]

I think you know I have always held and often expressed the opinion that whenever a mistake has been made in scientific work, it is best for the author of it to be the first to point it out. If he cannot be first he usually has the opportunity of being a good second. My main criticism of your bibliography is that you recommend in it, as an example of method, a paper in which the statistical methods and the numerical conclusions had been shown to be seriously in error.

This error arose from your attempt to use the experiment to study certain possible interactions which the experiment was designed to ignore and which had previously been ignored in previous experiments. It was obvious that it must be theoretically possible to use the data to examine these interactions and when you said that you had done so, and found that many of them were significant, I certainly supposed that you had done the work right and did not imagine that the supposed significance was due to your having greatly under-estimated the residual error.

To Yates belongs the credit of having read your paper with sufficient care to see where you went wrong in particular and principle and to show how it should have been carried out. If your mistake was a natural one in the current state of statistical knowledge then you should have given Yates the credit for having made a material advance in the subject, in addition to having ascertained the correct conclusion to be drawn from this particular experiment.

I think if you will re-read the two entries referring to your own paper and to that of Yates you will see why they should have made an unfavourable impression upon all who were aware of the circumstances.

[1] For background to this letter, see Fisher's letter of 25 June 1934 to Irwin (p.348). See also Fisher's letter of 27 December 1933 to Wilks (p.299).

Fisher to F. Yates: 7 February 1952

. . . Rothschild has been consulting me on the ethics and proprieties of co-operation between biometricians and others in scientific work and is concerned that the biometricians do not receive such formal and explicit acknowledgment as should weigh with the A.R.C. in assigning grades, etc., or with universities for higher degrees.

Evidently, what would suit him best would be for the Biometric Society to draw up formal recommendations which should make clear to directors of research stations, and not least, to the biometricians themselves, exactly what forms of recognition or acknowledgment their colleagues think appropriate to different cases.

Of course, the cases are quite diverse, and it is not to be assumed that participators in joint work always form a just opinion as to the originality or scientific worth of their several contributions. In fact, I think we must proceed partly by options, e.g. if A does a job of statistics which meets B's requirements, but which B had supposed (rightly) to be of a comparatively routine nature within the known techniques of the statistician's craft, while A imagines (mistakenly) that he has done something exceptionally bright, it might possibly be held that none the less, A should have the option of adding an appendix explaining or demonstrating his method under his own name.

I shall be very much interested to hear how you react to all this and how you think the Biometric Society should take it up.

F. Yates to Fisher: 8 *February* 1952

. . . Rothschild did mention the question of giving due credit to statisticians when I saw him last.

To dispose of the least important category first, I have always found that acknowledgements at the end of a paper mean very little. Firstly, one cannot stop authors acknowledging one's help, when in fact all they have done is to have a conversation and then go away and ignore everything that has been said. Secondly, different authors seem to have very different standards as to what merits acknowledgment. If, therefore, acknowlegments are really to mean anything we shall have to lay down some form of wording which will differentiate them from the casual acknowledgments which are thrown around by some authors.

More important, I think, are the other two categories. My own opinion is that if the statistician has contributed to the interpretation of the results as a whole and is prepared to make himself jointly responsible for the whole substance of the paper, then he should be a co-author. If he has contributed by developing new statistical techniques but is either unwilling or unable to associate himself with the work as a whole then these statistical techniques should be made the subject of an appendix or a separate paper, depending on their importance and the space required for their presentation. If an appendix is added to a paper then the responsibility for this appendix rests squarely on the shoulders of the writer and not the author of the original paper.

Editors of journals do, of course, adopt some control over such appendices, but this control is at present not very critical, and I think it might be suggested to them that they should ask the question: 'Is your appendix really necessary?'

These, I am afraid, are my immediate and unconsidered reactions; I certainly think it would be a good plan to have the whole subject ventilated by the Biometric Society. I do find that at present I have to tell some people who ought to know better that it would be appropriate to include the name of one of my youngsters as a joint author in their work.

The problem is, of course, not entirely one-sided. We have a particular headache here in the reporting of survey work and co-operative field experiments in cases where the field work is done by a number of members of the National Agricultural Advisory Service, but in which most of the planning and the critical analysis and consideration of the results is carried out by our people. The difficulty here is that there are too many names of people involved in the N.A.A.S. for them to be placed at the head of the paper, but it does, quite rightly, I think, create bad feeling if the only names are those of the Rothamsted staff. It has been suggested that such papers should have no name at the head, but should state the names of the associated workers in the introduction. This, however, creates great difficulties in indexing. I have suggested that the N.A.A.S. should pick one or two of their people who are particularly interested in the work to be co-authors and make themselves jointly responsible with the people here. This, however, is liable to slow things down as it is difficult to get their agreement in any reasonable time. The alternative would be to put the N.A.A.S. names in the form of 'A.B. Smith *et al.*, National Agricultural Advisory Service' with a footnote giving the names of the others.

All these matters I think require thorough ventilation; there is considerable ill-feeling engendered from time to time and it is important that young scientists should get a reasonable number of papers to their name, as, rightly or wrongly, their prospects of promotion are greatly influenced by this. Actually, I am one of those who believe that a man cannot be a good, or at least a useful, scientist unless he publishes his work. Unfortunately, however, the selection boards tend to read the papers in title only and frequently do not give sufficient consideration to their real merit.

Fisher to F. Yates: 15 *February* 1952

On the enclosed two sheets I have attempted classifications, first of types of publication commonly used in joint work and secondly of types of cooperation.

I suppose any recommendations which we should like the A.R.C. to make to editors or to research directors should be in the form that certain classes of one category are appropriate to certain classes of the other. Anyway, this seems to be the kind of classification which may help us to clear our heads if we are to give useful advice. . . .

[Sheet 1]

The various types of outcome of participating in scientific work which is later to be published can be classified as follows:–

a) Joint authorship with indistinguishable responsibility.

b) Joint authorship in which the authors are prepared to make a joint statement, dividing their responsibility either in respect of the techniques employed, e.g. a physicist looks after the electron microscope, or specifically by sections or paragraphs of the paper. In this case (b), editors should ask the authors to make their statement at the time of publication for the benefit of committees awarding higher degrees, increase of pay, changes of status, etc., who need to apportion credit.

c) A technical appendix for which one writer or group of writers is specifically responsible, and for which the other author or authors have no responsibility, even though it is published with and distributed in separates with the main paper.

d) The author or authors of the paper acknowledge specific aid from another. E.g. 'we are indebted to Professor Einstein for developing the relativity corrections of our formulation'.

e) Acknowledgment of a general character is usually given to the Head of Department or Professor for his kindness, encouragement, advice of an unspecific nature.

[Sheet 2]

Classification of types of co-operation.

1. B has been associated with A throughout the research period, perhaps doing work of a different kind, yet influencing the course of the research during its progress.

2. B, as a scientific employee in the same institution, has done certain work, e.g. chemical analyses required in A's research.

3. B has been consulted on A's initiative on a specific point which he has examined at his request, and the result of which is incorporated in the publication.

The intrinsic importance of the work done by B in these situations may vary very greatly, and its intrinsic importance may not be clearly and justly apprehended by the persons concerned.

F. Yates to Fisher: April 1952

I am enclosing a redraft of the authorship and acknowledgments paper. As you will see I am suggesting that we put this out under our joint names (an example of co-authorship!). When we have all approved of the text I propose, as indicated, to seek the approval of the other members of the Regional Committee.[1] This form of presentation will enable the document to be dealt with without further formality, but if you feel that some alternative form of presentation would be preferable please let me know. When we have the approval of all the biometricians I propose to send it to Lord Rothschild to see if he agrees to its circulation and possible publication in its present form. I think we might then see if *Nature* would be willing to publish it as a matter of general scientific interest.

The chief modification in the revised draft is the inclusion of specimen forms of acknowledgment which you felt were desirable. It might be worth including a suggested form of a note for 'Instructions to contributors'. If you feel this is worth while perhaps you might like to try your hand at drafting one.

[1] The British Regional Committee of the Biometric Society.

Fisher to F. Yates: 28 *April* 1952

I am returning your suggested draft with no modification save on the first page a reference to University degrees, where shared responsibility in joint publications is often a matter of great difficulty.

I suppose any instruction to contributors by editors must be exceedingly brief and might be no than 'In cases of joint authorship, authors should, so far as is possible, specify their respective contributions if individual credit is to be reserved in regard to the conferment of Doctorates, etc.. Acknowledgments also should be, so far as possible, specific.'

Notes on Correspondents

Aitken, Alexander C., FRS (1895–1967). Professor of Mathematics, University of Edinburgh, 1946–65. Author of *Statistical mathematics* (1939), etc.

Baker, John R., FRS (1900–84). Reader in Cytology, University of Oxford, 1955–67. Author of *Science and the planned state* (1945), etc.

Barnard, George A. (1915–). Reader and then Professor of Mathematical Statistics, Imperial College, London, 1948–66. Professor of Mathematics, University of Essex, 1966–75. A Vice-president of the Royal Statistical Society during Fisher's Presidency, 1952–4.

Bartlett, Maurice S., FRS (1910–). Assistant Lecturer in Statistics, University College London, 1933–4. Statistician, Imperial Chemical Industries Ltd., 1934–8. Professor of Statistics, University of Manchester, 1947–60, and University College London, 1960–7. Professor of Biomathematics, University of Oxford, 1967–75. Author of *Essays in probability and statistics* (1962), etc.

Behrens, Walter U. (1902–62). Research worker, Institute for Field and Plant Production, University of Königsberg, 1927–31, Institute for Agricultural Chemistry and Bacteriology, University of Berlin, 1932–5, and Agricultural Research Station, Kali-Chemie AG, Hannover, 1935–62. Author of *Mathematische Methoden für Versuchsansteller* (1933).

Bliss, Chester I. (1899–1979). Entomologist, US Department of Agriculture, 1926–33. Research worker in the Galton Laboratory, 1933–5, and Institute for Plant Protection, Leningrad, USSR, 1936–7. Biometrician, Connecticut Agricultural Experiment Station, 1938–71. Author of *Statistics in biology* (1967).

Box, George E.P., FRS (1919–). Statistician, Imperial Chemical Industries Ltd., 1948–56. Professor of Statistics, University of Wisconsin, 1960–. Co-author (with G.C. Tiao) of *Bayesian inference in statistical analysis,* (1973) etc.

Buckland, William R. (1914–). Head, Statistical Advisory Service, London Transport Executive, 1946–60. Manager, Statistical Advisory Service, The Economist Intelligence Unit, London, 1960–79. Visiting Lecturer and then Professor in Management Sciences, Royal Holloway College London, 1970–81. Co-author (with M.G. Kendall) of *A dictionary of statistical terms* (1955).

Buros, Oscar K. (1905–78). Assistant Professor and then Professor of Education, Rutgers University, New Brunswick, 1932–65.

Campbell, Norman R. (1880–1949). Physicist on research staff of General Electric Company, England. Author of *An account of the principles of measurement and calculation* (1928), etc.

Clark, *Sir* Wilfred E. Le Gros, FRS (1895–1971). Professor of Anatomy, University of Oxford, 1934–62. Author of *History of the primates* (1949), etc.

Cochran, William G. (1909–80). Member of Statistical Department at Rothamsted, 1934–9. Professor in Statistical Laboratory, Iowa State College, 1939–46, and in Institute of Statistics, University of North Carolina, Raleigh, 1946–9, Biostatistics Department, Johns Hopkins University, 1949–57, and Statistics Department, Harvard University, 1957–76. Co-author with G.M. Cox of *Experimental designs* (1957).

Corbière, Henri L.G. (1888–1980). Poet and scientific writer.

Cornish, E. Alfred (1909–73). Australian research worker in the Galton Laboratory, 1937–8. Chief, CSIRO Division of Mathematical Statistics, 1954–73. Professor of Mathematical Statistics, University of Adelaide, 1960–4. Co-author with Fisher of *CP* 148 and 281.

Daniels, Henry E., FRS (1912–). Statistician, Wool Industries Research Association, 1935–47. Lecturer in Mathematics, University of Cambridge, 1947–57. Professor of Mathematical Statistics, University of Birmingham, 1957–78.

Darmois, Georges (1888–1960). Director of the Institute of Statistics, University of Paris, 1945–60. Author of *Statistique mathématique* (1928), etc.

Deming, W. Edwards (1900–). Physicist, US Department of Agriculture, 1927–39. Professor of Statistics, New York University, 1946–76. Author of *Statistical adjustment of data* (1943).

Dufrénoy, H. Jean (1894–1972). Director, Plant Pathology Station, Pont-de-la-Maye, France, 1936.

Edelstam, Carl. A Swedish zoologist, 1948.

Eisenhart, Churchill (1913–). Statistician, University of Wisconsin Experiment Station, 1937–47, Statistical Research Group, Columbia University, 1944–5, and US National Bureau of Standards, 1946–83.

Fieller, Edgar C. (1907–60). Mathematical Assistant to Karl Pearson and then Lecturer in Statistics, University College London, 1929–34. Head of Statistics Section, National Physical Laboratory, 1946–52, and Ministry of Supply, 1952–60. A Vice-president of the Royal Statistical Society during Fisher's Presidency, 1952–4.

Finney, David J., FRS (1917–). Member of Statistical Department at Rothamsted, 1939–45. Reader and then Professor of Statistics, University of Aberdeen, 1954–66. Professor of Statistics, University of Edinburgh, 1966–84. Director, Agricultural Research Council Unit of Statistics, 1954–84. Author of *Statistical method in biological assay* (1952), etc.

Fisher, Arne (1886–1944). Actuary, Western Union Telegraph Company, USA, 1920–44. Author of *The mathematical theory of probabilities* (1915).

Fraser, Donald A.S. (1925–). Assistant Professor and then Professor of Statistics, University of Toronto, 1949–. Author of *The structure of inference* (1968), etc.

Fréchet, Maurice (1878–1973). Professor, Institut Henri Poincaré, University of Paris, 1929–49. Author of *Généralités sur les probabilités: eléments aléatoires* (1935), etc.

Frisch, Ragnar (1895–1973). Professor and Director of Research, Economic Insititute, University of Oslo, Norway, 1931–71.

Goossens, Leon (1890–1978). An engineer concerned with spinning and the textile industry, Societe Anonyme de la Lys, Gand, Belgium, 1923–57.

Grant, Robert (1878–1959). A partner in the publishing firm Oliver and Boyd Ltd., Edinburgh, 1898–1928, and director 1929–53.

Gray, Horace. American research worker in the Galton Laboratory, 1935–6. Co-author with Fisher of *CP* 110.

Greenwood, Major, FRS (1880–1949). Professor of Epidemiology and Vital Statistics, London School of Hygiene and Tropical Medicine, 1928–45.

Harrod, *Sir* Roy F., FBA (1900–78). Nuffield Reader in Economics, University of Oxford, 1952–67. Author of *The foundations of inductive logic* (1956), etc.

Heckstall-Smith, Hugh W. (1896–1973). Headmaster, Chippenham Grammar School, Wiltshire, 1935–9. Author of *Doubtful schoolmaster,* (1962), etc.

Hick, William E. (1912–74). Reader in Experimental Psychology, University of Cambridge, 1954–66.

Hotelling, Harold (1895–1973). American research worker at Rothamsted, 1929. Professor of Statistics, Institute of Statistics, University of North Carolina, Chapel Hill, 1946–61.

Immer, Forrest R. (1899–1946). American research worker at Rothamsted, 1930. Associate Professor and then Professor of Agronomy and Plant Genetics, University of Minnesota, 1935–46. Co-author (with H.K. Hayes) of *Methods of plant breeding* (1942). Co-author with Fisher of *CP* 96.

Irwin, J. Oscar (1898–1982). Member of Statistical Department at Rothamsted, 1928–31, and of scientific staff of Medical Research Council at London School of Hygiene and Tropical Medicine, 1931–65.

James, Alan T. (1924–). Visiting Lecturer and then Associate Professor of Mathematics, 1958–63, and Professor of Mathematical Statistics, 1963–5, Yale University. Professor of Mathematical Statistics, University of Adelaide, 1965–89.

James, Geoffrey S. (1924–). Lecturer in Statistics, University of Leeds, 1952–80.

Jeffreys, *Sir* Harold, FRS (1891–1989). Reader in Geophysics, 1931–46, and Plumian Professor of Astronomy and Experimental Philosophy, 1946–58, University of Cambridge. Author of *Scientific inference* (1931), etc.

Jordan, Charles (1871–1959). Professor in the Faculty of Political Economy, University of Budapest, 1933–40. Author of *The calculus of finite differences* (1939), etc.

Kendall, *Sir* Maurice G., FBA (1907–83). Statistician, Chamber of Shipping, 1941–9. Professor of Statistics, University of London, 1949–61. Author of *The advanced theory of statistics* (1943), etc.

Keyfitz, Nathan (1913–). Statistician, Dominion Bureau of Statistics, Ottawa, 1936–59. Professor of Sociology and Demography, Harvard University, 1972–80. Author of *Applied mathematical demography* (1977), etc.

Kislovsky, Dmitriy A. (1894–1957). Professor of Animal Breeding at Vogda Milk Institute, USSR, 1928–33, and at Moscow Zootechnical Institute and Timiryasev Agricultural Academy, 1933–57.

Koopmans, Tjalling (1910–85). Lecturer, Netherlands School of Economics, Rotterdam, 1936–8. Director of Research, Cowles Commission, University of Chicago, 1948–54. Professor of Economics, Yale University, 1955–81. Nobel Prize in Economics, 1975. Author of *Statistical inference in dynamic economic models* (1950), etc.

Kyburg, Henry E. (1928–). Research Associate, Rockefeller Institute, 1961–2. Professor of Philosophy, University of Rochester, New York, 1965–. Author of *Probability and the logic of rational belief* (1961), etc.

McGuigan, Frank J. (1924–). Professor of Psychology, Hollins College, Virginia, 1955–76, and US International University, San Diego, 1983–. Author of *Experimental psychology* (1960), etc.

McKie, Douglas (1896–1967). Lecturer and then Professor in the Department of the History and Philosophy of Science, University College London, 1934–64. Author of *Antoine Lavoisier; scientist, economist and social reformer* (1952), etc.

Maclean, John (1887–1964). Professor of Mathematics, Wilson College, Bombay, 1930–40. Author of *Descriptive mathematics* (1946), etc.

MacMahon, Percy A., FRS (1854–1929). Author of *Combinatory analysis* (1915), etc.

Mahalanobis, Prasanta C., FRS (1893–1972). Director and Secretary, Indian Statistical Institute, 1931–72. Author of *Experiments in sampling in the Indian Statistical Institute* (1961), etc.

Mather, *Sir* Kenneth, FRS (1911–90). Lecturer, Galton Laboratory, 1934–7. Head, Genetics Department, John Innes Horticultural Institution, 1938–48. Professor of Genetics, University of Birmingham, 1948–65. Vice-chancellor, University of Southampton, 1965–71. Author of *Statistical analysis in biology* (1943), etc. Co-author with Fisher of *CP* 145, 146, 192, and 196.

Michie, Donald (1923–). Research Associate, University of London, 1952–8. Professor of Machine Intelligence, University of Edinburgh, 1967–84. Author of *On machine intelligence* (1974), etc.

Needham, Joseph, FRS, FBA (1900–). Biochemist and historian of science. Master of Gonville and Caius College, Cambridge, 1966–76. Author of *Science and civilization in China* (1954), etc.

Nelder, John A., FRS (1924–). Statistician, National Vegetable Research Station, UK, 1950–68. Head of Statistical Department, Rothamsted Experimental Station, 1968–84. Author of *Computers in biology* (1974), etc.

Neyman, Jerzy (1894–1981). Lecturer and then Reader in Statistics, University College London, 1934–8. Professor and Head of Statistical Laboratory, University of California, Berkeley, 1938–55.

Nixon, W.L. Benedict (1920–). Research Assistant in Genetics Unit, Institute of Psychiatry, University of London, 1951–8. Senior Lecturer in Computer Science, University College London, 1970–85.

Prescott, James A., FRS (1890–1987). Professor of Agricultural Chemistry, 1924–55, and Director, 1938–55, Waite Agricultural Research Institute, University of Adelaide.

Quenouille, Maurice H. (1924–73). Research Fellow, Insitute of Statistics, University of Oxford, 1953–5. Professor of Statistics, University of Southampton, 1966–73. Author of *The design and analysis of experiment* (1953), etc.

Rao, C. Radhakrishna, FRS (1920–). Professor, 1949–83, Director and Secretary, 1972–6, Indian Statistical Institute. Author of *Linear statistical inference and its application* (1973), etc.

Richardson, Lewis F., FRS (1881–1953). Author of *Weather prediction by numerical process* (1922), etc.

Rider, Paul R. (1888–1984). American research worker in the Galton Laboratory, 1935–6. Professor of Mathematics, Washington University, St Louis, Missouri, 1939–54. Chief Statistician, Aeronautical Research Laboratory, Wright-Patterson Air Force Base, 1951–64. Author of *Introduction to modern statistical methods* (1938).

Rietz, Henry L. (1875–1943). Professor of Mathematics, University of Iowa, 1918–42. Author of *Mathematical statistics* (1943), etc.

Romanovsky, V. Ivanovich (1879–1954). Tashkent University, USSR. Author of *Discrete Markov chains* (1949), etc.

Rothschild, *Lord*, N.M. Victor, FRS (1910–90). Chairman of Agricultural Research Council, UK, 1948–58, etc.

Salaman, Redcliffe N., FRS (1874–1955). Director of Potato Virus Research Station, University of Cambridge, 1926–39. Author of *The history and social influence of the potato* (1949), etc.

Satakopan, V. Statistician in the Meteorological Office, India, 1939.

Savage, L. Jimmie (1917–71). Professor of Statistics, University of Chicago, 1954–60, and later in the University of Michigan, and Yale University. Author of *The foundations of statistics* (1954), etc.

Sayers, W.H. An English physicist, 1941.

Sheffield, Frederick D. (1914–). Associate Professor and then Professor of Psychology, Yale University, 1951–84.

Sheppard, Phillip M., FRS (1921–76). Research Officer, Zoology Department, University of Oxford, 1951–6. Professor of Genetics, University of Liverpool, 1963–76. Author of *Natural selection and heredity* (1958).

Shewhart, Walter A. (1891–1967). Physicist and statistician, Western Electric Company/Bell Telephone Laboratories, USA, 1918–56. Author of *Statistical method from the viewpoint of quality control* (1939).

Smith, Hugh Fairfield (1904–74). Research worker in the Galton Laboratory and at Rothamsted, 1935–7. Professor of Statistics, University of North Carolina, Raleigh, 1950–7, and University of Connecticut, Storrs, 1965–74.

Snedecor, George W. (1881–1974). Director of Statistical Laboratory, Iowa State College, 1933–47. Author of *Statistical methods applied to experiments in agriculture and biology* (1937).

Sprott, David A. (1930–). Associate Professor and then Professor of Statistics, University of Waterloo, Canada, 1958–.

Summerby, Robert F. (1886–1946). Canadian research worker at Rothamsted, 1931. Professor of Agronomy, McGill University, 1919–46.

Thom, Herbert C.S. (1910–). Statistician in the Weather Bureau, US Department of Commerce, Washington, 1941.

Thorpe, William H., FRS (1902–85). Lecturer in Entomology, 1932–59, and Professor of Animal Ethology, 1966–9, University of Cambridge. Author of *Science, man and morals* (1956), etc.

Tippett, Leonard H.C. (1902–85). Research worker at Rothamsted, 1927. Statistician, British Cotton Industries Association, 1924–65. A Vice-president of the Royal Statistical Society during Fisher's Presidency, 1952–4. Author of *The methods of statistics* (1931), etc. Co-author with Fisher of *CP* 63.

Tukey, John W. (1915–). Professor of Mathematics and then Professor of Statistics, Princeton University, 1950–. Author of *Exploratory data analysis* (1977), etc.

Villars, Donald S. (1900–). Physical chemist, US Rubber General Laboratories, 1938–46, and US Naval Ordnance Test Station/Naval Weapons Centre, 1948–70. Author of *Statistical design and analysis of experiments for development research* (1951).

Walker, *Sir* Gilbert T., FRS (1868–1958). Professor of Meteorology, Imperial College, London, 1924–34. Author of *Aberration* (etc).

Walker, Helen M. (1891–1983). Professor of Education, Teachers College, Columbia University, 1940–57. Author of *Studies in the history of statistical method* (1929), etc.

Wallis, W. Allen (1912–). Professor of Statistics and Economics, University of Chicago, 1946–62. Under Secretary for Economic Affairs, US Department of State, Washington, 1982–.

Wilks, Samuel S. (1906–64). Professor of Mathematical Statistics, Princeton University, 1944–64. Author of *Mathematical statistics* (1947), etc.

Wilson, Edwin B. (1879–1964). Professor in Department of Vital Statistics, Harvard University, 1922–45.

Wishart, John (1898–1956). Research assistant, Galton Laboratory, 1925–6. Member of Statistical Department at Rothamsted, 1927–31. Reader in Statistics, University of Cambridge, 1931–56. Co-author with Fisher of *CP* 56, 85, and 90.

Wood, Eric C. (1904–). Chief chemist, Virol Ltd., London, 1931–49.

Yates, Frank, FRS (1902–). Fisher's colleague at Rothamsted, 1931–3, and Head of Statistical Department at Rothamsted, 1933–68. Author of *Sampling methods for censuses and surveys,* 1949. Co-author with Fisher of *CP* 110 and *Statistical tables*.

Zeller, A. Statistician in a British research centre for pest control, 1957.

List of References to Collected Papers of R.A. Fisher

CP 1 (1912) On an absolute criterion for fitting frequency curves. *Messeng. Math.* **41**, 155–60.

CP 4 (1915) Frequency distribution of the values of the correlation coefficient in samples from an indefinitely large population. *Biometrika* **10**, 507–21.

CP 9 (1918) The correlation between relatives on the supposition of Mendelian inheritance. *Trans. R. Soc. Edinb.* **52**, 399–433.

*CP*11 (1919) The genesis of twins. *Genetics* **4**, 489–99.

*CP*12 (1920) A mathematical examination of the methods of determining the accuracy of an observation by the mean error, and by the mean square error. *Mon. Not. R. Astron. Soc.* **80**, 758–70.

*CP*14 (1921) On the 'probable error' of a coefficient of correlation deduced from a small sample. *Metron* **1**, 3–32.

*CP*15 (1921) Studies in crop variation. I. An examination of the yield of dressed grain from Broadbalk. *J. Agric. Sci.* **11**, 107–35.

*CP*16 (1921) Some remarks on the methods formulated in a recent article on the quantitative analysis of plant growth. *Ann. Appl. Biol.* **7**, 367–72.

*CP*18 (1922) On the mathematical foundations of theoretical statistics. *Phil. Trans.* **A 222**, 309–68.

*CP*19 (1922) On the interpretation of χ^2 from contingency tables, and the calculation of P. *J. R. Statist. Soc.* **85**, 87–94.

*CP*20 (1922) The goodness of fit of regression formulae, and the distribution of regression coefficients. *J. R. Statist. Soc.* **85**, 597–612.

*CP*21 (1922) (with W.A. Mackenzie) The correlation of weekly rainfall. *Q. J. R. Meteorol. Soc.* **48**, 234–42.

*CP*22 (1922) (with H.G. Thornton and W.A. Mackenzie) The accuracy of the plating method of estimating the density of bacterial populations. *Ann. Appl. Biol.* **9**, 325–59.

*CP*24 (1922) On the dominance ratio. *Proc. R. Soc. Edinb.* **42**, 321–41.

*CP*25 (1922) The systematic location of genes by means of crossover observations. *Am. Nat.* **56**, 406–11.

*CP*30 (1923) Note on Dr. Burnside's recent paper on errors of observation. *Proc. Camb. Phil. Soc.* **21**, 655–8.

*CP*31 (1923) Statistical tests of agreement between observation and hypothesis. *Economica* **3**, 139–47.

*CP*32 (1923) (with W.A. Mackenzie) Studies in crop variation. II. The manurial response of different potato varieties. *J. Agric. Sci.* **13**, 311–20.

*CP*34 (1924) The conditions under which χ^2 measures the discrepancy between observation and hypothesis. *J. R. Statist. Soc.* **87**, 442–50.

*CP*35 (1924) The distribution of the partial correlation coefficient. *Metron* **3**, 329–32.

*CP*36 (1924) On a distribution yielding the error functions of several well known statistics. *Proc. Int. Congr. Math., Toronto* **2**, 805–13.

*CP*37 (1924) Studies in crop variation. III. The influence of rainfall on the yield of wheat at Rothamsted. *Phil. Trans.* **B 213**, 89–142.

*CP*38 (1924) A method of scoring coincidences in tests with playing cards. *Proc. Soc. Psych. Res.* **34**, 181–5.

*CP*39 (1924) (with S. Odén) The theory of the mechanical analysis of sediments by means of the automatic balance. *Proc. R. Soc. Edinb.* **44**, 98–115.

*CP*42 (1925) Theory of statistical estimation. *Proc. Camb. Phil. Soc.* **22**, 700–25.

*CP*43 (1925) Applications of 'Student's' distribution. *Metron* **5**, 90–104.

*CP*44 (1925) Expansion of 'Student's' integral in powers of n^{-1}. *Metron* **5**, 109–20.

*CP*46 (1925) Sur la solution de l'équation intégrale de M.V. Romanovsky. *C.R. Acad. Sci. Paris,* **181**, 88–9.

*CP*47 (1925) The resemblance between twins, a statistical examination of Lauterbach's measurements. *Genetics* **10**, 569–79.

*CP*48 (1926) The arrangement of field experiments. *J. Min. Agric. G B* **33**, 503–13.

*CP*49 (1926) Bayes' theorem and the fourfold table. *Eugen. Rev.* **18**, 32–3.

*CP*50 (1926) On the random sequence. *Q. J. Roy. Meteorol. Soc.* **52**, 250.

*CP*51 (1926) On the capillary forces in an ideal soil: correction of formulae given by W.B. Haines. *J. Agric. Sci.* **16**, 492–503.

*CP*55 (1926) Periodical health surveys. *J. State Med.* **34**, 446–9.

*CP*56 (1927) (with J. Wishart) On the distribution of the error of an interpolated value, and on the construction of tables. *Proc. Camb. Phil. Soc.* **23**, 912–21.

*CP*57 (1927) (with T. Eden) Studies in crop variation. IV. The experimental determination of the value of top dressings with cereals. *J. Agric. Sci.* **17**, 548–62.

*CP*58 (1927) (with H.G. Thornton) On the existence of daily changes in the bacterial numbers in American soil. *Soil Sci.* **23**, 253–9.

*CP*59 (1927) On some objections to mimicry theory – statistical and genetic. *Trans. R. Ent. Soc. Lond.* **75**, 269–78.

*CP*61 (1928) The general sampling distribution of the multiple correlation coefficient. *Proc. R. Soc.* **A 121**, 654–73.

*CP*62 (1928) On a property connecting the χ^2 measure of discrepancy with the method of maximum likelihood. *Atti Congr. Int. Mat., Bologna* **6**, 95–100.

*CP*63 (1928) (with L.H.C. Tippett) Limiting forms of the frequency distribution of the largest or smallest member of a sample. *Proc. Camb. Phil. Soc.* **24**, 180–90.

*CP*64 (1928) Further note on the capillary forces in an ideal soil. *J. Agric. Sci.* **18**, 406–10.

*CP*65 (1928) (with T.N. Hoblyn) Maximum-and minimum-correlation tables in comparative climatology. *Geogr. Ann.* **10**, 267–81.

*CP*68 (1928) The possible modification of the response of the wild type to recurrent mutations. *Am. Nat.* **62**, 115–26.

*CP*69 (1928) Two further notes on the origin of dominance. *Am. Nat.* **62**, 571–4.

*CP*70 (1928) Triplet children in Great Britain and Ireland. *Proc. R. Soc.* **B 102**, 286–311.

*CP*71 (1928) (with B. Balmukand) The estimation of linkage from the offspring of selfed heterozygotes. *J. Genet.* **20**, 79–92.

*CP*72 (1928) (with E.B. Ford) The variability of species in the Lepidoptera, with reference to abundance and sex. *Trans. R. Ent. Soc. Lond.* **76**, 367–79.

*CP*74 (1929) Moments and product moments of sampling distributions. *Proc. Lond. Math. Soc.* Series 2 **30**, 199–238.

*CP*75 (1929) Tests of significance in harmonic analysis. *Proc. R. Soc.* **A 125**, 54–9.

*CP*77 (1929) A preliminary note on the effect of sodium silicate in increasing the yield of barley. *J. Agric. Soc.* **19**, 132–9.

*CP*78 (1929) (with T. Eden) Studies in crop variation. VI. Experiments on the response of the potato to potash and nitrogen. *J. Agric. Sci.* **19**, 201–13.

*CP*79 (1929) The statistical method in psychical research. *Proc. Soc. Psych. Res.* **39**, 189–92.

*CP*81 (1929) The evolution of dominance: a reply to Professor Sewall Wright. *Am. Nat.* **63**, 553–6.

*CP*83 (1930) The moments of the distribution for normal samples of measures of departure from normality. *Proc. R. Soc.* **A 130**, 16–28.

*CP*84 (1930) Inverse probability. *Proc. Camb. Phil. Soc.* **26**, 528–35.

*CP*85 (1930) (with J. Wishart) The arrangement of field experiments and the statistical reduction of the results. *Imp. Bur. Soil Sci. Tech. Comm.* **10**, 23pp.

*CP*86 (1930) The distribution of gene ratios for rare mutations. *Proc. R. Soc. Edinb.* **50**, 205–20.

*CP*87 (1930) The evolution of dominance in certain polymorphic species. *Am. Nat.* **64**, 385–406.

*CP*90 (1931) (with J. Wishart) The derivation of the pattern formulae of two-way partitions from those of simpler patterns. *Proc. Lond. Math. Soc.* Series 2 **33**, 195–208.

*CP*91 (1931) The sampling error of estimated deviates, together with other illustrations of the properties and applications of the integrals and derivatives of the normal error function. *Br. Assoc. Math. Tab.* **1**, xxvi–xxxv (3rd edn., xxviii–xxxvii (1951)).

*CP*93 (1931) The evolution of dominance. *Biol. Rev.* **6**, 345–68.

*CP*95 (1932) Inverse probability and the use of likelihood. *Proc. Camb. Phil. Soc.* **28**, 257–61.

*CP*96 (1932) (with F.R. Immer and O. Tedin) The genetical interpretation of statistics of the third degree in the study of quantitative inheritance. *Genetics* **17**, 107–24.

*CP*102 (1933) The concepts of inverse probability and fiducial probability referring to unknown parameters. *Proc. R. Soc.* **A 139**, 343–8.

*CP*104 (1933) On the evidence against the chemical induction of melanism in Lepidoptera. *Proc. R. Soc.* **B 112**, 407–16.

*CP*105 (1933) Selection in the production of the ever-sporting stocks. *Ann. Bot.* **47**, 727–33.

*CP*106 (1933) Number of Mendelian factors in quantitative inheritance. *Nature* **131**, 400–1.

*CP*108 (1934) Two new properties of mathematical likelihood. *Proc. R. Soc.* **A 144**, 285–307.

*CP*109 (1934) Probability, likelihood and quantity of information in the logic of uncertain inference. *Proc. R. Soc.* **A 146**, 1–8.

*CP*110 (1934) (with F. Yates) The 6 × 6 Latin squares. *Proc. Camb. Phil. Soc.* **30**, 492–507.

*CP*113 (1934) The effect of methods of ascertainment upon the estimation of frequencies. *Ann. Eugen.* **6**, 13–25.

*CP*114 (1934) The amount of information supplied by records of families as a function of the linkage in the population sampled. *Ann. Eugen.* **6**, 66–70.

*CP*115 (1934) The use of simultaneous estimation in the evaluation of linkage. *Ann. Eugen.* **6**, 71–6.

*CP*116 (1934) Some results of an experiment on dominance in poultry, with special reference to polydactyly. *Proc. Linn. Soc. Lond.* **147**, 71–81.

*CP*118 (1934) (with C. Diver) Crossing-over in the land snail *Cepaea nemoralis*. *Nature* **133**, 834–5.

*CP*119 (1934) Professor Wright on the theory of dominance. *Am. Nat.* **68**, 370–4.

*CP*123 (1935) The mathematical distributions used in the common tests of significance. *Econometrica* **3**, 353–65.

*CP*124 (1935) The logic of inductive inference. *J.R. Statist. Soc.* **98**, 39–54.

*CP*125 (1935) The fiducial argument in statistical inference. *Ann. Eugen.* **6**, 391–8.

*CP*130 (1935) On the selective consequences of East's theory of heterostylism in *Lythrum*. *J. Genet.* **30**, 369–82.

*CP*131 (1935) The detection of linkage with 'dominant' abnormalities. *Ann. Eugen.* **6**, 187–201.

*CP*132 (1935) The detection of linkage with recessive abnormalities. *Ann. Eugen.* **6**, 339–51.

*CP*133 (1935) The sheltering of lethals. *Am. Nat.* **69**, 446–55.

*CP*135 (1935) Dominance in poultry. *Phil. Trans.* **B 225**, 197–226.

*CP*137 (1936) Uncertain inference. *Proc. Am. Acad. Arts Sci.* **71**, 245–58.

*CP*138 (1936) The use of multiple measurements in taxonomic problems. *Ann. Eugen.* **7**, 179–88.

*CP*139 (1936) (with S. Barbacki) A test of the supposed precision of systematic arrangements. *Ann. Eugen.* **7**, 189–93.

*CP*141 (1936) 'The coefficient of racial likeness' and the future of craniometry. *J.R. Anthropol. Inst.* **66**, 57–63.

*CP*142 (1936) Heterogeneity of linkage data for Friedreich's ataxia and the spontaneous antigens. *Ann. Eugen.* **7**, 17–21.

*CP*143 (1936) Tests of significance applied to Haldane's data on partial sex linkage. *Ann. Eugen.* **7**, 87–104.
*CP*144 (1936) Has Mendel's work been rediscovered? *Ann. Sci.* **1**, 115–37.
*CP*149 (1937) Professor Karl Pearson and the method of moments. *Ann. Eugen.* **7**, 308–18.
*CP*151 (1937) On a point raised by M.S. Bartlett on fiducial probability. *Ann. Eugen.* **7**, 370–5.
*CP*153 (1937) The relation between variability and abundance shown by the measurements of the eggs of British nesting birds. *Proc. R. Soc.* **B 122**, 1–26.
*CP*156 (1938) Quelques remarques su l'estimation en statistique. *Biotypologie* **6**, 153–8.
*CP*162 (1939) The comparison of samples with possibly unequal variances. *Ann. Eugen.* **9**, 174–80.
*CP*164 (1939) A note on fiducial inference. *Ann. Math. Stat.* **10**, 383–8.
*CP*165 (1939) 'Student'. *Ann. Eugen.* **9**, 1–9.
*CP*181 (1941) The asymptotic approach to Behrens's integral, with further tables for the *d* test of significance. *Ann. Eugen.* **11**, 141–172.
*CP*183 (1941) The interpretation of experimental four-fold tables. *Science.* **94**, 210–1.
*CP*205 (1945) A new test for 2 × 2 tables. *Nature* **156**, 388.
*CP*249 (1953) Dispersion on a sphere. *Proc. R. Soc.* **A 217**, 295–305.
*CP*251 (1952) The expansion of statistics. *J. R. Statist. Soc.* **A 116**, 1–6.
*CP*256 (1954) The analysis of variance with various binomial transformations. *Biometrics* **10**, 130–9.
*CP*257 (1954) Contribution to a discussion of a paper on interval estimation by M.A. Creasy. *J. R. Statist. Soc.* **B 16**, 212–3.
*CP*261 (1955) Statistical methods and scientific induction. *J. R. Statist. Soc.* **B 17**, 69–78.
*CP*264 (1956) On a test of significance in Pearson's *Biometrika tables* (no. 11). *J. R. Statist. Soc.* **B 18**, 56–60.
*CP*265 (1956) (with M.J.R. Healy) New tables of Behrens' test of significance. *J. R. Statist. Soc.* **B 18**, 212–6.
*CP*267 (1957) The underworld of probability. *Sankhyā* **18**, 201–10.
*CP*272 (1958) The nature of probability. *Centennial Rev.* **2**, 261–74.
*CP*276 (1958) Cancer and smoking. *Nature* **182**, 596.
*CP*285 (1961) The weighted mean of two normal samples with unknown variance ratio. *Sankhyā* **23**, 103–14.
*CP*288 (1962) The simultaneous distribution of correlation coefficients. *Sankhyā* **24**, 1–8
*CP*289 (1962) Some examples of Bayes' method for the experimental determination of probabilities *a priori*. *J. R. Statist. Soc.* **B 24**, 118–24.

Name Index

An asterisk indicates a correspondent, the page references to the letters being shown in italics.

Subject Index

The manufacturer's authorised representative in the EU for product safety is Oxford University Press España S.A. of El Parque Empresarial San Fernando de Henares, Avenida de Castilla, 2 - 28830 Madrid (www.oup.es/en or product.safety@oup.com). OUP España S.A. also acts as importer into Spain of products made by the manufacturer.
Printed and bound by CPI Group (UK) Ltd, Croydon, CR0 4YY
14/09/2026
02216286-0008